Electric Machines: Principles, Applications, and Control

Electric Machines: Principles, Applications, and Control

Editor:

Nelson Bolívar, Ph.D

ArclerPress

www.arclerpress.com

Electric Machines: Principles, Applications, and Control

Editor: Nelson Bolívar, Ph.D

708 3rd Avenue, 6th Floor
New York
NY 10017
United States of America
www.arclerpress.com

ISBN: 978-1-68094-404-4

Library of Congress Control Number: 2016951061

About the Editor

Nelson Bolívar, Ph.D

Nelson Bolivar has a PhD. in physics from the University of Lorraine in France finished in 2014. His expertise is in quantum systems and condensed matter. His interest includes spintronic devices and correspondences between general relativity and condensed matter. He is currently an associate professor at the Central University of Venezuela

Contents

List of Contributors

M. J. Kramer
Department of Materials Science and Engineering, Iowa State University, Ames, IA 50011, USA.
Ames Laboratory, U.S. Department of Energy, Iowa State University, ISU Campus, Ames, IA 50011, USA.

R. W. McCallum
Department of Materials Science and Engineering, Iowa State University, Ames, IA 50011, USA.
Ames Laboratory, U.S. Department of Energy, Iowa State University, ISU Campus, Ames, IA 50011, USA.

I. A. Anderson
Department of Materials Science and Engineering, Iowa State University, Ames, IA 50011, USA.
Ames Laboratory, U.S. Department of Energy, Iowa State University, ISU Campus, Ames, IA 50011, USA.

S. Constantinides
Arnold Magnetic Technologies Corporation, Rochester, NY 14625, USA.

Arash A. Samimi
Applied Magnetics Group, Queen's University, Kingston, ON K7L 3N6, Canada

Thomas W. Krause
NDE Development, Royal Military College of Canada, Kingston, ON K7K 7B4, Canada

Lynann Clapham
Applied Magnetics Group, Queen's University, Kingston, ON K7L 3N6, Canada

Matthew Gallaugher
Department of Mining and Materials Engineering, McGill University, Montreal, QC H3A 0C5, Canada

Yaoyao Ding
Department of Mining and Materials Engineering, McGill University, Montreal, QC H3A 0C5, Canada

Pampa Ghosh
Department of Mining and Materials Engineering, McGill University, Montreal, QC H3A 0C5, Canada

Richard Chromik
Department of Mining and Materials Engineering, McGill University, Montreal, QC H3A 0C5, Canada

Andrew M. Knight
Department of Electrical and computer Engineering, University of Alberta, Edmonton, AB T6G2V4, Canada

Ran Zhang
North Carolina A & T State University, Greensboro, USA.

Marwan Bikdash
North Carolina A & T State University, Greensboro, USA.

Mohammed Zeki Al-Faiz
Department of Computer Engineering, Al-Nahrain University, Baghdad, Iraq

Mohammed S. Saleh
Department of Electronics Engineering, Diyala University, Diyala, Iraq

Ahmed A. Oglah
Department of Control & Systems Engineering, University of Technology, Baghdad, Iraq

Radosław Kępiński
Department of Automation, Biomechanics and Mechatronics, Lodz University of Technology, Stefanowskiego 1/15, 90-924 Lodz, Poland

Jan Awrejcewicz
Department of Automation, Biomechanics and Mechatronics, Lodz University of Technology, Stefanowskiego 1/15, 90-924 Lodz, Poland

Donat Lewandowski
Department of Automation, Biomechanics and Mechatronics, Lodz University of Technology, Stefanowskiego 1/15, 90-924 Lodz, Poland

Chin Kat Meng
Department of Engineering Sciences & Mathematics, College of Engineering, Universiti Tenaga Nasional, Kajang, Selangor, Malaysia

Nasri A. Hamid
Department of Engineering Sciences & Mathematics, College of Engineering, Universiti Tenaga Nasional, Kajang, Selangor, Malaysia

Mohd. Yusri Abd. Rahman
Department of Engineering Sciences & Mathematics, College of Engineering, Universiti Tenaga Nasional, Kajang, Selangor, Malaysia

Shazia Baloch
Electrical Engineering Department, Balochistan University of Engineering and Technology, Khuzdar, Pakistan

Mukhtiar Ahmed Mahar
Electrical Engineering Department, Mehran University of Engineering and Technology, Jamshoror, Pakistan

Raza Haider
Electrical Engineering Department, Balochistan University of Engineering and Technology, Khuzdar, Pakistan

Shakeel Ahmed
Civil Engineering Department, Balochistan University of Engineering and Technology, Khuzdar, Pakistan

Chengjun Xia
School of Electric Power, South China University of Technology, Guangzhou, China

Yun Zhou
School of Electric Power, South China University of Technology, Guangzhou, China

Kun Men
Electric Power Research Institute of CSG, China Southern Power Grid, Guangzhou, China

Yinggeng Xie
School of Electric Power, South China University of Technology, Guangzhou, China

K. Prakasam
Department of EEE, K.S.R. College of Engineering, Tiruchengode, India

S. Ramesh
Department of EEE, K.S.R. College of Engineering, Tiruchengode, India

Hussein Sarhan
Faculty of Engineering Technology, Al-Balqa' Applied University, Salt, Jordan

Ben Hamed Mouna
Department of Electrical Engineering, National Engineering School of Gabes, Tunisia

Sbita Lassaâd
Department of Electrical Engineering, National Engineering School of Gabes, Tunisia

Mohammad Abdul Mannan
Department of Electrical & Electronic Engineering, American International University, Bangladesh (AIUB), Dhaka, Bangladesh

Asif Islam
Department of Electrical & Electronic Engineering, Bangladesh University of Engineering & Technology (BUET), Dhaka, Bangladesh

Mohammad Nasir Uddin
Department of Electrical & Electronic Engineering, American International University, Bangladesh (AIUB), Dhaka, Bangladesh

Mohammad Kamrul Hassan
Department of Electrical & Electronic Engineering, American International University, Bangladesh (AIUB), Dhaka, Bangladesh

Toshiaki Murata
Department of Electrical Engineering, Kitami Institute, Hokkaido, Japan

Junji Tamura
Department of Electrical Engineering, Kitami Institute, Hokkaido, Japan

Nader N. Barsoum
Department of Electrical and Computer Engineering, Curtin University Sarawak Campus, Miri, Malaysia

Pin Rui Chin
Department of Electrical and Computer Engineering, Curtin University Sarawak Campus, Miri, Malaysia

M. L. Elhafyani
Laboratory of Electrical Engineering and Maintenance (LEEM), University Mohammed 1, Oujda, Morocco

S. Zouggar
Laboratory of Electrical Engineering and Maintenance (LEEM), University Mohammed 1, Oujda, Morocco

M. Benkaddour
Department of Physics, Faculty of Science, University Mohammed 1, Oujda, Morocco

A. Aziz
Laboratory of Electrical Engineering and Maintenance (LEEM), University Mohammed 1, Oujda, Morocco

Arie Shenkman
Electrical and Electronics Engineering Department, Shamoon College of Engineering, Beer Sheva, Israel

Saad Tapuchi
Electrical and Electronics Engineering Department, Shamoon College of Engineering, Beer Sheva, Israel

Dmitry Baimel
Electrical and Electronics Engineering Department, Shamoon College of Engineering, Beer Sheva, Israel

Boel Ekergård
The Division for Electricity, Uppsala University, Uppsala, Sweden

Rafael Waters
The Wave Energy Converter Department, The Wave Energy Company Seabased Industry AB, Uppsala, Sweden.

Mats Leijon
The Division for Electricity, Uppsala University, Uppsala, Sweden

Ahmad M. Alkandari
College of Technological Studies, Department of Electrical Technology Sheweikh, Kuwait, Kuwait

S. A. Soliman
Electrical Power and Machines Department, Misr University for Science and Technology, Giza, Egypt

Mansour H. Abdel-Rahman
Department of Electrical Engineering, Mansoura University, Mansoura, Egypt

Deng-Chern Sue
Pacific Engineers & Constructors, Ltd., Taipei, China

V. Chandrasekaran
Department of Electrical and Electronics Engineering, Sri Eshwar College of Engineering, Coimbatore, India

D. Nagarajan
Department of Electrical and Electronics Engineering, Karpagam College of Engineering, Coimbatore, India

B. Nagaraj
Department of Electronics and Communication Engineering, Karpagam College of Engineering, Coimbatore, India

Masahide Hojo
Institute of Technology and Science, The University of Tokushima, Tokushima, Japan

Ryo Ikeshita
Institute of Technology and Science, The University of Tokushima, Tokushima, Japan

Yoshinobu Ueda
Power Utility Sector Business Unit, Meidensha Corporation, Tokyo, Japan

Toshihisa Funabashi
EcoTopia Science Institute, Nagoya University, Nagoya, Japan

Omessaad Elbeji
Department of Electric Automatic, National Engineering School of Gabes, Gabes, Tunisia

Mouna Ben Hamed
Department of Electric Automatic, National Engineering School of Gabes, Gabes, Tunisia

Lassaad Sbita
Department of Electric Automatic, National Engineering School of Gabes, Gabes, Tunisia

Preface

Electric machines are often synonyms for electric motors and electric generators. The term refers to those devices that has certain mobility but furthermore, to those that transform electric currents into motion or motion into electric currents. The modern society depends upon energy to subsist, this energy comes from a number of different sources, some very polluting, like burning fossil fuels, others very low polluting - at least during energy harvesting – like solar cells, and in between, lies the inductive generators. This machines take advantage of mechanical energy, for instance water sources and wind power, and convert part of this energy in electric currents.

The synchronization and control of these machines is complex, not only from the mechanical point view, but electronically. The engineer accounts for duty cycles, torque potential, cost, efficiency and magnetic core assembly among others, and the advances in these fields is growing fast. Electric Machines Principles, Applications, and Control is intended for students and engineers, as well as materials scientists will find this book to be a good comprehensive resource for learning about the fundamental control and functioning of these machines, focusing on specific examples that are also top research material in the area, suitable for the interested reader. Content coverage of the book is as follows. First chapter provides an overview of non-rare earth permanent magnets for traction motors and generators. The purpose of second chapter is to investigate the feasibility of using surface magnetic Barkhausen noise for the evaluation of AC core loss, and further, to examine potential origins of magnetic loss in non-oriented electrical steel. Third chapter presents the fault-tolerant control problem of DC motors. In fourth chapter, the application of type-2 fuzzy logic concepts to the position control of a simple DC motor were illustrated. Fifth chapter covers dynamical simulation of a nonlinear stepper motor system. Sixth chapter describes the performance of DI-BSCCO superconducting electric motor. In seventh chapter, work power factor and total harmonic distortion of separately excited DC motor controlled with AC-DC converter has been analyzed at-load. Eighth chapter proposes a method for the identification of inductive motors load partition based on coherence and ninth chapter aims to present dissimilar stator faults which are classified under electrical faults using motor current signature analysis (MCSA) and the comparison of simulation and hardware results. The aim of tenth chapter is to provide energy efficient control of three-phase induction motor drive and eleventh chapter deals with performance analysis and implementation of a three phase inverter fed induction motor (IM) drive system. The purpose of twelfth chapter is to propose a fuzzy logic speed controller of induction motor where flux and torque decoupling strategy is decoupled in terms of magnetizing current instead of stator current to alleviate the effects of core loss. Thirteenth chapter focuses on Ethernet control AC motor via programmable logic controller (PLC) using LabVIEW. Fourteenth chapter presents the capacity variation effect, speed of drive and load as function of the wind system without and with analogical regulator. Fifteenth chapter proposes a new type of the synchronous capacitive machine operated on a principle of the electric field effect. In sixteenth chapter calculations of the varying inductances profile for a synchronous linear surface mounted permanent magnet generator has been presented and seventeenth chapter explores on the steady state analysis of a double-fed induction generator (DFIG) adopted for wind power generation. Eighteenth chapter deals with high efficiency double–fed induction generator applied to wind power generator technical analyses and nineteenth chapter highlights new development in the performance improvement synchronous motor. Twentieth chapter describes a converter control to contribute to enhancing the synchronizing torque. In last chapter, a model of a variable speed wind turbine using a permanent magnet synchronous generator (PMSG) is presented and the control schemes are proposed.

Editor
Nelson Bolívar, Ph.D

INTRODUCTION

An electric machine is a device that produces electricity or does work using electricity. It usually has a motor that drives a set of gears and belts. An electric machine may not have any moving parts at all. One example of an electric machine is the electrical generator, which produces electricity using a source of mechanical energy, such as an engine running on a fossil fuel, a turbine, an electric motor, steam, wind, or water.

Electrical generators may be either alternating current (AC) or direct current (DC) generators. Power utilities produce AC voltages in several thousands of volts that are then distributed and stepped down using a transformer. Cars and other means of transportation use DC generators to generate about 13.8 volts direct current (VDC) to power up built-in equipment, such as ignition systems, stereos, two-way radios, and laptops.

An electric machine may use either an AC or DC motor. AC motors are very popular where AC power is available. In homes, they are used in fans, air conditioners, washing machines, dishwashers, and other appliances. DC motors are popular in smaller equipment and where only DC power is available. In cars, DC motors are used in power windows, wipers, engine starters, and blower fans.

Electric motors are used in all industries, and range from simple machines that shape materials to machines that create miniaturized high-density circuit boards. Either humans or computers act as the controller. Computers are able to repeat processes under tight tolerances, whereas humans are normally unable to work in such conditions.

Special machines are capable of storing energy as kinetic energy (KE) in satellites. Using a magnetic bearing, a KE storage device is able to bear a 331 lb (150 kg) rotating flywheel at about 3,600 revolutions per minute. The flywheel spins while suspended in a vacuum. When the satellite in orbit is in view of the sun, solar photovoltaic panels generate the needed electricity for the satellite and to maintain the spin of the flywheel. During dark times, the battery system takes over. At prolonged darkness, the battery voltage drops below cut off, which is where the flywheel is engaged to generate electricity.

New electric machines are also foreseen in the nanotechnology field. Very small machines are thought to be one day capable of closed surgical processes, for example. The impact is that patients may undergo major surgery from outside the body, and it would be like nothing is happening.

Generator

A generator is an electrical machine which converts mechanical form of energy into electrical form. Generator works on the principle that whenever a conductor moves in a magnetic field, an emf gets induced in the conductor. This principle is called as generator action.

Generators have generally two basic parts named "Stator" and "Rotor". Mechanical energy is provided to the rotor of a generaotor by means of a prime mover (i.e. a turbine). Turbines are of different types like steam turbine, water turbine, wind turbine etc. Mechanical energy can also be provided by IC engines or similar other sources.

To learn more about how generators work, read the following articles.

- AC Generator (converts mechanical energy into Alternating Current (AC) electricity)
- DC Generator (converts mechanical energy into Direct Current (DC) electricity)

Motor

A motor is an electrical machine which converts electrical energy into mechanical energy. When a current carrying conductor is placed in a magnetic field, the conductor experiences a mechanical force and this is the principle behind motoring action.

Just like generators, motors also consists two basic parts, stator and rotor. In a motor, we give electric supply to both the stator and rotor windings which causes a mechanical force between the stator and rotor. This force causes the rotor to rotate.

To learn more about electric motors, read the following articles.

- AC motors: (i) Induction motors and (ii) Synchronous motor
- DC motors: (i) Brushed DC motor and (ii) Brushless DC motor

Transformers

Transformers do not actually make conversion between mechanical and electrical form, but they transfer electric power from one circuit to another circuit. They can decrease or increase the voltage while transferring the power without changing the frequency, but with the corresponding increase or decrease in the current. Transformers also belong to electrical machines.

Step up transformers increases the voltage level from primary to secondary but with the corresponding decrease in the current and vice versa.

Classification of Electric Machines

There are several methods of classifying electric machines:

Electric Power Supply

Electric machines are classified as D.C. and A.C. machines as well as according to their stator and rotor constructions as shown in Figure 1.

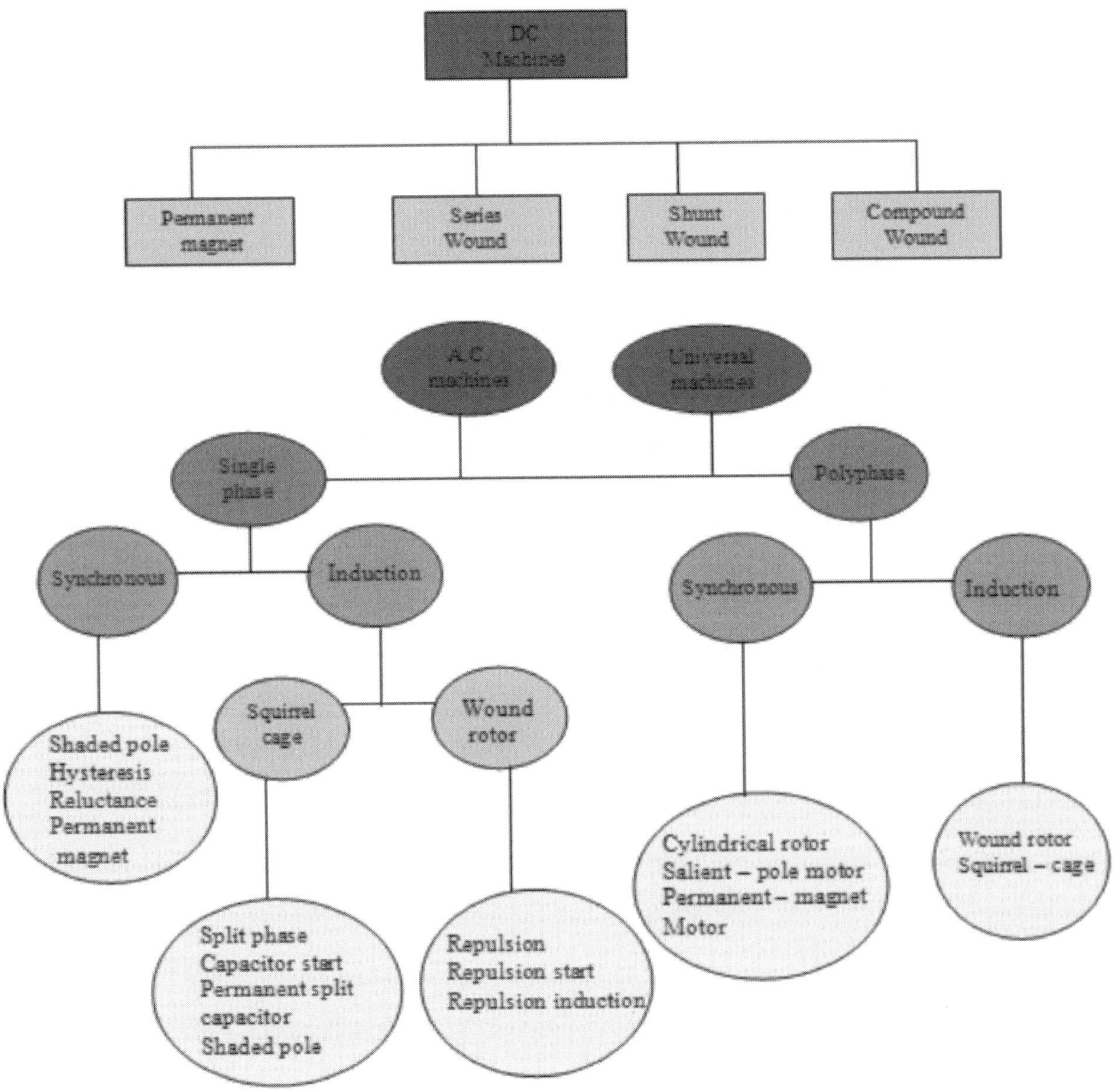

Figure 1: Classification of Electric Machines.

National Electric Manufacturers' Association (NEMA) Standards

NEMA standards are voluntary standards of the National Electric Manufacturers Association and represent general practice in industry. They define a product, process, or procedure with reference to nomenclature composition, construction, dimensions, tolerances, operating characteristics, performance, quality, rating, and testing. NEMA classifications of Electric Machines are summarized in Table 1.

Table 1: NEMA Classifications.

Types	Features
A. Open:	
i. Drip-proof	Operate with dripping liquids up to 15^0 from vertical
ii. Guarded	Guarded by limited size openings (less than 0.75 inch)
iii. Externally ventilated	Ventilated with separate motor driven blower, can have other types of protection
B. Totally enclosed:	
i. Non-ventilated (TENV)	Not equipped for external cooling
ii. Fan-cooled (TEFC)	Cooled by external integral fan
iii. Water-cooled	Cooled by circulating water

Basic Features of Electric Machines

The basic structural features of a D.C. machine are:

- *Stator*: The stator carries the field winding. The stator together with the rotor constitutes the magnetic circuit or core of the machine. It is a hollow cylinder.
- *Rotor*: It carries the armature winding. The armature is the load carrying member. The rotor is cylindrical in shape.
- *Armature Winding*: This winding rotates in the magnetic field set up at the stationary winding. It is the load carrying member mounted on the rotor. An armature winding is a continuous winding; that is, it has no beginning or end. It is composed of a number of coils in series as is shown in Figure 2.

Depending on the manner in which the coil ends are connected to the commutator bars, armature windings can be grouped into two: lap windings and wave windings. Wave winding gives greater voltage and smaller current ratings while the lap winding supplies greater current and smaller voltage ratings.

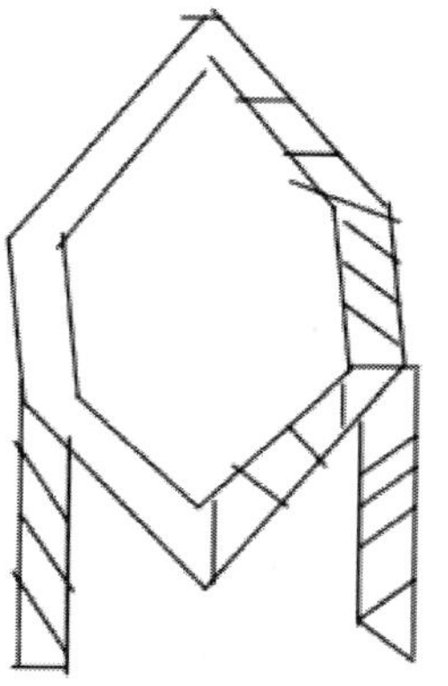

Figure 2: An Armature Coil.

- *Field Winding*: This is an exciting system which may be an electrical winding or a permanent magnet and which is located on the stator.
- *Commutator*: The coils on the armature are terminated and interconnected through the commutator which comprised of a number of bars or commutator segments which are insulated from each other. The commutator rotates with the rotor and serves to rectify the induced voltage and the current in the armature both of which are A.C.
- *Brushes*: These are conducting carbon graphite spring loaded to ride on the commutator and act as interface between the external circuit and the armature winding.
- *Poles*: The field winding is placed in poles, the number of which is determined by the voltage and current ratings of the machine.

- *Slot/Teeth*: For mechanical support, protection from abrasion, and further electrical insulation, non-conducting slot liners are often wedged between the coils and the slot walls. The magnetic material between the slots is called teeth. Figure 3 shows a cross-sectional views of slot/Teeth geometry

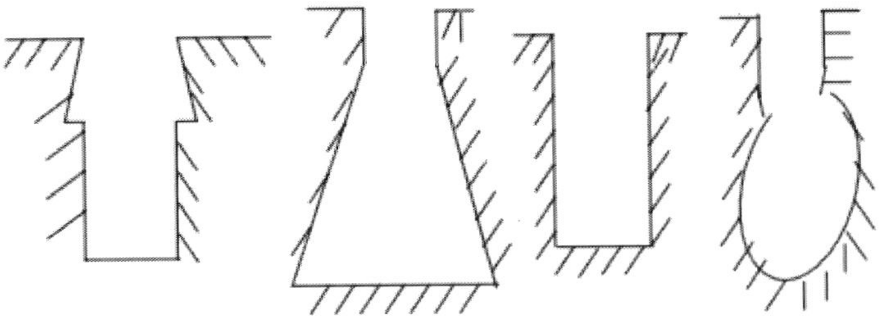

Figure 3: Typical Armature Slot Geometry.

On the other hand, the basic constructional features of an A.C. machine (e.g induction machine) are:

- *Rigid Frame*: The whole construction ensures compact and adaptable design at low weight and low vibration level in all operating conditions and throughout the whole speed range [8]. Figure 4 shows a basic ABB Rigid Frame Design.

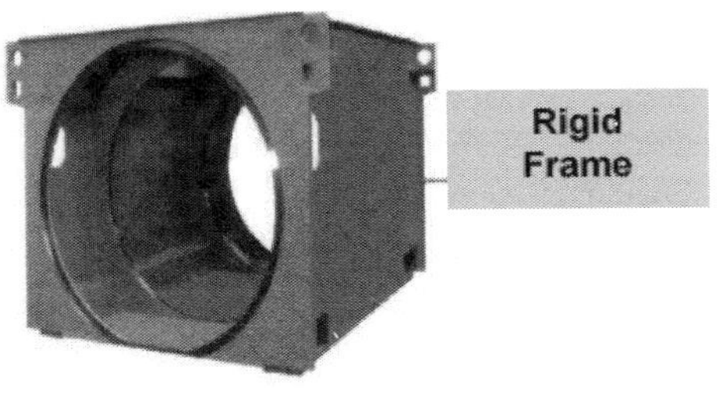

Figure 4: ABB Rigid Frame.

- *Stator Package*: The stator core is a stack of thin electrical sheet steel laminations insulated by a heat resistant inorganic resin. The radial cooling ducts ensure uniform and efficient cooling. The stator package, shown in Figure 5 forms a solid block which retains its rigidity throughout the long lifetime of the machine.

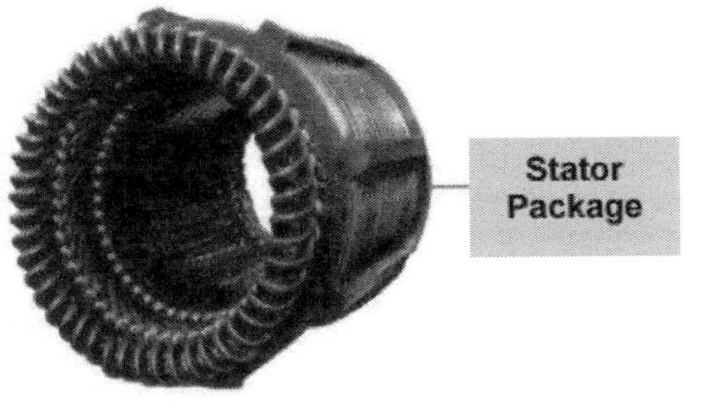

Figure 5: ABB Stator Package.

- *Rotor Construction*: The rotor of A.C. machines can be of wound type or squirrel cage type. A typical squirrel cage rotor is shown in Figure 6. Depending on the number of poles and whether the shaft is of the spider or cylindrical type, the rotor core is shrunk onto the shaft and the conductor bars tightly caulked into the slots to prevent bar vibration.

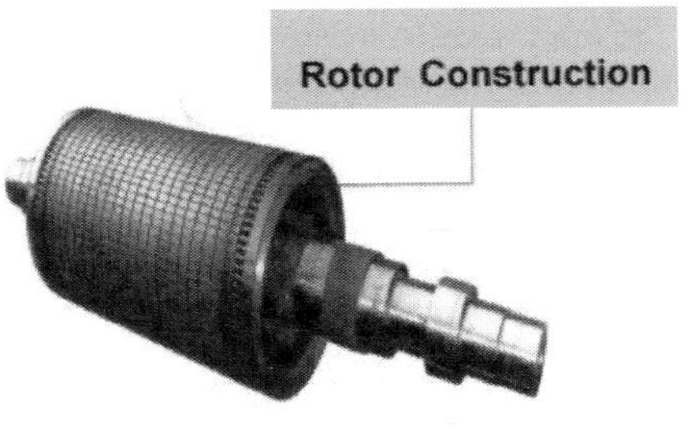

Figure 6: ABB Rotor Construction.

- *Rugged Bearing Assemblies*: The bearings are designed for reliable continuous operation and ease of maintenance. Depending on the rated power either spherical seated self-aligning sleeve bearings or anti-friction bearings with a life time of over 100,000 hours are available. See Figure 7 for ABB rugged bearing assemblies.

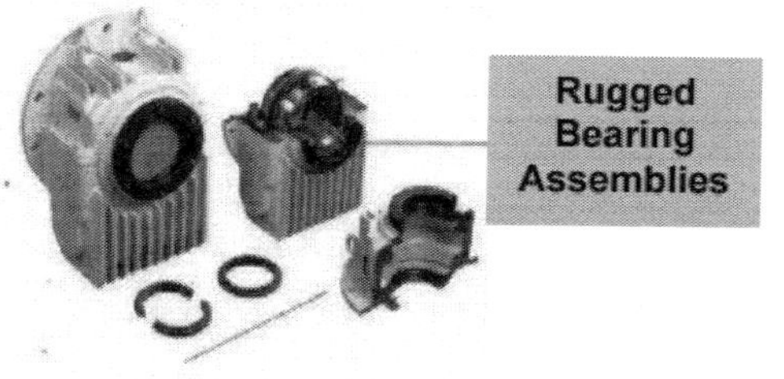

Figure 7: ABB Rugged Bearing Assemblies.

Basic Principals of Operation

Electric motors and generators are a group of devices used to convert mechanical energy into electrical energy, or electrical energy into mechanical energy, by electromagnetic means. A machine that converts mechanical energy into electrical energy is called a generator, alternator, or dynamo, and a machine that converts electrical energy into mechanical energy is called a motor.

Two related physical principles underlie the operation of generators and motors. The first is the principle of electromagnetic induction discovered by Michael Faraday in 1831. If a conductor is moved through a magnetic field, or if the strength of a stationary conducting loop is made to vary, a current is set up or induced in the conductor. The converse of this principle is that of electromagnetic reaction, first discovered by Andre' Ampere in 1820. If a current is passed through a conductor located in a magnetic field, the field exerts a mechanical force on it.

Both motors and generators consist of two basic units, the field, which is the electromagnet with its coils, and the armature (the structure that supports the conductors which cut the magnetic field and carry the induced current in a generator or the exciting current in a motor). The armature is usually a laminated soft-iron core around which conducting wires are wound in coils.

Determination of Machine Parameters

The nameplate gives sufficient information on the rated current, power, frequency, voltage, winding temperature, and stator winding c n. However, it may be necessary to determine the winding resistances and reactance as well as the mechanical properties of the machine in order to evaluate the machine performance under both steady and dynamic conditions. For instance, the following tests are usually carried out to determine the parameters of an asynchronous machine.

No-Load Test - The aims of the no – load test are to determine:

- Stator ohmic/copper losses
- Stator core losses due to hysteresis and eddy current
- Rotational losses due to friction and windage
- Magnetizing inductance.

The test is carried out at rated frequency and with balanced polyphase voltages applied to the stator terminals. Readings are taken at rated voltage, after the motor has been running for a considerable period of time necessary for the bearings to be properly lubricated. At no—load, the machine slip and the rotor current are very small thereby resulting to a negligible no—load rotor loss.

Blocked—Rotor Test - The blocked—rotor test provides information necessary to determine:

- The winding resistances
- The leakage reactances

In this test, the rotor is blocked by external means to prevent rotation. In the blocked—rotor test, the slip is unity $(s \cong 1)$ and the mechanical load resistance, RM is zero; thereby resulting in a very low input impedance of the equivalent circuit.

- Retardation Test: The retardation test is carried out to determine the test motor moment of inertia. In this test, a no—load is carried out with and without additional standard induction machines can be obtained from manufacturer's data as well as from the Finite—Element—Analysis (FEA) calculation results [11] such standard curves are shown in Figure 8 and Figure 9.

Electric Machines Protection and Maintenance

Electrical and mechanical faults can impose unacceptable conditions and protective devices are therefore provided to quickly disconnect the machine from grid. In order to ensure that electrical machines receive adequate protection, extensive testing is performed to verify the high quality of assembly. After a machine of a particular type has been type tested for electrical characteristics, all subsequent machines of the same type undergo a routine test programme.

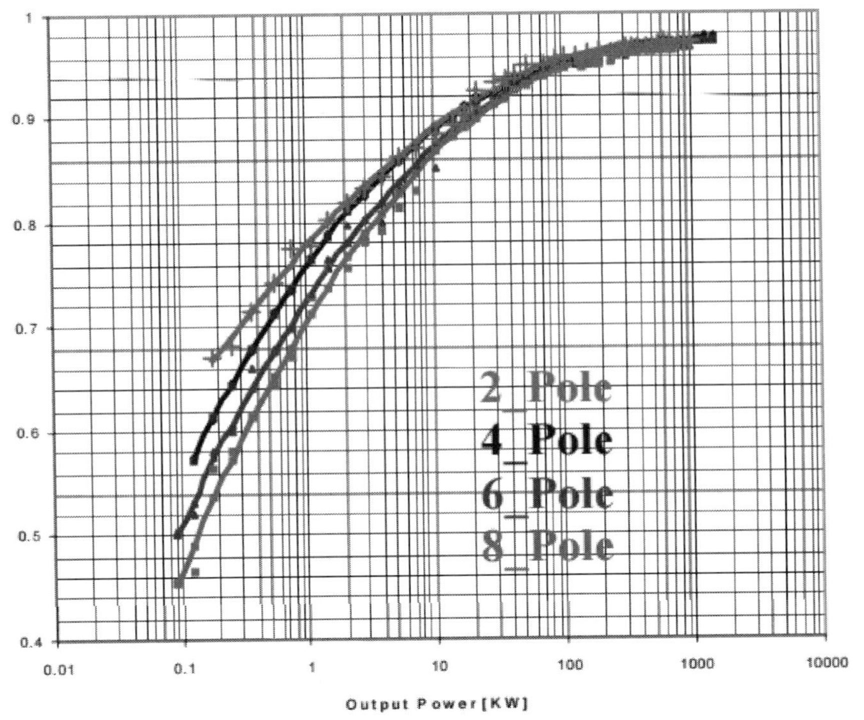

Figure 8: Efficiency of Induction Machine.

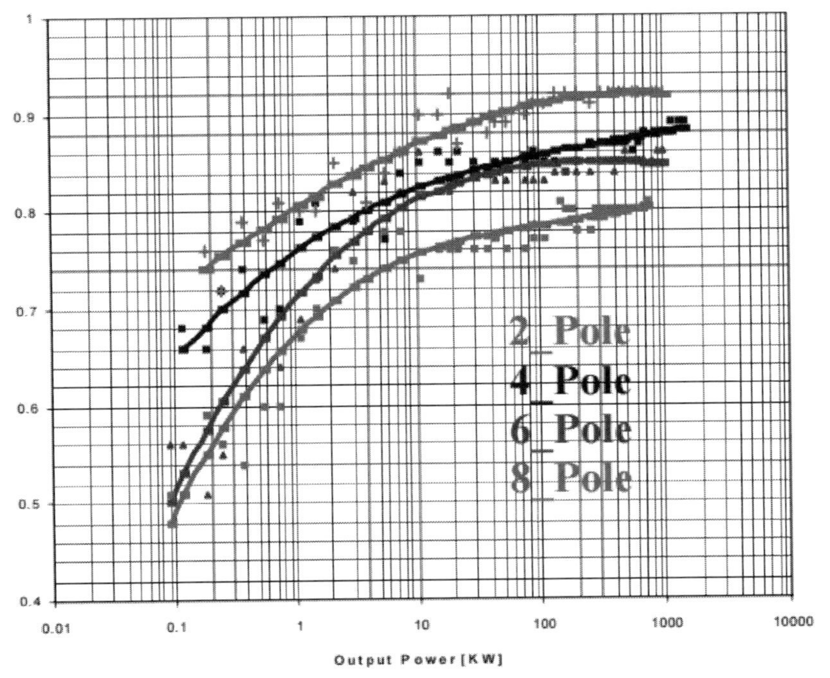

Figure 9: Power factor of Induction Machine.

Routine and type test programmes can take these form:

- Routine test programme
    - Bearing control
    - Control of the insulation
    - Ohmic resistance measurement

- Vibration measurement
- Short circuit test
- No- load test
- High voltage test

- Type test programme
  - Routine test
  - No-load curve
  - Load point
  - Heat run test

After the type test programme, the electrical machine is identified with a protective symbol. Degrees of protection by enclosures for electrical machines are quoted with the letters IP and two characteristic numerals. The first numeral designates the degree of protection for persons against contact with live or moving parts inside the enclosure and of machines against ingress of solid foreign bodies. The second numeral designates the degree of protection against harmful ingress of water. Some of the degrees of protection are IP23, IP54, IP55, and IP56. For instance, in IP23, the first numeral means operation against contact by a finger with live or moving parts inside the enclosure while the second numeral denotes protection against water.

Generally, when deciding on a particular type of motor protection, this should be done with actual operating conditions in mind. Motors are protected by current—dependent motor protection circuit breakers and/or over current relays. These are particularly effective in cases like locked rotor or interrupted run-up, and are therefore indispensable in large machines with thermally critical rotor.

The temperature-dependent devices serve to protect the motor against the effects of excessive winding heating due to overload, increased ambient temperature, impaired cooling, intermittent operation, high switching frequency, and phase failure.

Unscheduled downtime and resultant high repair cost reduce profits. There is a need to set objectives to manage maintenance, schedule repair, adjustments, and control cost. When carrying out servicing or repairing electric machines, like power generation equipment, do the following:

- Make sure the unit is off-line
- Make sure the generator engine is stopped
- Make sure all batteries are disconnected
- Make sure all capacitors are discharged.

The generator is a component of the generator set and should be tested with the entire system. The service manual for the voltage regulator provide tests to determine if the generator is the cause of a generator set malfunction.

The following procedure should be used to help identify and define the problem:

1. Perform visual checks to help identify the problem.
2. If previous tests were performed from the voltage regulator service manual, use the test results to help identify the problem.
3. Check troubleshooting knowledge-bases to help identify the problem.
4. Perform the Generator Functional Test to help identify the problem.

After the work has been carried out the machine is to be marked by an addition name plate with the following data:

- Date
- Operative firm
- If necessary, mode of repair
- If necessary, signature of the expert

Electric Machines Applications

Figure 10, Figure 11, and Figure 12 show some of the practical applications of electric machines.

Figure 10: SCHORCH Motor 90KW, 400V, 2975rpm, Compressor drive.

Figure 11: SCHORCH Motors 160KW, 400V, 1485rpm, Driving pumps.

Figure12: SCHORCH Hydrogenerator 45KW, 400V, 1025rpm.

Asynchronous Machines

- Petroleum and chemical pumps
- Cooling towers
- Air-handling equipment

- Compressors
- Process machinery
- Blowers and fans
- Drilling machines
- Grinders
- Lathes
- Conveyors
- Crushers, etc.

Synchronous Machines

- Power generation
- Wind energy turbines
- Power factor correction
- Voltage regulation improvement of transmission lines
- Electric clock drives
- Gasoline engine drives
- Servo drives
- Compressors, etc

D.C. Machines

- Rolling mills
- Elevators
- Conveyors
- Electric locomotives
- Rapid transit systems
- Cranes and hoists
- Lathes
- Machine tools
- Blowers and fans, etc

CHAPTER 1

NON-RARE EARTH PERMANENT MAGNETS FOR TRACTION MOTORS AND GENERATORS

M. J. Kramer, R. W. McCallum, I. A. Anderson, and S. Constantinides

INTRODUCTION

Transportation accounts for 28% of energy consumption in the United States,[1] of which 93% comes from petroleum. Options for replacing fossil fuels for transportation are limited. While biofuels and hydrogen can replace fossil fuels for use in internal combustion engines, electric motors in combination with an internal combustion engine or all-electric vehicles are seen as the primary technologies capable of reducing dependence on petroleum for the transportation sector. Since the advent of high-flux density permanent magnets based on rare earth elements such as neodymium (Nd) or samarium (Sm), permanent magnet-type electric machines[1] are preferred over induction machines when weight and size are factors, such as in hybrid electric vehicles.[2] For electric generation, wind is one of the more cost-effective renewable energy sources. Permanent magnet-based-systems have several advantages over induction machines that are heavier, noisier and require more maintenance due to their high-ratio gear boxes.[3,4] However, the advantages of the permanent magnet-based electric machines maybe overshadowed by the current limited availability and price of their key constituent, rare earth elements.[5–9] We will present an overview of the past, present, and future prospects for magnet alloys used in permanent magnet-based electric machines in light of current market pressures on rare earth elements and prospects for replacing alloys based on rare earth elements with more abundant and less strategically critical elements.

Permanent magnets are arguably one of the earliest functional materials. Lodestones, comprised mainly of magnetite, were known in ancient times.[10] The magnetic properties of carbon steel were first reported in the 1600s by Sir William Gilbert. With the discovery of electromagnetism in the early 19th century by André-Marie Ampère and the formalism of Maxwell's equations, the stage was set for the linking of mechanical work to electromagnetism so that by the end of the 1800s, simple electric motors and generators were in production. In fact, all-electric and hybrid automobiles were commercially available more than 100 years ago with the first hybrid built in 1900 by Ferdinand Porsche. The drawback to electric cars was their short range due to low battery capacity and low efficiency of the electric motors. There have been significant advances in both these areas with the most efficient motors for vehicle applications being permanent magnet motors made possible by high-performance permanent magnets. A permanent magnet must have a large magnetic induction (B) and the ability to maintain B in a fixed direction relative to the body of the magnet in the presence on an opposing magnetic field (H). B as a function of H is given by B(H) = M(H) + H, where M(H) is the magnetization of the material as a function of H. The energy product of a magnet is the product of B(H) and H The largest negative value of this product when B(H) and H are opposed is $(BH)_{max}$, which is used as a figure of merit. It is a measure of the maximum amount of work that can be performed by a permanent magnet in units of megagauss oersteds (MGOe, or kJ/m^3in SI units). The ability of a magnet to maintain its magnetization in the presence of an opposing magnetic field is referred to as its intrinsic coercivity. It is characterized by the value of H required to reduce M(H) to zero, H_{cJ} is referred to as the intrinsic coercivity.

Through the first half of the 1900s, most permanent magnets were Fe-based alloys. The first significant improvement in the maximum energy product was achieved in columnar grown alnico (Fig. 1). The improvement came about from the grain alignment of the cubic phases and the application of a magnetic field during cooling through the spinodal decomposition into the nominally FeCo body-centered cubic (bcc) and the primitive cubic β-AlNi phases.[11] This produces what is essentially an aligned array of nanoscaled compass needles in a nonmagnetic matrix. These Fe-based alloys have sufficient

Citation: M. J. Kramer, R. W. McCallum, I. A. Anderson, and S. Constantinides. Non-Rare earth permanent magnets for traction motors and generators, JOM, July 2012; Volume 64, Issue 7, pp 752–763. DOI: 10.1007/s11837-012-0351-z.

intrinsic magnetization from the Fe and Co atoms, which results in measurable residual induction (B_r), whereas the shape of the needles makes it favorable for the magnetization to point along the axis of the needle, resulting in shape anisotropy. Improving the resistance of an alloy to demagnetizing fields, i.e., H_{cJ}, was paramount to improving energy product in Fe-based alloys. While PtCo and ferrites were significant discoveries in the mid-20th century, alnico continued to be the industry standard for high-end permanent magnets.[12]

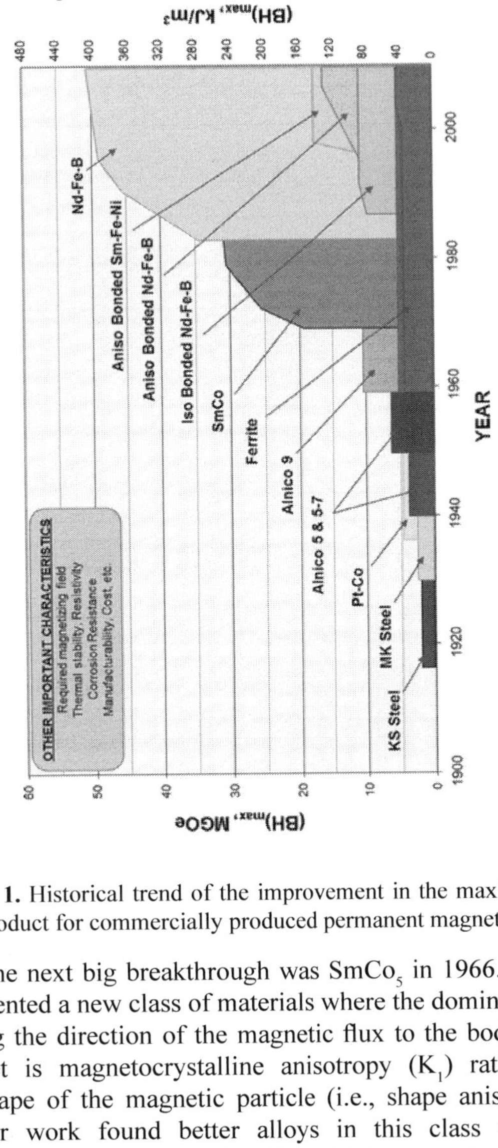

Figure 1. Historical trend of the improvement in the maximum energy product for commercially produced permanent magnets.

The next big breakthrough was $SmCo_5$ in 1966.[13,14] This represented a new class of materials where the dominant force linking the direction of the magnetic flux to the body of the magnet is magnetocrystalline anisotropy (K_1) rather than the shape of the magnetic particle (i.e., shape anisotropy).[2] Further work found better alloys in this class with the highest $(BH)_{max}$ being the Sm_2TM_{17} (TM = transition metals, typically Co and Fe) compound, but high Co content is required for use in commercial applications.[15,16] The next big jump in energy density came in 1981, after the price for Co increased precipitously due to political unrest in what is now the Democratic Republic of the Congo, the world's leading producer for high-quality, low-cost cobalt. This uncertainty in the Co market spurred researchers in Japan and the United States to investigate alternatives, and it resulted in the discovery of a new compound based on Nd, Fe, and B.[17–19] This stoichiometric, layered tetragonal compound ($Nd_2Fe_{14}B$, space group P 42/m n m) has a nearly ideal separation of the Fe atoms resulting in strong exchange coupling of the Fe and Nd sublattices resulting in the highest theoretical $(BH)_{max}$ of any compound to date at 64 MGOe. This compound and alloys derived from it has enabled the miniaturization of many of the more common consumer products such as ear buds for portable audio devices and hard disk drives. The estimated market demand for Nd-based magnets in 2012 in disk drives, CD-ROMs, and DVD-ROMs alone is forecast to exceed 13,100 tons.[3]

With the discovery of the rare earth element-based magnet alloys, using primarily Nd and Sm, the demand for rare earth element metals also increased throughout the 1980s. At the same time, new rare earth element resources were becoming available in China, causing a significant reduction in prices. The major producer in the world, the Molycorp's Mountain Pass mine in California, was experiencing problems with aging infrastructure, which required significant investment. With the low rare earth element prices, improvements to the Mountain Pass mine were not economically viable, and in a few short years in the 1990s, China became the main producer of rare earth elements (Fig. 2).[7] By 2000, China was dominating the worldwide rare earth element production. During the discovery period of rare earth element-based magnets, the major producers of rare earth element-based magnet alloys were in Japan and the United States. Over the period 2001 to 2005, the majority of the rare earth element-based magnet companies either ceased production or relocated to China to take advantage of lower manufacturing costs. Concurrently, numerous domestic Chinese manufacturers initiated the production of Nd-based magnets. While the demand for rare earth element-based magnets for disk drives, other electronics, and power hand tools had been increasing steadily, the rare earth supplies were more than capable of meeting that demand. Growing concern for air quality, carbon emissions, and rising cost of petroleum spurred the demand for alternatives to traditional production and for more efficient utilization of energy. This resulted in a rapid escalation of demand during the past decade for hybrid electric cars and wind-powered electric generators. This increased demand resulted in political pressures with regard to supply. In 2009, China announced that it would begin to curtail its export of the rare earths europium, terbium, and dysprosium.[20,21] In March 2012, the European Union, Japan and United States announced that they would be filing a complaint for unfair trading practices against China's export quotas on rare earth elements to the World Trade Organization.[22]

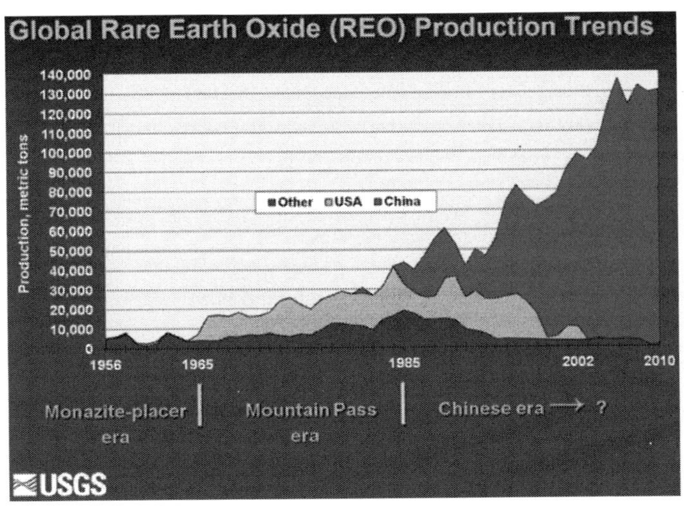

Figure 2. Timeline showing the worldwide production of rare earth elements on an oxide basis by major producers[7].

Regardless of the politics surrounding rare earth elements, increase in demand, driven primarily by electric machines, will be a major factor in both cost and availability of rare earth elements in the near future. The higher temperature performance requirement of generators and traction motors requires an adjustment to the Nd-based permanent magnets alloy to improve its high temperature resistance to demagnetization. This is accomplished with the addition of heavy rare earth elements, primarily Dy. Dy, while reducing B_r, provides a moderate increase in T_c and a significant increase in the anisotropy field, which helps maintain a useful H_{cJ} at temperatures up to ~200°C. The availability, hence cost, of the rare earth elements is not uniform. Dy, being one of the more "rare" of the not-so-rare rare earths, was trading near $100/kg in early 2009 to as high as $3400/kg in August of 2011, and it remains near $2000/kg as of April 2012. The amount of Dy added can be as high as 12 wt.% for the more demanding electric machine applications performing at 220°C (Fig. 3).[23] There appears to be sufficient resources for historical demands. However, projected demands for electric vehicles and wind turbines through the next 5–10 years shows that the availability of Nd, while a concern is not as critical as Dy if Nd-based permanent magnets are used in these applications. Even with projected increase in rare earth element production from new mines coming online (Table I), this demand will not be met. Switching to Sm-based alloys does not alleviate the shortage of resources since Sm comprises less than 4% of the rare earth elements. Samarium is not available in adequate quantities to provide the volume of magnet product needed for a wholesale replacement of Nd-based magnets. It does represent a complementary product, especially for products expected to perform at temperatures above 160°C. The trade-offs between rare earth element-based magnets and potential substitutes will be discussed in more detail below.

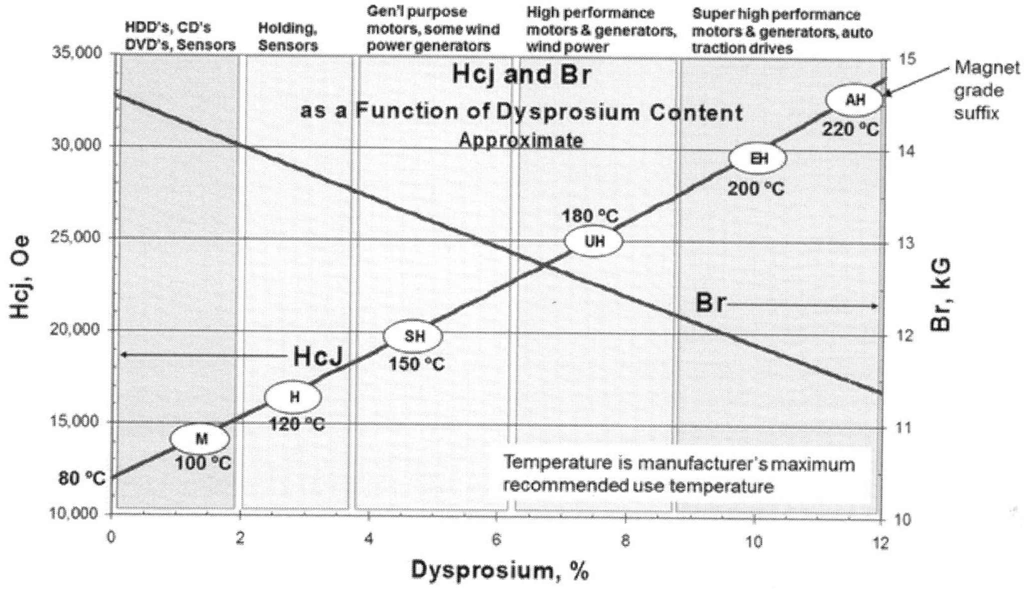

Figure 3. A comparison of the Dy content (in wt.% of the alloy) in various grades of Nd-based permanent magnets and how the Dy content affects the intrinsic coercivity (H_{cJ}) and induction (B_r). The general uses for the various grades of magnets is shown along the top columns. Data are supplied by Arnold Magnetic Technologies, www.arnoldmagnetics.com.

Table I. Potential sources for additional rare earth element (REE) resources through 2015

Select REE	2010 Production	Potential Sources of Additional Production								2015 Capacity	% Supply Increase
		Mountain Pass	Australia			Vietnam	South Africa	Russia and Kazakhsistan	India		
			Mt. Weld	Nolans Bore	Dubbo Zirconium						
Ce	42,000	18,100	10,300	4800	960	1500	2300	290	1200	81,000	93
Pr	5900	1550	1200	590	110	120	250	20	140	9900	74
Nd	20,000	4300	4100	2200	370	320	830	44	460	33,000	65
Sm	2800	290	510	240	56	27	125	5	68	4000	43
Dy	1600	19	22	30	53		34	1		1700	6

Note that many of these localities do not have large proven reserves of Dy.[6]

Given the current state of technology, the options for replacing permanent magnet machines all have shortcomings. For wind turbines, the options are to continue to install induction-based generators or hybrid drive generators with a simpler 2-stage gearbox and 1/3 the weight of permanent magnets required for a direct drive generator. For larger systems, 10 MW and up, superconducting generators are being researched. For automotive traction drive motors, the size and weight considerations are more demanding, especially for hybrid vehicles due to limited space between the internal combustion engine and the wheels. While induction motors are an option for all electric vehicles, power conversion efficiency and mass-to-power ratio clearly favor a permanent magnet-based motor.[2] However, the advantages of permanent magnet-based machines disappear if lower energy product non-rare earth element-based magnets are used. Even though most non-rare earth element permanent magnet materials developed prior to the invention of rare earth element permanent magnets are still commercially produced, the development of these older materials has been neglected. In this article, we will discuss a few of the more promising non-rare earth element systems that have the potential for replacing rare earth element-based permanent magnet alloys in electric machines and challenges to improving their properties to meet the demanding performance criteria for permanent magnet machines.

ALTERNATIVE COMPOUNDS

A high energy density is a necessary, but not a sufficient, condition for a commercially viable magnet alloy. It must be stable enough for the application so that the irreversible losses are minimal over the expected lifetime of the device, which can be 25 years for generators, and 10 or more years for traction drive motors. The intrinsic coercivity must be sufficient to resist demagnetization under maximum load and temperature. It must be made to be thermodynamically and physically stable (minimal flux change with temperature and resistance to corrosion, decomposition or oxidation) in the environment the device will operate. For instance, rare earth element-based alloys are prone to rapid breakdown in wet, warm, and corrosive environments if not adequately coated.[24,25] Since the moments couple along specific crystallographic axes, it is necessary to grain-align the compound in order to attain the maximum energy product. But this texturing must be done to exacting tolerances. Ideally, the grains should be under the single domain limit, the size at which the energy of a domain wall in a grain exceeds the reduction in magnetostatic energy resulting from the domain wall. In practice, other factors that limit the nucleation of reverse domains and pin domain walls once the reverse domains nucleate allow the grain size to be about an order of magnitude larger. For ferrites, that is ~1 μm equivalent spherical diameter, ~3 μm for Nd-based magnets, and 4–5 μm for SmCo. Highly irregular grain boundaries are also detrimental to domain-wall pinning, thus reducing coercivity. This extreme sensitivity to processing can greatly affect manufacturing costs. For instance, isotropic Nd-based magnets are produced by rapid quenching techniques resulting in grain sizes of a few tens of nanometers. These alloys are used in bonded magnets that have an energy product of ~10 MGOe while fully dense isotropic magnets can have energy products up to 16 MGOe and are less expensive to produce than sintered alloys. Figure [4] shows a scale of approximate cost as a function of maximum energy product for a wide range of permanent magnets. Since there are various grades and chemistries for many permanent magnet alloys, the boxes are approximate ranges of performance and cost. Part of the higher cost in the sintered rare earth element-based magnets is that powder metallurgical production results in large amounts of waste: grind swarf and saw kerf, as well as breakage. An additional concern for permanent magnet-based electric machines is the low electrical resistivity of the metal alloys that enhances eddy current losses unlike most permanent magnet ferrites, which are nonmetallic and nonconductive. Therefore, to minimize cost, the size and shape of the magnets must be closely matched to the motor design. Injection molding would accomplish this, but achieving the necessary energy density needed for generators and traction motors has been a challenge even in Nd-based magnets. This is due to the dilution effect of the nonmagnetic binder, which occupies about 35% of the magnet volume in injection molded magnets and 20–25% in compression bonded magnets.

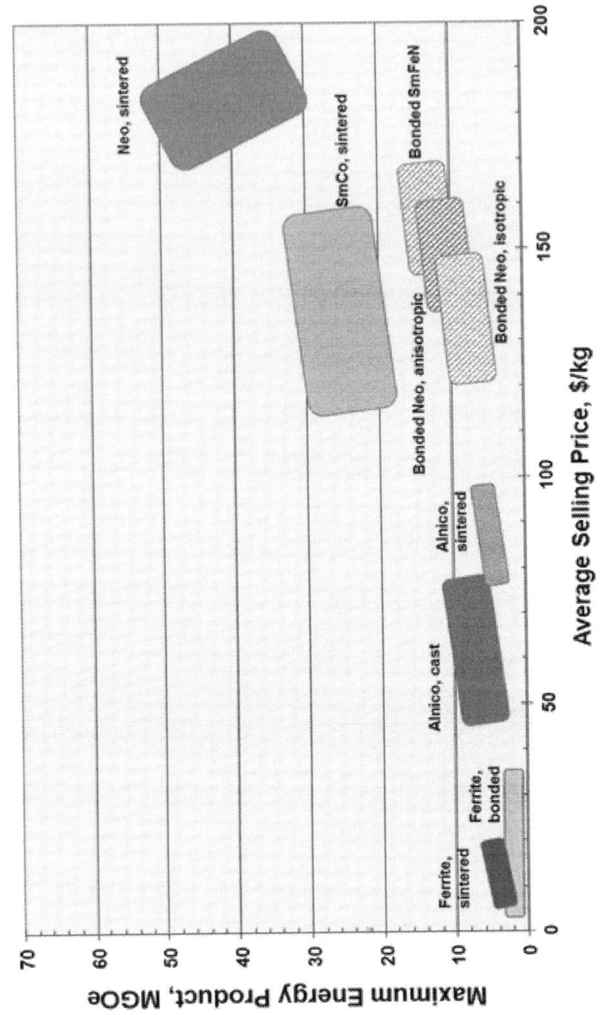

Figure 4. A composite of the maximum energy product for various commercially produced permanent magnets and their relative costs. Many factors such as shape, complexity, and size contribute to a magnet's selling price. The values shown are fair estimates of standard shapes and sizes. The selling price is that observed in the United States and Europe.

While a large maximum energy product does provide electric machine designers the greatest flexibility, there are considerations beyond $(BH)_{max}$. These include the T_c (Curie temperature) and the temperature dependence of the magnetic properties (Fig. 5). In the following discussion, we will focus on those compounds that show some promising characteristics or pathways to improving the compounds' magnetic properties that would allow for use in generators and traction motors.

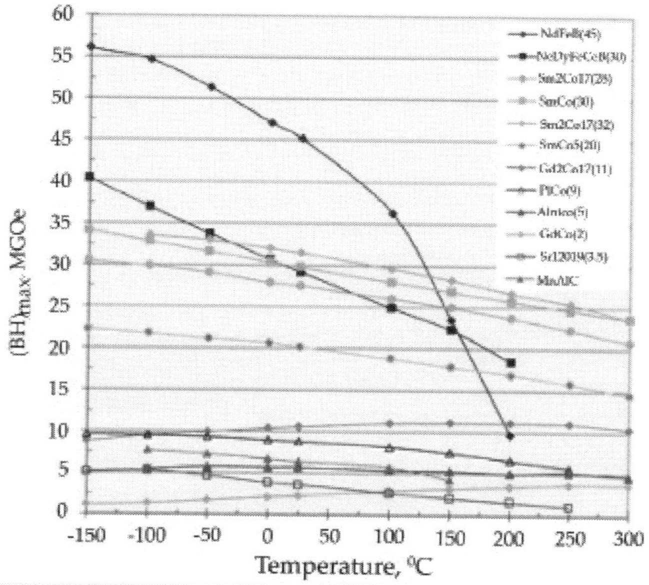

Figure 5. The temperature performance of the maximum energy product as a function of temperature for many common commercial magnets and some promising alternatives. While the Nd-based magnets have superior maximum energy product at ~100°C or below, the Sm-based alloys do better above 200°C. Currently, Dy is added to Nd-based alloys to boost its performance in generators and traction motors. Many alternatives to Nd-based alloys will need to boost their maximum energy product by a factor or two to be comparable with current rare earth element-based magnets

Alnico.

Alnico has a number of very promising characteristics: a high T_c, low thermal coefficient for coercivity and induction, and the ability to magnetize in nonplanar orientations. Significant strides in improving the energy density were made in the 1950s and 1960s based primarily on processing.[26,27] The highest energy product alnico grades are 5–7 and 9. These are both grain aligned and spinodally decomposed[28] as they are cooled in a magnetic field.[29] The improvements in properties were based primarily on empirical studies. Computational and characterization tools to understand and predict optimal nanostructure of the spinodal domains were three decades in the future. A number of minor alloying elements, primarily Cu, Ti, and Nb, are added to the base alloy of 8–13 wt.% Al, 13–28 wt.% Ni, 0–42 wt.% Co, with the balance Fe, to promote columnar growth and enhance H_{cJ}. Key questions remaining to be addressed in further optimization of this class of alloys include: To what extent has the spinodal decomposition been optimized? More specifically, how does the dimension, uniformity of shape, and degree of chemical segregation affect H_{cJ}? For example, doubling of the H_c of alnico 5 to 7 from 750 to 1500 oersteds without sacrificing B_r would result in an increase in $(BH)_{max}$ from 7.5 to ~14 MGOe. Determining how to achieve this without a better understanding of the fundamental mechanisms for domain pinning in these compounds would be difficult.

Ferrites

By weight, ferrites are by far the largest single class of permanent magnets produced (567,330 tons in 2010). They are made with inexpensive elements and, being oxides, have excellent chemical stability for electric machine applications. While significant improvements have been made in preceding decades in the energy density for ferrites,[30] there are a number of significant intrinsic limitations for this class of compounds. First, they have a very high mass-to-magnetization ratio so that the size of a permanent magnet-based electric machine would be too large for a hybrid vehicle. For wind turbine applications, the cost of the additional structure to support the more massive nacelle is more than the incremental cost of the rare earth element permanent magnets. The low magnetization is due to the fact the ferrites are actually ferrimagnets with two different Fe lattices with unequal magnetizations that are coupled antiferromagnetically. Therefore, the net magnetization is the difference between the two lattice magnetizations rather than the sum. As a result, the magnetization is approximately a factor of 4 less than that of Nd-based permanent magnets. Unlike Nd-based permanent magnets, the low temperature performance is a problem. The two magnetic sublattices have different temperature dependences resulting in large temperature dependences of the magnetic properties. In the ferrites, this results in the net magnetization increasing much more rapidly than the anisotropy constant with decreasing temperature. The anisotropy field and, hence, the coercivity is determined by the ratio of these two with the net result that H_{ci} decreases with decreasing temperature, a unique property of these materials. At −40°C, the coercivity is reduced to the minimum practical value, making electric machines based on these materials marginally inoperable in higher latitudes during the coldest part of the winter. Significant improvements in B_r are difficult to achieve simply because the magnetic coupling is based on antiferromagnetic exchange between the magnetic sublattices across oxygen atoms. Doping of selective Fe sites with Co has resulted in significant enhancements of B_r and H_{cJ} when paired with La substitution on the Sr site, but the extent of doping is limited by phase stability. While the magnetization can be increased by lowering the magnetization of the sublattice with the lower magnetization, this inevitably decreases the exchange and lowers T_c.

Other Fe-Based Compounds without Rare Earth Elements

The challenge to designing a high-performance permanent magnet is getting enough density of the elements with significant moments, the transition elements (TM) Mn, Cr, Fe and Ni, in a configuration where their moments can align ferromagnetically. Fe is preferred since it has the largest moment, is abundant, and is inexpensive. Unfortunately, there is an optimum TM-TM spacing for each element; closer spacing tends toward antiferromagnetic alignment while larger spacings reduce the exchange and lowers T_c and the volume magnetization. The optimum spacing is achieved using nonmagnetic elements to stabilize suitable structures. Additionally, the compound structure should form a lower symmetry compound than cubic, which would have the additional benefit of large magnetocrystalline anisotropy to provide coercivity. In fact, FePt forming the L10 structure is an ideal compound, except that cost and availability of Pt renders it impractical for any high-volume application.

TM compounds with N have shown some promise[31–33] as they have a large moment and have the potential to be produced in volume at low cost. The main challenge has been that the magnetic phases of interest are either metastable compounds or have a moderately low decomposition temperatures,[34–36] and these phases have yet to demonstrate significant coercivity.[37] There are two metastable compounds of interest, the tetragonal α′ with about 10 at.% N and the more ordered α″ $Fe_{16}N_2$.[38] For instance, nitriding Sm_2Fe_{17} provides a high-energy product compound but has yet to be fabricated into an oriented, fully dense material.[31,39] Compounds with even moderate decomposition temperatures (in the range of ~400°C) have two significant drawbacks for electric machines; first, they may have unacceptable irreversible losses over time, and more importantly, in order to obtain high energy density in a magnet, the magnet must approach 100% density. Therefore, being able to synthesize fine powders with a theoretical $(BH)_{max}$ comparable with Nd-based alloys is not sufficient. These compounds will need to be compacted into grain-aligned, dense monolithic parts. Typical processing conditions for consolidation of powders into dense compacts results in decomposing the compound or in excessive grain growth, either of which significantly degrades magnetic performance. The challenge to producing usable permanent magnets out of this class of alloys will be to stabilize their metastable structures so that they can be consolidated into dense monolithic parts with long-term stability.[40]

The search for new or modified materials may be aided by modern computer codes based on density functional theory (DFT). These allow for a fairly rapid assessment of "what if" scenarios for either known or postulated compounds.[41] This can be as straightforward as evaluating substitutions to known compounds and determining their thermodynamic stability or as sophisticated as performing genetic searches for new compounds.[42] Critical metrics for promising compounds are the significant number of ferromagnetic moments and high magnetocrystalline anisotropy.[43] Since magnetocrystalline anisotropy is a combination of the spin–orbit and crystal field effects, these are all parameters that can be calculated using a number of DFT programs.[44,45] Small concentrations of impurities of C and N in bcc Fe were shown to modify the hybridization of the Fe 3d orbitals, even though the anisotropy contribution of the C and N were small.[46–48] More important could be the strain effects from the interstitials themselves. Theoretical calculations have suggested that compounds containing mixtures of 3d transition elements and 4d and 5d elements can have large magnetocrystalline anisotropy via large spin–orbit coupling.[49,50]

One interesting approach is to look at marginally stable materials. Fe-Ni meteorites contain a magnetically hard FeNi compound having the L10 structure, called tetrataenite.[51] This compound has a relatively high magnetocrystalline anisotropy and a magnetization equal to that of $Nd_2Fe_{14}B$. Unfortunately, these phases do not form under normal laboratory conditions but rather require the extremely slow cooling rates which have occurred in meteorites. The challenge is to develop a kinetic pathway to form the hard magnetic FeNi L10 structure. The existence of this phase suggests that all the possible permutations of Fe-based systems have not been fully exploited. Thermodynamic calculations, more in-depth first-principles studies and more comprehensive screen tools, such as combinatorial analysis, will be required to more fully investigate this phase space. Advanced far-from-equilibrium processing will be needed to explore new phase space for compounds that possess anisotropy with reasonable magnetization by chemical substitutions and other processing methods to induce chemical ordering.

Mn-Based Compounds

There are two possible non-Fe-based compounds that have shown significant coercivity: MnBi and MnAl. While Mn-containing compounds are typically antiferromagnetic, the moments can be induced to align if the separation of the Mn atoms is greater than 2.96 Å.[52] However, the total moments in these compounds tend to be low as the large spacing translates to a small volume fraction of magnetic atoms which results in a low energy density. MnBi is unusual in that its coercivity actually increases with temperature. Its T_c is limited by the decomposition of the magnetic phase through a peritectic reaction at 355°C.[53] Reported $(BH)_{max}$ is in the range of ~6–7 MGOe.[54] The magnetic phase is hexagonal, having the NiAs structure.[55] The low melting temperature of Bi and the high vapor pressure of Mn have posed some challenges to processing, but the compound can be formed using rapid solidification.[56] However, this results in an isotropic structure greatly reducing the energy product. Like most permanent magnet compounds, maintaining a fine grain structure is crucial to retaining its coercivity.

MnAl: With a theoretical $(BH)_{max}$ of ~12 MGOe and with its low density of 5.2 g/cm³, MnAl has an attractive energy density per unit mass.[57–59] The metastable ferromagnetic phase is of the L10-type and termed the τ phase. It is difficult to synthesize due to its sluggish reactions at low temperatures where it is stable.[52,60] Mechanical milling has been shown to be a viable route to producing this metastable phase while retaining a fine grain structure. As with rapid solidification, this method results in an isotropic grain structure reducing the magnetic properties. While the addition of C increases the stability of the τ phase, it significantly decreases the anisotropy field (55–39 kOe) and T_c (380–285°C).[61] This lower T_c greatly reduces the potential for energy applications.

The challenge to both of these systems is to increase the operating temperature capability and magnetization. Increasing the operating temperature in MnBi is particularly problematic since it is the stability of the phase rather than the strength of the exchange interaction that limits high-temperature operation. Thus, a large temperature excursion would not only demagnetize the magnet but also result in the formation of a liquid phase within the machine. For most generators and traction motors, the operating temperature is limited to well below 355°C by the temperature stability of other components, in particular the insulating materials, so 355°C is not an engineering limit. However, ternary additions that could possibly increase M_s tend to reduce the peritectic temperature, limiting choices for alloy improvement.[62,63] On the plus side, the coercive field for MnBi is nearly five times that of Nd-based compounds at 127°C.[64]The temperature stability of the MnAl τ-phase is also of concern since prolonged annealing has been shown to decompose the binary phase.[65] While additions of C can stabilize the magnetic phase, the resulting degradation to the magnetic properties is unacceptable. Finding an alloying element that increases the stability of the τ-phase without degrading the magnetic properties will be the key making this a viable permanent magnet.

Nonstrategic Rare Earth Elements

The resource availability and demand for the various rare earth elements are not correlated; hence, the strategic concern is not uniform across the Lanthanide series. Only a few of the rare earth elements currently have high market demand, both in absolute terms and more importantly, relative to their natural abundances in commercially viable mines. The most critical supply risks in the next 5 years are Dy, Eu, Nd, Tb, and Y while Ce, La, and to some extent Pr are less so.[6] In terms of resources, Ce comprises nearly 40% of all rare earth elements in the Earth's crust,[7] and in light rare earth element deposits like Mountain Pass, the ratio can be as high as 50%. Conversely, Dy, which is critical to providing the increase in high temperature capability for Nd-based alloys, is less than 2% of all the rare earth elements and is found in abundance in only a few localities. While a wide variety of magnetic compounds can be synthesized from nearly every lanthanide element, few have the high energy density of the Nd-, Pr-, and Sm-based compounds. Heavy rare earths, for instance, are unsuitable for permanent magnets even though they have large moments since these moments couple antiparallel to the transition-metal sublattice. Furthermore, the rare earth anisotropy contribution (easy-axis or easy-plane) changes sign in the middle of each of the two rare earth half-series. This means, for example, that Sm yields a positive anisotropy contribution but not Nd.[12] Furthermore, the de Gennes factor causes the rare earth elements from the middle of the series to yield the largest Curie temperature. The T_c for $Gd_2Fe_{14}B$ is nearly 100°C higher than $Nd_2Fe_{14}B$ but its anisotropy is too low to make is it a useful magnet.[66] Substitutions, even at low levels, can have a profound effect on the magnetic properties.[66,67] While considerable efforts have gone into exploring a large number of possible minor alloying elements, the majority of the work has been in Nd- and Sm-based compounds. Much less research has been performed on compounds based on the more abundant Ce and La. Ce and La comprise up to about 85% of the ore in most light rare earth mines on an oxide basis. Alloys which utilize some of these more abundant rare earth elements would help to mitigate excessive demand on Nd and more importantly Dy, and at the same time increase market demand for the more abundant Ce and La providing a higher return on investment for the light rare earth element mines. However, La does not have a magnetic moment and, hence, functions only as a spacer in the crystal structure, and Ce is not well behaved due to its unusual electronic structure. Unfortunately, in intermetallic compounds with Fe, Co, or Ni, Ce has a strong tendency to give up its single 4f electron and become a 4+ or a mixed valence ion rather than the 3+ typical of most of the other rare earths. While 4+ Ce might be expected to behave like the nonmagnetic rare earth elements La, Lu, and Y with little effect on the T_c and only having magnetocrystalline anisotropy from the TM lattice, it appears that Ce is mixed valence with a resulting hybridization of the 4f electrons with the transition metal 3d electrons, which results in the lowering of T_c. If means can be found to compensate for the detrimental properties of Ce, then the availability of rare earth element magnets from existing sources could be greatly enhanced.

DISCUSSION

While a large number of permanent magnet compounds has been discovered over the last 50 years,[68] the number that have the necessary characteristics to be viable candidates for the demanding application of generators and traction motors is limited. The key factors that need to be considered when investigating new or modified compositions are T_c, K_1, and M_s. For generators and traction motors, the weight-to-power ratio is of prime importance. While Curie temperature in and of itself is not that important provided it is above the operating window for the electric machines, it places an absolute limit on the temperature dependence of B_r and H_{cJ}. Large negative temperature coefficients for these parameters are problematic. First, it requires that considerably more material be used than that required based on normal operating temperatures since the motor must be designed to maximum load conditions. Alternatively, additional cooling systems must be employed. Second, power control is more complicated since motor power output is not linear with temperature. Operating too close to the minimum load line for the motor can also result in large irreversible losses, reducing motor lifetime. The magnetocrystalline anisotropy also provides a key metric to the ultimate H_{cJ} that can be achieved in the compound while M_s is a metric for the B_r. How closely a magnet based on a given compound can achieve these theoretical limits is dependent on processing. Here, it is well established that the grain size must approach the single domain particle size limit where the formation of a domain wall inside the grain is unfavorable. Since the total magnetization of a magnet is the vector sum of the individual grains along the direction of the bulk magnetization, the degree of grain alignment has a significant role to play. In fact, the maximum energy product of an aligned magnet is 2.5 times that of an isotropic magnet. In addition, irregularities at grain boundaries can also provide regions of high flux concentration and reduce the area of the domain wall required to nucleate a reverse domain. This can result in easy nucleation of reversal of magnetic domains. Therefore, being able to reproducibly process new materials to such exacting parameters sets a high bar to introduce new materials into the market.

In the $Nd_2Fe_{14}B$ magnet, there are two functional parts of the structure: the Fe layers provide the magnetization and T_c while the Nd layers provide the magnetocrystalline anisotropy. At room temperature, there is more anisotropy than is actually required for most applications, but the compound lacks sufficient moments. Additional moments can be provided by nanostructuring of a high-magnetization soft phase and a high-anisotropy hard phase.[69] Here, the mixture of a hard and soft magnetic phase at the appropriate length scales, typically on the order of a few nm, gives rise to uniform magnetic response as a single domain extends over both hard and soft phases providing a high magnetization while their small dimension limits easy reversal of the soft phase since their moments are pinned to the hard phase. While modeling has suggested that energy products of such nanostructured alloys could exceed 100 MGOe in an anisotropic magnet, nothing close to this has been achieved in practice.[70] The primary difficulty is that processing methods which provide good texturing tend to require high temperatures, which result in coarser grained alloys than required for effective exchange coupling. While this is being actively pursued in the rare earth element-based compounds,[39]this approach is just as valid for other ferromagnetic compounds and is particularly attractive for Mn-based alloys, which have high H_{cJ} but poor B_r.

Compounds based primarily on Fe have the opposite problem. Here, the challenge will be to induce H_{cJ} without sacrificing B_r. For most of the Fe-based magnets, H_{cJ} is dependent on shape rather than magnetocrystalline anisotropy. For those compounds utilizing shape anisotropy, such as alnico, the key factor is how to improve the shape anisotropy. Theoretical studies suggest that increasing the aspect ratio will increase H_{cJ}.[71] Recent calculations also suggest that the sharpness of the interface between the magnetic and nonmagnetic phase can also have a measurable effect. Another promising approach discussed above is to induce magnetocrystalline anisotropy. The most promising approach is to judiciously introduce either interstitials or substitutional atoms that result in degeneracy of the normal bcc Fe structure, as seen in the Fe-N compounds. These substitutions need to be thermodynamically stable and easily produced on industrial scales. However, while nanostructuring is an excellent pathway to produce novel metastable compounds, engineering materials must be near theoretical density to be economical.

For a mass market, the materials of choice for the alloy development must be plentiful, secure, and relatively inexpensive. The electric vehicle market is expected to grow to 20 million vehicles per year, or approximately 20% of the global vehicle market by 2020. Current technology uses 12.3–13.3 g of rare earth per kW for motors with peak power 60 and 110 kW.[72] The increased demand for rare earth elements can be estimated to be between 6400 and 11,200 metric tons of which ~25 wt.% is Dy. Current direct drive wind generators utilize approximately 600 kg of Nd-magnets per MW. In 2010, China installed 18,928 MW of wind power generation or approximately 50% of all wind power installed globally that year. Of that total, ~25% was permanent magnet generators using 2840 metric tons of Nd-magnets. The rest of the world's wind generators up to this point have been primarily induction machines, and the small number of permanent magnet generators used only 751 metric tons of Nd magnets. If the rest of the world were to install permanent magnet generators to meet greenhouse gas reduction, the demand for Nd magnets would increase by nearly 50% by 2020 from 2010 levels even while assuming about a 30% increase in the efficiency of the newer generators and that only 75% newly installed capacity would use permanent magnet generators. However, these numbers belie the real problem, which is an adequate supply Dy (Table II). Of particular importance is the increased demand for Dy relative to its natural abundance. While demand for the Nd is growing rapidly under the most aggressive scenario for permanent magnet-based electric machines, increased production will generally keep pace (Table I). The most significant bottleneck in supplying higher demand for permanent magnet generators and traction motors is the upper limit temperature performance and the associated cost and availability of the Dy. It is here where the technological breakthroughs are going to be most important.

Table II. Projection of Dy supply and demand, in tons, and looming shortage based on projected increased demand for Nd-based permanent magnets for traction motors and wind generators

	2005	2010	2011	2012	2013	2014	2015	2020
Oxide								
Demand	1015	1597	1836	2020	2181	2356	2544	3568
Supply	1020	1600	1610	1625	1650	1675	1700	
Shortage			–226	–395	–531	–681	–844	
Metal								
Demand	810	1280	1470	1620	1750	1880	2040	2850
Supply	816	1280	1290	1300	1320	1340	1360	
Shortage			–180	–320	–430	–540	–680	

Assume 80% of the oxide is made into metal.

While the Nd-based magnets possess the highest energy product near room temperature and down to 140 K, at higher temperatures, especially the upper operating temperature range for hybrid vehicle motors and wind turbines, a number of other alloys are close contenders (Fig. 5). For instance, one of the advantages of alnico alloys is that they have a very flat energy density over a broad range of temperatures. While alnico's Co content is high and the availability of Co can also be a concern, the resources for Co are better distributed than rare earth elements. A modest gain in the energy density can make this alloy a viable alternative. Prospects for other Fe plus transition metal systems are longer term. While theoretical models have shown promise for this general class of alloys, it is not clear yet whether they can be synthesized and if so, would they have the thermodynamic stability to be used in a permanent magnet electric machines. The potential payback for an inexpensive high energy density magnet made from relatively abundant materials would be revolutionary.

Mn-based alloys also have promising high-temperature magnetic properties, in particular the MnBi class of compounds. These compounds pose a different set of materials optimization problems since they lack significant magnetization. While theoretical pathways have been identified with which to dramatically enhance magnetization in such compounds, it has yet to be demonstrated in a bulk compound.

CONCLUSIONS

The push to reduce greenhouse gasses will require a restructuring of both energy production and utilization. Electric machines for both power generation and traction motors can figure prominently in this new economy. While Nd-based permanent magnet electric machines provide obvious power-to-weight advantage over induction machines, the availability and price of the rare earth metals are making permanent magnet-based electric machines less favorable. Of particular concern is the cost and availability of Dy, which is a key rare element with limited availability that is required to improve the high-temperature performance of Nd-based alloys for use in generators and traction motors. Alternatives to rare earth-based alloys do exist, but at the current time, they lack the necessary energy density to replace Nd-based magnets. Some of these potential replacements actually exceed performance of the Nd-based alloys above 200°C. Many of these compounds have been known for decades, but serious interest in their optimization waned once the rare earth element-based compounds were discovered. Fortunately, researchers have new tools with which to characterize and understand the mechanisms for coercivity and magnetization in these compounds. Computational tools can provide detailed descriptions of the relationship between atomic composition, order, and interaction of the magnetic spins. These can even be used to discover potentially new compounds. Experimental methods for exploring wide areas of phase space are also being utilized to identify new magnetic compounds. Significant research efforts are underway to exploit these possibilities. Like the early 1980s, a shortage of critical materials today can be a boon for materials discovery.

ACKNOWLEDGEMENTS

We would like to thank Ralph Skomski, David Sellmyer, and Jeff Shield for their helpful suggestions. This work was supported by the U.S. Department of Energy, Office of Energy Efficiency and Renewable Energy (EERE), under its Vehicle Technologies Program, through the Ames Laboratory. Ames Laboratory is operated by Iowa State University under contract DE-AC02-07CH11358.

FOOTNOTES

[1]Electric machines, a classification that encompasses both motors and generators, are ubiquitous in the modern industrial economy spanning power output from a few milliwatts to megawatts.

[2]While PtCo was the first compound discovered with high-magnetocrystalline anisotropy, its cost rendered it impractical for commercial use in bulk.

[3]Projected market demand is a compilation of industry and government sources.

REFERENCES

1. Center for Transportation Analysis, U.S. Department of Energy, Transportation Energy Data Book, 30th ed. (Washington, DC: Center for Transportation Analysis, U.S. Department of Energy, 2011), http://cta.ornl.gov/data/tedb30/Edition30_Full_Doc.pdf.
2. K.R. Weeber, M.R. Shah, K. Sivasubramaniam, A. El-Refaie, Q. Ronghai, C. Stephens, and S. Galioto, IEEE Power and Energy Society General Meeting, 2010, pp. 1–6.
3. A.M. Bradshaw and T. Hamacher, Chemsuschem 5, 550 (2012).
4. E. Peeters, J. Van Bael, and P. Van Tichelen, Proc. 2nd IASME/WSEAS International Conference on Energy & Environment (www.worldses.org: WSEAS, 2007), pp. 154–158.
5. Critical Materials Strategy, U.S. DOE, 2010,http://energy.gov/sites/prod/files/edg/news/documents/criticalmaterialsstrategy.pdf.
6. Critical Materials Strategy, U.S. DOE, 2011,http://energy.gov/sites/prod/files/DOE_CMS2011_FINAL_Full.pdf.
7. U.S. Geological Survey, Rare Earth Elements—Critical Resources for High Technology, USGS Fact Sheet 087-02, 2002.
8. M. Humphries, Rare Earth Elements: The Global Supply Chain (Washington, DC: C.R. Service, 2011).
9. R. Moss, E. Tzimas, H. Kara, P. Willis, and J. Kooroshy, Critical Metals in Strategic Energy Technologies—Assessing Rare Metals as Supply-Chain Bottlenecks in Low-Carbon Energy Technologies (Brussels, Belgium: Joint Research Center, Institute for Transuranium Elements, 2011),http://publications.jrc.ec.europa.eu/repository/handle/111111111/22726.
10. H.P. Schlemmer, Radiologe 45, 356 (2005).
11. A.J.J. Koch, M.G. Van der Steeg, and K.J. de Vos, AIEE Spec. Publ. T-91, 173 (1957).
12. R. Skomski and J.M.D. Coey, Permanent Magnetism (Bristol, U.K: Institute of Physics, 1999).
13. W. Ostertag and K.J. Strnat, Acta Crystallogr. 21, 560 (1966).
14. K. Strnat, G. Hoffer, W. Ostertag, and J.C. Olson, J. Appl. Phys. 37, 1252 (1966).
15. M.G. Benz and D.L. Martin, Appl. Phys. Lett. 17, 176 (1970).
16. K.J. Strnat, Proc. IEEE 78, 923 (1990).
17. N.C. Koon, C.M. Williams, and B.N. Das, J. Appl. Phys. 52, 2535 (1981).

18. J.J. Croat, B Am. Phys. Soc. 26, 239 (1981).
19. M. Sagawa, W. Yamagishi, and Z. Henmi, J. Appl. Phys. 52, 2520 (1981).
20. P.C. Dent, Adv. Mater. Process. 16, 277 (2009).
21. D. Normile, Science 330, 1598 (2010).
22. T. Orlik and C.-W. Yap, "China's Rare Earth Recoil," Wall Street Journal (14 March 2012).http://online.wsj.com/article/SB10001424052702304450004577278981461660606.html.
23. S. Constantinides (Paper presented at DOE Vehicle Technologies Program APEEM R&D FY12 Kickoff Meeting, November 2–4, 2011, Oak Ridge National Laboratory, Oak Ridge, TN, 2011).
24. M. Katter, L. Zapf, R. Blank, W. Fernengel, and W. Rodewald, IEEE Trans. Magn. 37, 2474 (2001).
25. H. Bala, N.M. Trepak, S. Szymura, A.A. Lukin, V.A. Gaudyn, L.A. Isaicheva, G. Pawlowska, and L.A. Ilina, Intermetallics 9, 515 (2001).
26. R.B. Campbell and C.A. Julien, J. Appl. Phys. 32, S192 (1961).
27. C.A. Julien and F.G. Jones, J. Appl. Phys. 36, 1173 (1965).
28. J.W. Cahn, Acta Metall. Mater. 9, 795 (1961).
29. J.W. Cahn, J. Appl. Phys. 34, 3581 (1963).
30. V.A.M. Brabers and K.H. Buschow, ed., Handbook of Magnetic Materials, vol. 8 (Amsterdam, the Netherlands: North-Holland, 1995), p. 189.
31. J.M.D. Coey and P.A.I. Smith, J. Magn. Magn. Mater. 200, 405 (1999).
32. N. Ji, M.S. Osofsky, V. Lauter, L.F. Allard, X. Li, K.L. Jensen, H. Ambaye, E. Lara-Curzio, and J.P. Wang, Appl. Phys. Lett. 98, 092506 (2011). doi:10.1063/1.3560051.
33. N. Ji, L. Allard, E. Lara Curzio, and J.-P. Wang, Appl. Phys. Lett. 98, 092506 (2011).
34. K.H. Jack, Proc. R. Soc. Lond. Ser.-A 208, 200 (1951).
35. K.H. Jack, Proc. R. Soc. Lond. Ser.-A 208, 216 (1951).
36. B.J. Kooi, M.A.J. Somers, and E.J. Mittemeijer, Metall. Mater. Trans. A 27, 1063 (1996).
37. H. Takahashi, H. Shoji, and M. Takahashi, J. Magn. Magn. Mater. 174, 57 (1997).
38. M.Q. Huang, W.E. Wallace, S. Simizu, and S.G. Sankar, J. Magn. Magn. Mater. 135, 226 (1994).
39. G.C. Hadjipanayis, J.F. Liu, A. Gabay, and M. Marinescu, J. Iron. Steel Res. Int. 13, 12 (2006).
40. K.H. Jack, Nitrides Oxynit. 325–3, 91 (2000).
41. J. Hafner, C. Wolverton, and G. Ceder, MRS Bull. 31, 659 (2006).
42. M.C. Nguyen, X. Zhao, M. Ji, C.-Z. Wang, B. Harmon, and K.-M. Ho, J. Appl. Phys. 111, 07E338 (2012).
43. R. Skomski, A. Kashyap, and A. Enders, J. Appl. Phys. 109, 07E143 (2011). doi:10.1063/1.3562445.
44. K.D. Belashchenko, V.P. Antropov, and N.E. Zein, Phys. Rev. B 73, 073105 (2006). doi:10.1103/Physrevb.73.073105.
45. V.P. Antropov, M. van Schilfgaarde, S. Brink, and J.L. Xu, J. Appl. Phys. 99, 08F507 (2006).
46. N.I. Medvedeva, D. Van Aken, and J.E. Medvedeva, J. Phys.-Condens. Mat. 22, 316002 (2010). doi:10.1088/0953-8984/22/31/316002.
47. G. Rahman, I.G. Kim, H.K.D.H. Bhadeshia, and A.J. Freeman, Phys. Rev. B 81, 184423 (2010). doi:10.1103/Physrevb.81.184423.
48. M. Kogachi, N. Tadachi, H. Kohata, and H. Ishibashi, Intermetallics 13, 535 (2005).
49. Kashyap, P. Manchanda, P.K. Sahota, R. Skomski, J.E. Shield, and D.J. Sellmyer, IEEE Trans. Magn. 47, 3336 (2011).
50. V. Sharma, P. Manchanda, R. Skomski, D.J. Sellmyer, and A. Kashyap, J. Appl. Phys. 109, 07A727 (2011). doi:10.1063/1.3562256.
51. E.R.D. Scott and R.S. Clarke, Nature 287, 255 (1980).
52. T. Ohtani, N. Kato, S. Kojima, K. Kojima, Y. Sakamoto, I. Konno, M. Tsukahara, and T. Kubo, IEEE Trans. Magn. 13, 1328 (1977).
53. H. Okamoto, Binary Alloy Phase Diagrams, ed. T.B. Massalski (Materials Park, OH: ASM, 1990).
54. D.T. Zhang, S. Cao, M. Yue, W.Q. Liu, J.X. Zhang, and Y. Qiang, J. Appl. Phys. 109, 07A722 (2011). doi:10.1063/1.3561784.
55. J.B. Yang, K. Kamaraju, W.B. Yelon, W.J. James, Q. Cai, and A. Bollero, Appl. Phys. Lett.79, 1846 (2001).
56. X. Guo, X. Chen, Z. Altounian, and J.O. Stromolsen, Phys. Rev. B 46, 14578 (1992).
57. R.B. Campbell and C.A. Julien, J. Appl. Phys. 32, S346 (1961).
58. Z.A. Abdelnour, H.F. Mildrum, and K.J. Strnat, IEEE Trans. Magn. 17, 2651 (1981).
59. Q. Zeng, I. Baker, J.B. Cui, and Z.C. Yan, J. Magn. Magn. Mater. 308, 214 (2007).
60. C. Muller, H.H. Stadelmaier, B. Reinsch, and G. Petzow, Z. Metallkd. 88, 620 (1997).
61. L. Pareti, F. Bolzoni, F. Leccabue, and A.E. Ermakov, J. Appl. Phys. 59, 3824 (1986).
62. X.Z. Li, P. Kharel, V.R. Shah, and D.J. Sellmyer, Philos. Mag. 91, 3406 (2011).
63. P. Kharel, R. Skomski, and D.J. Sellmyer, J. Appl. Phys. 109, 07B709 (2011). doi:10.1063/1.3537952.
64. J.B. Yang, W.B. Yelon, W.J. James, Q. Cai, M. Kornecki, S. Roy, N. Ali, and P. l'Heritier, J. Phys.-Condens. Matter 14, 6509 (2002).
65. Y. Du, J. Wang, J.R. Zhao, J.C. Schuster, F. Weitzer, R. Schmid-Fetzer, M. Ohno, H.H. Xu, Z.K. Liu, S.L. Shang, and W.W. Zhang, Int. J. Mater. Res. 98, 855 (2007).
66. J.F. Herbst, Rev. Mod. Phys. 63, 819 (1991).
67. C.D. Fuerst and J.F. Herbst, J. Appl. Phys. 69, 7727 (1991).
68. R.J. Parker, Advances in Permanent Magnetism (New York: Wiley-Interscience, 1990).
69. R. Skomski and J.M.D. Coey, Phys. Rev. B 48, 15812 (1993).
70. B. Balamurugan, D.J. Sellmyer, G.C. Hadjipanayisc, and

R. Skomskia, Scripta Mater. 67 (2012). doi:10.1016/j.scriptamat.2012.03.034.

71. R. Skomski, J. Phys.-Condens. Matter 15, R841 (2003).

72. Oak Ridge National Laboratory, Evaluation of the 2010 Toyota Prius Hybrid Synergy Drive System, ORNL/TM-2010/253 (Oak Ridge, TN: Oak Ridge National Laboratory, 2011), http://info.ornl.gov/sites/publications/files/Pub26762.pdf

CHAPTER 2 CORRELATION BETWEEN AC CORE LOSS AND SURFACE MAGNETIC BARKHAUSEN NOISE IN ELECTRIC MOTOR STEEL

Arash A. Samimi, Thomas W. Krause, Lynann Clapham, Matthew Gallaugher, Yaoyao Ding, Pampa Ghosh, Richard Chromik, and Andrew M. Knight

Keywords: Barkhausen noise, Flux control, Magnetic non-destructive testing, Core loss, Electrical steel

INTRODUCTION

This work reports the results from a collaborative study on the correlation between core loss and Barkhausen noise in non-oriented electrical steel (NOES) laminates, with applications to electric motors. The motivation of this work is the development of magnetic Barkhausen noise (MBN) as a non-destructive means of identifying factors that may affect core loss in NOES, which is of interest to an industrial partner. Furthermore, this work contributes to a better understanding of magnetic loss mechanisms, which can lead to improvement in the production of energy efficient laminates.

NOES or non-oriented Silicon–Iron steel is the most widely used soft ferromagnetic material, with applications in rotary machines [1]. Among the significant properties of NOES laminates are high permeability, low magnetic loss and low magnetostriction, which is achieved by an optimum combination of various factors including texture, grain size, composition and thickness [2, 3]. The demand for energy efficiency in electric motors and rotary machines has driven the growth of NOES in the world market, with the characterization of magnetic loss as the main target in NOES research [4]. Nevertheless, loss measurement and its evaluation is limited by: (1) Deficiencies in the traditional loss measurement techniques, i.e. the Epstein square or a single sheet tester. These are destructive, off-line and typically provide only an average of the magnetic properties in the rolling or transverse direction [5–8]. Therefore, development of a nondestructive evaluation technology could provide benefits for condition monitoring at various stages of manufacture. (2) The understanding of the Epstein technique is also limited by the absence of a 3-D model, which properly incorporates domain wall dynamics into total loss calculations [1]. Furthermore, a standard approach is to express total loss as the sum of hysteresis, bulk eddy current and an excess term, which compensates for the observed total loss [1, 9]. However, this approach ignores the influence of domain wall dynamics on magnetization processes and loss mechanisms [1, 10, 11]. Information about domain wall processes can be obtained from MBN emissions [12], which may be incorporated into a physical model to improve loss calculations. Therefore, MBN may provide the "missing link" in magnetic loss characterization.

MBN reflects clustering of magnetization discontinuities, which are temporally and spatially correlated jumps of domain walls [12–15]. The Barkhausen noise spectrum is affected by material microstructure [16, 17], texture [18], magnetic easy axis direction in which domain magnetization vectors tend to align [19], and stress state of the material [20], which in turn influence the magnetic loss. Therefore, it is expected that Barkhausen emissions contain information about the magnetic loss within ferromagnetic laminates [1]. The generation of Barkhausen noise is influenced by domain nucleation and annihilation processes, which typically occur at grain boundaries [11, 14]. The average time duration of a clustered pulse of Barkhausen emission is directly proportional to grain size [14], which also correlates with the magnetic loss. Domain wall motion is accompanied by localized eddy currents (micro-eddy currents) [10]. The resulting inhomogeneous distribution in permeability due to micro-eddy currents is related to higher losses in silicon iron steels [21]. The correlation between

Citation: Arash A. Samimi, Thomas W. Krause, Lynann Clapham, Matthew Gallaugher, Yaoyao Ding, Pampa Ghosh, Richard Chromik, and Andrew M. Knight. Correlation between AC core loss and surface magnetic barkhausen noise in electric motor steel, Journal of Nondestructive Evaluation, December 2014; Volume 33, Issue 4, pp 663–669. DOI:10.1007/s10921-014-0260-x.

MBN and core loss in grain oriented electrical steel laminates has been investigated in a limited number of earlier papers where the MBN technique is applied under field or current control [22–24]. The present work uses a flux-controlled MBN technique to further investigate the applicability of MBN for non-destructive monitoring of core loss in NOES. Flux control has been achieved by using pole feedback control of magnetic flux waveforms, to improve MBN reproducibility. An in-house designed and built flux controller combines an analog circuit with a digital algorithm for rapid surface measurements with minimum error in the control loop [25]. Control of flux reduces the likelihood of potential artifact effects such as those introduced by air gaps, and increases the reproducibility of measurement results [25–27].

MATERIALS AND METHODS

Samples

The samples were eight semi-processed and fully-processed NOES laminates having standard widths of 30 mm. An insulating coating with a measured thickness of 1 μm was uniformly distributed over the laminate surface.

Magnetic Measurements

Bulk magnetic characteristics and core loss in the samples were measured along the rolling direction (RD) under ac excitation at 1 T flux density at 30 Hz using a single sheet tester. These measurements were performed at University of Alberta. The samples had associated core loss values (W) as listed in Table 1, along with sample thickness (T), grain diameter (D) and measured B–H parameters.

Table 1. Measured parameters for non-oriented electrical steel laminates with increasing ac core loss (W) at 1 T at 30 Hz

Sample	D	T	W	μ_r	H_c	B_R
#	**(μm)**	**(mm)**	**(W/kg)**		**(A/m)**	**(T)**
1	198	0.35	0.3223	13141.42	23.78	0.7468
2	177	0.35	0.3178	15604.88	24.03	0.7885
3	186	0.35	0.3151	15233.12	23.47	0.7552
4	167	0.35	0.3083	16410.95	23.05	0.7857
5	129	0.30	0.3031	17703.06	22.10	0.8178
6	130	0.30	0.2976	17081.01	21.99	0.8049
7	128	0.25	0.2871	19654.62	20.92	0.8389
8	123	0.25	0.2861	19293.19	20.99	0.8364

D (grain diameter), T (thickness), μ_r (permeability), H_c (coercivity), B_R (remanence)

Local magnetic measurements were performed using a flux-controlled Barkhausen noise system at Queen's University. The MBN apparatus consisted of a surface dual-core probe (tetrapole) [25, 28] in combination with an in-house designed magnetic flux controller, which controlled the flux waveform to obtain consistency and repeatability in local MBN measurements [25, 26]. Figure 1 is a schematic diagram of one of the flux-controlled cores of the tetrapole probe. The poles of the U-shaped Supermendur core had cross sections of 4.2 mm × 9.5 mm and were 21.2 mm apart. Each pole was wrapped with a 500 turn excitation coil and a 50 turn feedback coil, which monitors the flux at the closest distance to the sample surface. As shown in the schematic in Fig. 1, a reference voltage, V_{ref} is generated by a digital-to-analog converter and fed into an analog feedback amplifier. The excitation coil voltage, V_{ex}, and the measured feedback coil voltage, V_F, are sampled by an analog-to-digital converter and processed through a personal computer. V_{ref} is adjusted in each iteration and the control loop repeats until V_F converges to the target feedback coil voltage. The convergence time (time to measure) is less than 10 s for frequencies between 10 to 100 Hz, which are typically used in NDT measurements [25]. The tetrapole probe uses flux superposition to achieve a net magnetization in the plane of the laminate in different orientations. This permitted angular Barkhausen measurements on two of the samples characterized by the highest and lowest core loss values, respectively.

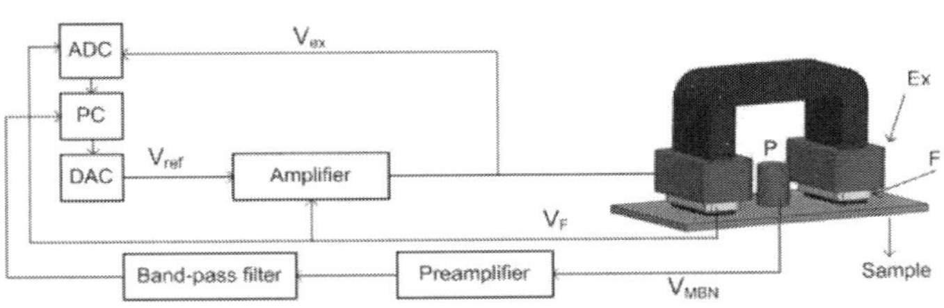

Figure 1. Schematic diagram of the flux-controlled MBN system showing the feedback loop and the probe with the excitation (Ex), feedback (F) and peak-up (P) coil.

A spring-loaded solenoidal pick-up assembly, shown in green in the center of the pole pieces in Fig. 1, used a ferrite core, which is wrapped with 100 turns of AWG 44 copper coil, surrounded by a permeable sheath and assembled into a conductive brass shield to minimize the sensing radius to about 2.4 mm at 30 Hz [28]. Between the electromagnet poles, immediately beneath the small pick-up coil, finite element modeling has shown that the flux is uniform [28]. The high spatial resolution pick-up coil couples to the vertical magnetization components projected out of the surface of the sample, which in turn induce a voltage in the pick-up coil. The high frequency Barkhausen emissions (>1 kHz) were fed to a preamplifier and the low frequency excitation signal was filtered out. The rms envelope of the Barkhausen signal is produced and the total Barkhausen noise energy (BN_E) is then calculated as:

$$BN_E = \int (BN_{env})^2 d\phi, \quad (1)$$

in which ϕ is the feedback signal phase corresponding to time within the cycle. Similar to the single sheet B–H measurements, the flux-controlled MBN measurements were also taken at 30 Hz excitation frequency. This was the lowest frequency with which the current system could still maintain flux control. The peak flux density used for MBN measurements was 100 mT measured at the excitation magnet poles, which results in approximately four to five times higher flux density level within the laminates, taking into account the effect of sample thickness.

Grain Size and Texture Measurements

Preparation for texture and grain size measurements involved coating removal, polishing with 1 μm oil-based diamond suspension and 3 % Nital etching to reveal the grain structure. Images were captured at 50× magnification using a Nikon Epiphot 200 optical microscope equipped with Clemex Image Analysis software. The grain size measurements were performed according to ASTM E112 [29].

Texture measurements were performed at McGill University using X-ray diffraction on the top surface (rolling plane) of the samples. A Bruker X-Ray Diffraction Goniometer operating at 35 kV and 45 mA was used to capture diffraction data over a 6 mm × 6 mm scanned area. Three incomplete pole figures [(110), (200), and (211)] were collected, and used to calculate the orientation distribution function (ODF) using Tex-Tools software. To quantitatively correlate magnetic properties with texture, the ODF was then used to calculate the magneto-crystalline anisotropy energy (E_a) [30], i.e. the energy required to magnetize the sample along the RD. In addition to the texture information, the calculation for E_a incorporated the aluminum and silicon content. Table 2 includes the Al and Si contents, which were obtained from the ICP-AES analysis and combustion techniques, as well as the calculated value of E_a obtained using the first order anisotropy constant K1 [30]:

$$K1 = 4.77 - (0.21256)(\text{wt\%Si}) - (0.03816)(\text{wt\%Al}). \quad (2)$$

Table 2. Chemical composition in wt%, and calculated magneto-crystalline energy (E_a) along RD

Sample #	Si	Al	E_a (J/m^3)
1	3.27	0.54	0.411
2	2.81	0.26	0.402
3	3.26	0.53	0.405
4	2.78	0.25	0.402
5	3.51	0.62	0.361
6	3.19	0.58	0.344
7	3.61	0.57	0.301
8	3.24	0.60	0.293

The chemistry was found using ICP-AES and combustion techniques

RESULTS

Correlations with Core Loss

Figure 2 indicates an inverse relationship between core loss and MBN energy measured along the RD. This correlation implies that the optimum magnetic performance is achieved when higher MBN emissions are measured, which is qualitatively in agreement with earlier studies [22, 23] on grain oriented Si–Fe laminates. The most recent paper [24] also observed a strong correlation between MBN and core loss in both oriented and non-oriented Si–Fe laminates, however in that case the correlation was a direct, rather than an inverse one. The difference in behaviour from the present study is likely due to a smaller grain size relative to the previous samples [24] and a different level of excitation field. Figure 2 also demonstrates the relatively high sensitivity of the MBN technique to only a small range of variation in core loss.

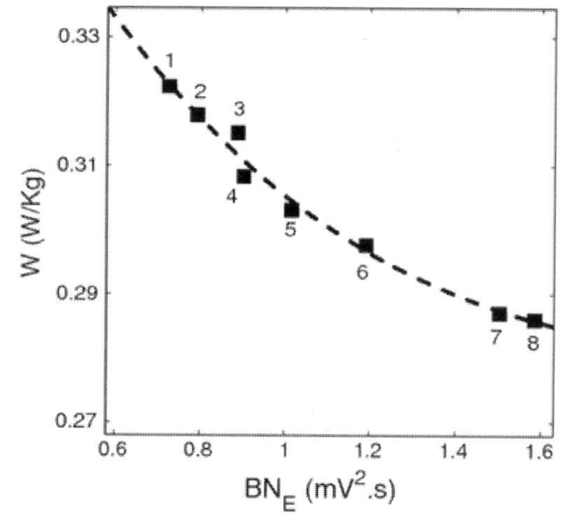

Figure 2. The core loss as a function of Barkhausen energy (BNEE) along RD. The *dotted line* is the best polynomial fit to the data.

Correlations with Grain Size and Texture

Figure 3 suggests a linear relationship between core loss and the inverse square root of grain diameter. This result agrees with earlier observations for laminates with grains larger than the optimum size at which the lowest loss and highest permeability is achieved [2, 3]. The observed inverse relationship between Barkhausen energy and grain size, agrees with the Hall–Petch relationship as given by [31]:

$$BN_E = C_g.d_g^{-\frac{1}{2}}, \qquad (3)$$

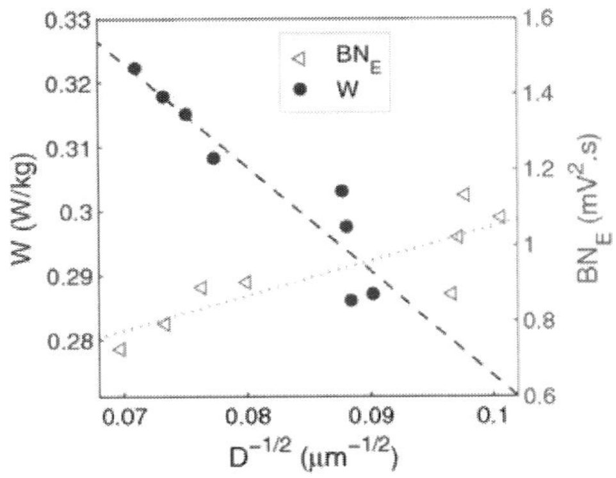

Figure 3. Barkhausen energy (BN_E) and core loss (W) as a function of the inverse square root of the grain diameter (D).

in which C_g is a constant and d_g is the grain size. The correlation R-squared factor associated with the observed Hall–Petch relationship is 0.71, which is statistically significant.

In fine grained steels, each grain surrounds a region of highly correlated 180° domain walls, known as magnetic objects [11]. The primary energy barrier, which defines the region of a magnetic object is, therefore, the grain boundary, at which nucleation and pinning of domain walls occur. As grain size increases, the grain boundary density per unit volume decreases [17] and Barkhausen energy is reduced according to a Hall–Petch relation for ferrite grains. As mentioned in Sect. 2.2, the thickness variations can result in approximately 25 % increase in sample flux density from the largest to the smallest laminate thickness, which in turn modifies the MBN generation. Therefore, in Fig. 3, Barkhausen energy was normalized to thickness, as performed elsewhere [23], to compensate for the thickness variations, and hence to more clearly examine the relationship of grain size with BN_E. To normalize MBN energy values, a dimensionless factor was introduced by normalizing the laminate thicknesses with respect to the largest thickness of 0.35 mm. The MBN energy values were then multiplied by this factor. The remaining scatter in the data may be due to the very small window of variations in other microstructural features of the laminates, which affect their magnetic properties.

Figure 4 plots linear correlations between core loss (W) and Barkhausen energy (BN_E) versus the magneto-crystalline energy (E_a) determined along the RD. In this plot E_a and BN_E were both normalized to compensate for thickness variations. Lower E_a is associated with more magnetic easy axes aligned close to the RD, i.e. the energetically favorable direction of magnetization, and therefore, a higher MBN energy [32]. Furthermore, less energy is required to magnetize the samples with lower E_a, which in turn reduces the core loss [33, 34] as observed in Fig. 4.

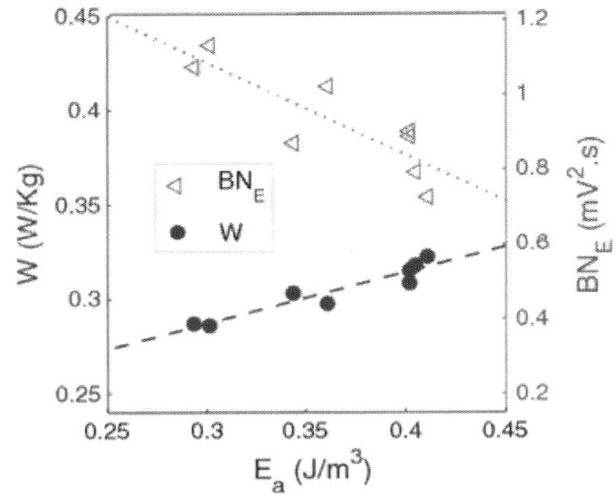

Figure 4. Barkhausen energy (BN_E) and core loss (W) as a function of magneto-crystalline energy (E_a) along RD.

Correlations with B–H Characteristics

The variations in grain size and texture of the Si–Fe laminates modify the configuration of domain walls within the samples, which in turn influence the magnetic characteristics of the B–H loop, the core loss and the generation of Barkhausen noise. Figure 5 shows the interplay among these parameters. The variation of core loss with these three parameters, as shown in Fig. 5, is qualitatively in agreement with earlier observations for laminates with grains larger than the optimum size [2, 3].

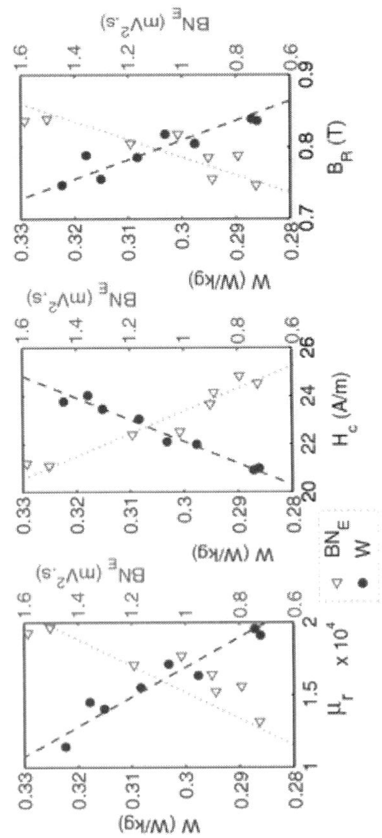

Figure 5. Correlations between Barkhausen energy (BN_E) and core loss (W) with hysteresis parameters, i.e. permeability (μ), coercivity (H_c) and remanence (B_R).

DISCUSSION

This work investigated the interplay and correlations among MBN emissions, core loss, grain size, texture and B–H characteristics of the NOES laminates. The significant aspect of this work is the inverse correlation between Barkhausen energy and core loss, as shown in Fig. 2. In the following discussion, this correlation is interpreted in terms of the localized nonuniform magnetization changes within the ferromagnetic laminates as a result of abrupt domain wall motion.

The observation of correlations between Barkhausen noise energy and core loss at 30 Hz suggests that the same parameters are present in both effects. These parameters are most likely associated with the influence of anisotropy energy on domain configuration and by extension the anisotropic Barkhausen noise energy as observed in Fig. 4. Therefore, the precise frequency is not a factor, but the particular domain interactions are. Reference [22], for example, correlates MBN and core loss at totally different magnetization frequencies of 12 and 60 Hz, respectively. Furthermore, at higher frequencies, macroscopic eddy current effects make a progressively larger contribution, substantially complicating the magnetization in the sample.

The discontinuous movement of domain walls is accompanied by local micro-eddy currents, which are known to be a source of magnetic loss in ferromagnetic materials along with macroscopic eddy currents and hysteresis loss [10]. Micro-eddy currents circulate around the moving domain wall to oppose the local changes in magnetization, and thereby damp the velocity of domain walls and generate resistive heating [10, 21]. Localized microscopic eddy currents add vectorily to bulk eddy current components to contribute to the total power loss (P) over the unit volume of the laminate sample given as [18, 22, 35]:

$$P = \rho \int (\overline{I}_{ma} + \overline{I}_{mi})^2 dv, \tag{4}$$

in which ρ is the resistivity of the material, $\overline{I}_{mi}$ is the micro-eddy current vector associated with domain wall movements, and $\overline{I}_{ma}$ is the homogeneous macroscopic eddy current component. Equation (4) indicates that the power loss is smaller for samples in which a larger amount of their micro-eddy current vector ($\overline{I}_{mi}$) is projected out of the surface of the laminate, with a consequently larger voltage induced into the surface Barkhausen pick-up coil [22]. Therefore, samples with lower loss exhibit larger Barkhausen energy as shown in Fig. 2.

An array of simultaneously moving (jumping) domain walls interacts with each other, and thereby cancels out their adjacent micro-eddy currents [10], which further reduces the total power loss [22]. It is therefore expected that a higher degree of micro-eddy current cancellation can further reduce the magnetic loss. To investigate this, Ref. [22] introduced a phenomenological parameter, α′, which indicates the strength of the eddy current field interactions between simultaneously moving domain walls that contribute to the magnetic easy axis of the laminate. α′ is found by fitting an energy expression (E) to the angular Barkhausen measurements and is given by [22]:

$$E = \alpha cos^2(\theta - \phi) + \alpha' cos(\theta - \phi) + \beta, \tag{5}$$

in which θ is the angle of the field with respect to the sample magnetic easy axis, ϕ is the angle of the magnetic easy axis with respect to the RD. β is the isotropic background contribution to Barkhausen energy and is associated with the domain walls oriented randomly with respect to the magnetic easy axis. α and α′ are fitting parameters associated with angular variations of MBN signal. The parameter of interest is α′, which is directly proportional to the local eddy current field interaction (H_{E_i}) given by the sum over the mth neighboring domain wall as [36]:

$$H_{E_i} = C_i \sum_m v_m e^{-X_{im}\pi/d}, \tag{6}$$

where C_i is a constant, v_m is the wall velocity, X_{im} is the distance between adjacent domain walls and d is the laminate thickness. Equation (6) indicates that the micro-eddy current interaction is proportional to wall velocity (v), and exponentially decreases with the distance between the two adjacent walls. An increase in α′ with decreasing power loss was observed in grain-oriented silicon laminates [22]. This was attributed to a finer spacing of 180° domain walls in samples with lower loss, which increases the interactions between micro-eddy currents by decreasing the distance between adjacent walls according to Eq. (6) [36]. The increase in micro-eddy current interactions, therefore results in a reduction in the contribution of micro-eddy currents to the loss process.

Following the same approach as Ref. [22], angular Barkhausen measurements were taken, as indicated in Sect. 2.2, on two of the samples under study with the lowest and highest core loss values, i.e. sample 8 and 1, respectively. Then the energy variations were plotted with respect to the angle of excitation as a polar plot shown in Fig. 6. The energy expression from Eq. (5) was used to calculate the α′ values for sample 8 and 1, which were found to be 0.05 and 0.03 ($mV^2.s$) respectively, when taking into account the laminate thickness. A larger interaction parameter for the sample with the lowest core loss, i.e. sample 8, is in agreement with Ref. [22].

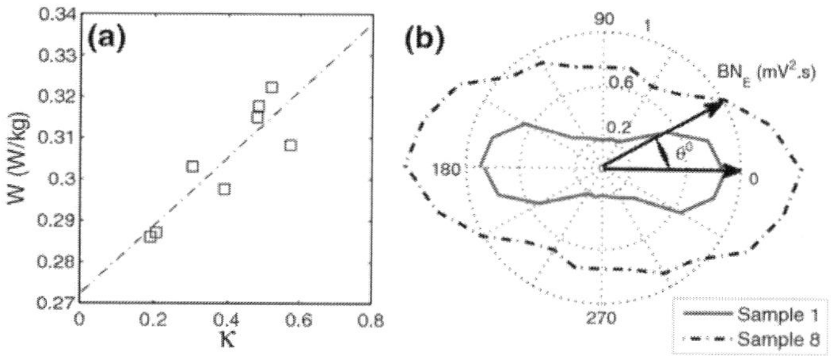

Figure 6. a Core loss (W) as a function of the magnetic anisotropy factor (κ), b comparison of the angular Barkhausen measurements between samples with the highest and lowest core loss values, i.e. sample # 8 and #1 respectively.

Figure 6 also demonstrates a more pronounced magnetic anisotropy for sample 8 with the higher loss. A dimensionless anisotropy parameter, κκ, was introduced to evaluate the degree of angular variations as [32]:

$$\kappa = \frac{BN_E(R) - BN_E(T)}{BN_E(R) + BN_E(T)}, \tag{7}$$

in which R and T corresponds to rolling and transverse direction respectively. Figure 6 shows an increase in core loss with increasing magnetic anisotropy, κ, which is attributed to a more inhomogeneous magnetization process [23]. Therefore, the smaller anisotropy and larger interaction parameter of sample 8 was correlated with its lower core loss relative to sample 1. Furthermore, the constantly higher Barkhausen noise energy of sample 8, as shown in Fig. 6, was attributed to its smaller grain size.

CONCLUSION

The present work investigated the applicability of surface Barkhausen noise measurements as a complementary, nondestructive method to the standard techniques for loss evaluation in NOES laminates. An in-house designed and built flux controller was used to effectively control the flux waveforms for local repeatable measurements at the surface of the laminates. A strong correlation was found between Barkhausen emissions and core loss for uni-directional magnetization applied along the RD. This correlation was established despite the very small variations in core loss values. The results were qualitatively explained in terms of micro-eddy currents, with higher interactions and therefore cancellation, likely for the samples having lower core losses. MBN was further correlated with grain size, texture and B–H curve parameters, which influence the core loss in the laminates. The angular MBN measurements showed that MBN may be used for local non-destructive evaluation of the magnetic anisotropy and magnetic easy axis, which influences the core loss in the laminates. The results suggest that flux-controlled MBN has the potential to be a valuable inspection technology for monitoring of NOES laminate quality.

ACKNOWLEDGMENTS

This research was supported by General Motors (GM) and Collaborative Research and Development Grant from Natural Sciences and Engineering Research Council of Canada (NSERC-CRD).

REFERENCES

1. Moses, A.J.: Energy efficient electrical steels: magnetic performance prediction and optimization. Scr. Mater. 67(6), 560–565 (2012)
2. Shiozaki, M., Kurosaki, Y.: The effects of grain size on the magnetic properties of nonoriented electrical steel sheets. J. Mater. Eng. 11(1), 37–43 (1989)
3. Stephenson, E.T., Marder, A.R.: The effects of grain size on the core loss and permeability of motor lamination steel. IEEE Trans. Magn. 22(2), 101–106 (1986)
4. Gutfleisch, O., Willard, M.A., Brück, E., Chen, C.H., Sankar, S.G., Liu, J. P.: Magnetic materials and devices for the 21st century: stronger, lighter, and more energy efficient. Adv. Mater. 23(7), 821–842 (2011)
5. Moses A.J.: Possible future trends and research challenges related to 1 & 2 d magnetic properties of soft magnetic materials. Przegl. Elektr, pp. 11–16, (2011)
6. Zemánek, I., Havlíček, V.: Universal control and measuring system for modern classic and amorphous magnetic materials single/on-line strip testers. J. Magn. Magn. Mater. 304(2), e577–e579 (2006)
7. Marketos, P., Zurek, S., Moses, A.J.: Calculation of the mean path length of the epstein frame under non-sinusoidal excitations using the double epstein method. J. Magn. Magn. Mater. 320(20), 2542–2545 (2008)
8. Tumanski, S.: Handbook of Magnetic Measurements. CRC Press, Boca Raton (2011)
9. Zirka, S.E., Moroz, Y.I., Marketos, P., Moses, A.J.: Loss separation in nonoriented electrical steels. IEEE Trans. Magn. 46(2), 286–289 (2010)
10. Graham Jr, C.D.: Physical origin of losses in conducting ferromagnetic materials. J. Appl. Phys. 53(11), 8276–8280 (1982)
11. Bertotti, G.: General properties of power losses in soft ferromagnetic materials. IEEE Trans. Magn. 24(1), 621–630 (1988)
12. Jiles, D.C.: Dynamics of domain magnetization and the Barkhausen effect. Czechoslov. J. Phys. 50(8), 893–924 (2000)
13. Alessandro, B., Beatrice, C., Bertotti, G., Montorsi, A.: Phenomenology and interpretation of the Barkhausen effect in ferromagnetic materials. J. Appl. Phys. 64(10), 5355–5360 (1988)
14. Rautioaho, R., Karjalainen, P., Moilanen, M.: Coercivity and power spectrum of barkhausen noise in structural steels. J. Magn. Magn. Mater. 61(1), 183–192 (1986)
15. Bertotti, G., Fiorillo, F., Sassi, M.P.: Barkhausen noise and domain structure dynamics in Si–Fe at different points of the magnetization curve. J. Magn. Magn. Mater. 23(2), 136–148 (1981)
16. Ranjan, R., Jiles, D.C., Rastogi, P.: Magnetic properties of decarburized steels: an investigation of the effects of grain size and carbon content. IEEE Trans. Magn. 23(3), 1869–1876 (1987)
17. Yamaura, S., Furuya, Y., Watanabe, T.: The effect of grain boundary microstructure on barkhausen noise in ferromagnetic materials. Acta Mater. 49(15), 3019–3027 (2001)
18. Krause, T.W., Szpunar, J.A., Atherton, D.L.: Anisotropic flux density dependence of magnetic barkhausen noise in oriented 3 % Si–Fe steel laminates. IEEE Trans. Magn.39(1), 562–566 (2003)
19. Clapham, L., Heald, C., Krause, T.W., Atherton, D.L., Clark, P.: Origin of a magnetic easy axis in pipeline steel. J. Appl. Phys. 86(3), 1574–1580 (1999)
20. Iordache, V.E., Hug, E., Buiron, N.: Magnetic behaviour versus tensile deformation mechanisms in a non-oriented Fe-(3 wt%) Si steel. Mater. Sci. Eng.: A 359(1), 62–74 (2003)
21. Williams, H.J., Shockley, W., Kittel, C.: Studies of the propagation velocity of a ferromagnetic domain boundary. Phys. Rev. 80(6), 1090 (1950)
22. Krause, T.W., Szpunar, J.A., Birsan, M., Atherton, D.L.: Correlation of magnetic barkhausen noise with core loss in oriented 3 % Si–Fe steel laminates. J. Appl. Phys. 79(6), 3156–3167 (1996)
23. Birsan, M., Szpunar, J.A., Krause, T.W., Atherton, D.L.: Correlation between the Barkhausen noise power and the total power losses in 3 % Si–Fe. J. Appl. Phys. 79(8),

6042–6044 (1996)

24. Stupakov, O.: Local non-contact evaluation of the AC magnetic hysteresis parameters of electrical steels by the Barkhausen noise technique. J. Nondestruct. Eval. 32(4), 405–412 (2013)
25. White, S., Krause, T.W., Clapham, L.: A multichannel magnetic flux controller for periodic magnetizing conditions. IEEE Trans. Instrum. Meas. 61(7), 1896–1907 (2012)
26. White, S., Krause, T.W., Clapham, L.: Control of flux in magnetic circuits for Barkhausen noise measurements. Meas. Sci. Technol. 18(11), 3501 (2007)
27. Patel, H.V., Zurek, S., Meydan, T., Jiles, D.C., Li, L.: A new adaptive automated feedback system for Barkhausen signal measurement. Sensors Actuators A 129(1), 112–117 (2006)
28. White S.: A Barkhausen noise NDE system for CANDU ◯R CANDU ◯R feeder pipes. Queen's University (2009)
29. ASTM Standard. E112, 2010, standard test methods for determining average grain size astm international, West Conshohocken, 2010, doi: 10.1520/e0112-10
30. Yonamine, T., Landgraf, F.J.G.: Correlation between magnetic properties and crystallographic texture of silicon steel. J. Magn. Magn. Mater. 272, E565–E566 (2004)
31. Sakamoto, H., Okada, M., Homma, M.: Theoretical analysis of Barkhausen noise in carbon steels. IEEE Trans. Magn. 23(5), 2236–2238 (1987)
32. Krause, T.W., Clapham, L., Atherton, D.L.: Characterization of the magnetic easy axis in pipeline steel using magnetic Barkhausen noise. J. Appl. Phys. 75(12), 7983–7988 (1994)
33. Matsumura, K., Fukuda, B.: Recent developments of non-oriented electrical steel sheets. IEEE Trans. Magn. 20(5), 1533–1538 (1984)
34. Cullity, B.D., Graham, C.D.: Introduction to Magnetic Materials. Wiley, Hoboken (2011)
35. Chikazumi, S., Charap, S.H.: Physics of Magnetism. Wiley, New York (1964)
36. Carr Jr, W.J.: Energy loss resulting from domain wall motion. J. Appl. Phys. 30(4), S90–S91 (1959)

CHAPTER 3

FAULT TOLERANT NEURO-ROBUST POSITION CONTROL OF DC MOTORS

Ran Zhang and Marwan Bikdash

Keywords: Fault-Tolerant, Neuro-Robust, Position Control, DC Motors

INTRODUCTION

In this paper, we proposed a neuro-robust position control of DC motors incorporating a fault-tolerant control model with the condition of parameter uncertainty and unknown load torque. DC motors are widely used in industry such as mechanics, robotics, and aerospace engineering. In recent years, some methods are applied for the position control of DC motors. For example, an adaptive PID control tuning was proposed to cope with the control problem for a class of uncertain chaotic systems with external disturbance [1]. Sliding mode control (SMC) is one of the popular strategies to deal with uncertain control systems [2-4]. Based on certain assumptions, various control approaches, such as model-based control, nonlinear inverse control, and neural network based control (and others) have been proposed [5-6].

This work is concerned with position control of DC motors by neuro-robust control. A nonlinear model is considered incorporate with fault-tolerant model in which modelling uncertainties and nonlinearities are explicitly addressed. The NN unit is designed to cope with the disturbances and coupled dynamics. Unlike traditional NN based method, there is no need to analytically estimate the upper bound of the reconstruction error and uncertainties in the proposed approach. Both theoretical analysis and simulation studies verify the effectiveness of the control algorithms.

PROBLEM FORMULATION

Modeling of DC Motors

The DC motors dynamics is given by

Citation: R. Zhang and M. Bikdash. "Fault Tolerant Neuro-Robust Position Control of DC Motors," Journal of Electromagnetic Analysis and Applications, Vol. 3 No. 10, 2011, pp. 412-415. DOI: 10.4236/jemaa.2011.310065.

$$\frac{K_m}{rR}V - \frac{J_m}{r^2}\ddot{q} - \frac{B_m + K_b K_m/R}{r^2}\dot{q} - Q(t) = 0 \tag{1}$$

where q is the position of link, K_m is the torque constant, J_m is the motor inertia, B_m is the damping coefficient, K_b is the back emf constant, R is the armature resistance, r is the gear ratio, Q(t) is the load torque, and V is the control input in voltage [7]. Our control objective is to make the position q tracking to the desired position q_d. From (1), we have

$$\ddot{q} = \frac{rK_m}{J_m R}V - \frac{B_m + K_b K_m/R}{J_m}\dot{q} - \frac{r^2}{J_m}Q(t) \tag{2}$$

For later development, we can express (2) in the following compact form with $x = q$ as

$$\ddot{x} = bV + L(\bullet) \tag{3}$$

where

$$b = \frac{rK_m}{J_m R} \tag{4}$$

$$L(\bullet) = -\frac{B_m + K_b K_m/R}{J_m}\dot{q} - \frac{r^2}{J_m}Q(t) \tag{5}$$

A direct measurement of the load torque is difficult because high cost equipment is required. Thus, the load torque Q(t) is usually considered to be unknown. The control objective is to design V so that the actual position trajectory q tracks the desired q_d asymptotically.

Fault Tolerant Control Model

As actuator failures represent the most typical and devastate faults, we should explicitly address such faults in control design. The preceding control failures can be modeled for each control by the following model

$$U_a = \Delta\left(t, t_1^{L1}, t_2^{L1}, \cdots, t_n^{L1}\right) U_c + D\left(t, t_1^{L2}, t_2^{L2}, \cdots, t_n^{L2}\right) \quad (6)$$

with

$$\Delta\left(t, t_1^{L1}, t_2^{L1}, \cdots, t_n^{L1}\right) = \begin{bmatrix} \delta_1\left(t - t_1^{L1}\right) & & & \\ & \delta_2\left(t - t_2^{L1}\right) & & \\ & & \ddots & \\ & & & \delta_n\left(t - t_n^{L1}\right) \end{bmatrix} \quad (7)$$

$$D\left(t, t_1^{L2}, t_2^{L2}, \cdots, t_2^{L2}\right) = \begin{bmatrix} d_1\left(t - t_1^{L2}\right) \\ d_2\left(t - t_2^{L2}\right) \\ \vdots \\ d_n\left(t - t_n^{L2}\right) \end{bmatrix} \quad (8)$$

where U_a is the applied controller, U_c is the demanded control to be designed, $0 \le \delta_i(.) \le 1$ and $|d_i(.)| \le d_0 < \infty$ are the actuation health indicator and un-adjustable portion of the actuator. In general, $t_i^{F1} \ne t_j^{F1}, t_i^{F2} \ne t_j^{F2}\ (\forall i \ne j)$ and $t_i^{F1} \ne t_i^{F2}$. It should be mentioned that the proposed fault model includes the commonly used fault model as a special case [8,9]. With the proposed actuator fault model (6), we can model all the possible actuation faults as mentioned earlier with δ_i and d_i defined properly, as follows:

Fault Case 1) $\{\delta_i(.) = 1;\ d_i(.) = 0\}$: healthy Fault Case 2) $\{\delta_i(.) < 1 \text{ but } \delta_i(.) \ne 0;\ d_i = 0\}$: fading power but partially controllable Fault Case 3) $\{\delta_i(.) = 0;\ d_i(.) = 0\}$: completely shut down Fault Case 4) $\{\delta_i(.) = 0;\ d_i(.) = 0\}$: completely uncontrollable Fault Case 5) $\{\delta_i(.) < 1 \text{ but } \delta_i(.) \ne 0;\ d_i(.) \ne 0\}$: fading power but partially controllable with uncontrollable portion.

Most existing fault-tolerant control methods such as [8,9] are based on constant Δ(.) and D(.) which is practically restrictive. In our design, we treat both Δ(.) and D(.) as completely unknown time-varying. Presently, there is no technique available to anticipate/predict such faults precisely. We design a control scheme that does not explicitly use any information about δ_i, d_i, t_i^{F1}, t_i^{F2}. Namely, we assume that not only the type and the magnitude of the faults are unknown, the time profile of fault occurrence are also unpredictable.

Let us define $u_a = V$, and substitute for u_a from (6), we get

$$\ddot{x} = bu_a + L(\bullet)$$
$$= b\left(\Delta u_c + d\right) - \frac{B_m + K_b K_m / R}{J_m}\dot{q} - \frac{r^2}{J_m}Q(t) \quad (9)$$

which can be expressed as

$$\ddot{x} = gu_c + \eta(\bullet) \quad (10)$$

where

$$g = b\Delta$$
$$\eta(\bullet) = bd - -\frac{B_m + K_b K_m / R}{J_m}\dot{q} - \frac{r^2}{J_m}Q(t) \quad (11)$$

Thus, the model of DC motor has been constructed, and the u_c is the control input. We will consider the case that all the actuator partially loses its effectiveness in that $\delta_i(.) < 1$ but $\delta_i(.) \ne 0$ in the following control design.

CONTROL DESIGN

As the first step, we define the position acking error

$$e = x - x_d \quad (12)$$

where x_d is the desired position, and the filtered error variable

$$s = \dot{e} + \beta e(\beta > 0). \quad (13)$$

Because of (7), it is seen that the control objective is realized as long as s is driven to zero as time goes by. Therefore, we focus on designing u to stabilize s. From (3) and (5) we have

$$\dot{s} = gu_c + \eta(\bullet) - \ddot{x}_d + \beta\dot{e} \quad (14)$$

Due to the existence of the uncertainties, as lumped by $\eta(\bullet)$, traditional model-based control is not applicable. In this work, we design the neuro-robust control scheme based on the unavailable part $\eta(\bullet)$, which is to compensated by the following neural network unit,

$$\eta(\bullet) = f_{NN} + \varepsilon = w^T\varphi + \varepsilon \quad (15)$$

where $w^T \in R^{3\times l}$, $\varphi(\bullet) \in R^{l\times 1}$ are the optimal weight matrix and basic function vector of the neural network, respectively, and $\|\varepsilon\| \le E_0 < \infty$ is the reconstruction error with unknown upper bound. For this case, we construct the following controller

$$u_c = g^{-1}\left(-\beta\dot{e} - ks + \ddot{x}_d + u_n\right) \quad (16)$$

where $k > 0$ is control parameter chosen by designer. With the control scheme (16), it is seen that the closedloop dynamics become

$$\dot{s} = -ks + \eta + u_n \quad (17)$$

the NN based compensation is given as follows,

$$u_n = -\left[\hat{w}^T\varphi + u_{rb}\right] \quad (18)$$

with

$$\hat{w} = \int_0^t \varphi(\cdot) s^T(t) \mathrm{d}\tau,\ u_{rb} = \left[\int_0^t \|s\| \mathrm{d}\tau\right] \mathrm{sgn}(s(t))$$

$$\varphi_i = \frac{1-e^{-\alpha_i(\|X\|)}}{1+e^{-\alpha_i(\|X\|)}},\ \left(X = \left[r\ \dot{r}\ \theta\ \dot{\theta}\ \phi\ \dot{\phi}\right], i = 1,2,\cdots,l\right)$$

The overall control block diagram is shown in **Figure 1**. To address the stability, we consider the following Lyapunov function candidate,

$$V = \frac{1}{2} s^T s + \frac{1}{2} tr\left(\tilde{w}^T \tilde{w}\right) + \frac{1}{2}\left(E_0 - \int_0^t \|s\| \mathrm{d}\tau\right)^2 \tag{19}$$

where $\tilde{w} = w - \hat{w}$, it can be shown that

$$\begin{aligned}
\dot{V} &= s^T \dot{s} - tr\left(\tilde{w}^T \dot{\hat{w}}\right) + \left(E_0 - \textstyle\int_0^t \|s\| \mathrm{d}\tau\right)\left(-\|s\|\right) \\
&\le -ks^T s + s^T\left(w^T \varphi(\cdot) + \varepsilon\right) - s^T\left(\hat{w}^T \varphi(\cdot)\right) - s^T u_{rb} \\
&\quad + tr\left(w - \hat{w}\right)^T\left(-\dot{\hat{w}}\right) - \left(E_0 - \textstyle\int_0^t \|s\| \mathrm{d}\tau\right)\left(\|s\|\right) \\
&= -ks^T s + tr\left[\left(w - \hat{w}\right)^T\left(\varphi(\cdot)s^T - \dot{\hat{w}}\right)\right] + s^T\left(\varepsilon - u_{rb}\right) \\
&\quad - \left(E_0 - \textstyle\int_0^t \|s\| \mathrm{d}\tau\right)\left(\|s\|\right)
\end{aligned} \tag{20}$$

Note that $s^T(\varepsilon - u_{rb}) \le \left(E_0 - \int_0^t \|s\| \mathrm{d}\tau\right)\left(\|s\|\right)$, it follows that

$$\dot{V} \le -k_2 s^T s \le 0$$

Thus

$$s \in L_\infty \cap L_2 \cap L_1,\ \hat{w} \in L_\infty$$

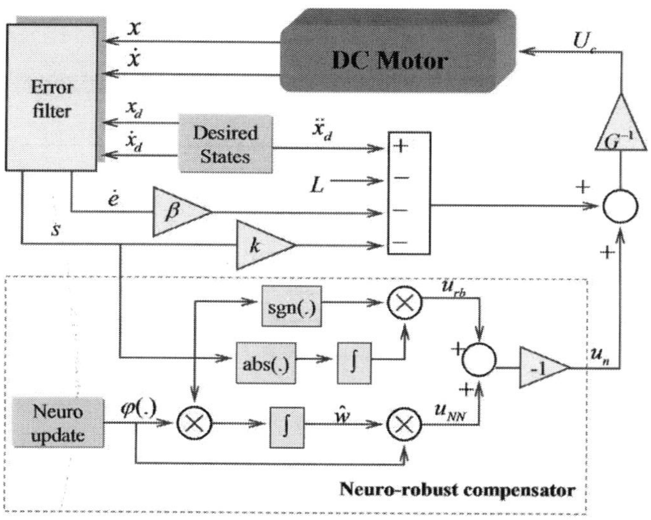

Figure 1. Block diagram of the proposed neuro-robust controller.

We can also show that s is uniformly continuous because $\dot{s} \in L_\infty$. By Babalart lemma, we conclude that $s \to 0$ as $t \to \infty$, then $e \longrightarrow 0$ and $\dot{e} \to 0$, as $t \longrightarrow \infty$, i.e: $q \longrightarrow q_d$ and $\dot{q} \to \dot{q}_d$ as $t \longrightarrow \infty$.

Remark 1. The control scheme, consisting of (16) and (18), has simple structure, and does not involve analytical estimation of the upper bound on the reconstruction error, making the design process simple and easy for implementation easy.

SIMULATION RESULTS

To verify the effectiveness of the controller, a numerical simulation study was conducted. The following functions were used in the simulation,

$$B_M = 2.68 \times 10^{-5}\text{Nms};\ K_b = 0.0603\ \text{Vs};$$
$$K_m = 0.060438586\ \text{Nm/A};\ R = 1.16\ \Omega;$$
$$J_m = 0.0000134\ \text{Kgm}^2;\ r = 1$$

Simulation results are given in Figures 2 and 3, where Figure 2 is a plot of the position tracking. The initial values are all set 0 at the very beginning for the two figures. The actual position is tracking to the desired position very quick and smooth in both cases. In first case, a one time position tracking is set, and in the second case, there is one more desired position set after the 4th second. Figure 3 is the tracking error for both position tracking. Note that although precise descriptions of system dynamics and external disturbances are unavailable, good tracking precision via the proposed neurorobust adaptive control still maintains with smooth control action and satisfactory position tracking

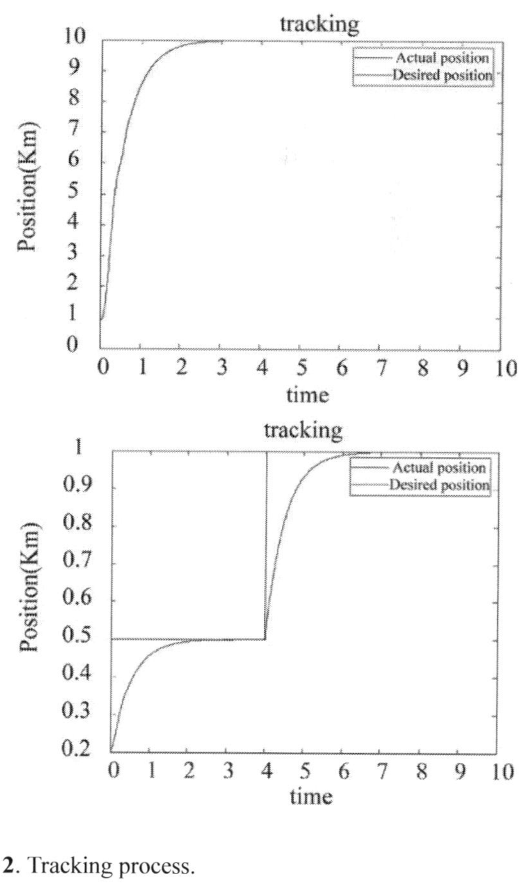

Figure 2. Tracking process.

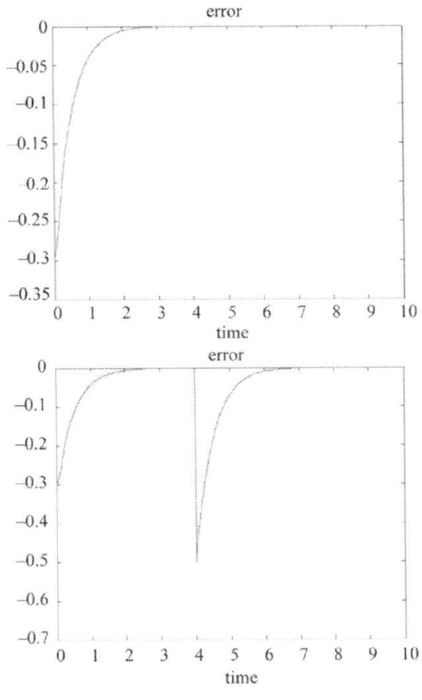

Figure 3. Tracking error.

CONCLUSIONS

The paper explored the fault-tolerant control problem of DC motors. Using neuro-robust approach, control algorithms achieve high precision position tracking of DC motors under uncertainties and disturbances. The salient feature of the proposed method lies in its flexibility and simplicity in design and implementation. There is no need to redesign or reprogram the control scheme for the different conditions, and no need of analytical estimation on the upper bound of the reconstruction error and uncertainties. A numerical example was simulated as a verification of the effectiveness of the proposed control algorithms.

REFERENCES

1. W. D. Chang and J. J. Yan, "Adaptive Robust PID Controller Design Based on a Sliding Mode for Uncertain Chaotic Systems," Chaos Solitions & Fractals, Vol. 26, No. 1, 2005, pp. 167-175. doi:10.1016/j.chaos.2004.12.013 [Paper reference:1]
2. J. Y. Hung, W. Gao and J. C. Hung, "Variable Structure Control: A Survey," IEEE Transactions on Industrial Electronics, Vol. 40, No. 1, 1993, pp. 2-22. doi:10.1109/41.184817
3. K. D. Young, V. I. Utkin and Ü. Özgüner, "A control Engineer'S Guide to Sliding Mode Control," IEEE Transactions on Control Systems Technology, Vol. 7, No. 3, 1999, pp. 328-342. doi:10.1109/87.761053
4. A. S. I. Zinober, "Variable Structure and Lyapnuov Control," Springer-Verlag, Berlin, 1994. doi:10.1007/BFb0033675
5. A. Calise and R. Rysdyk, "Nonlinear Adaptive Flight Control Using Neural Networks," Control Systems Magazine, Vol. 18, No. 6, 1998, pp. 14-25. doi:10.1109/37.736008
6. F. Lewis, "Nonlinear Network Structures for Feedback Control," Asian Journal of Control, Vol. 1, No. 4, 1999, pp. 205-228. doi:10.1111/j.1934-6093.1999.tb00021.x
7. H.-L. Choi, Y.-H. Chang, Y. Oh and J.-T. Lim, "On Robust Position Control DC Motor by PID Controller and Its Application to Humanoid Robot Arms," Proceedings of the 17th IFAC World Congress, Seoul, 6-11 July 2008. [Paper reference:1]
8. J. D. Boskovic, R. K. Prasanth and R. K. Mehra, "Reconfigurable Fault-tolerant Flight Control: Algorithms, Implementation and Metrics," AIAA Guidance, Navigation and Control Conference and Exhibit, Keystone, 21-24 August 2006.
9. M. Bodson and J. Groszkiewicz, "Multivariable Adaptive Algorithms for Reconfigurable Flight Control," IEEE Transactions on Control Systems and Technology, Vol. 5, No. 2, 1997, pp. 217-229. doi:10.1109/87.556026

CHAPTER 4

TYPE-2 FUZZY LOGIC CONTROLLERS BASED GENETIC ALGORITHM FOR THE POSITION CONTROL OF DC MOTOR

Mohammed Zeki Al-Faiz, Mohammed S. Saleh, and Ahmed A. Oglah

Keywords: Type-2 Fuzzy Logic Controller, Genetic Algorithm, DC Motor

INTRODUCTION

Fuzzy logic controllers (FLCs) are usually constructed using type-1 fuzzy sets [1], referred to as type-1 FLCs. Such FLCs have been applied to many areas, especially for the control of complex nonlinear systems that are difficult to model analytically [2,3]. Despite their popularity, research has shown that type-1 FLCs may have difficulties in modeling and minimizing the effect of uncertainties [4]. This limitation may restrict the usefulness of design methods that tune the FLCs using the genetic algorithm (GA) and a model of the controlled process. Since it is impossible for a model to capture all the characteristics of the actual plant, the performance of a controller designed using a model will inevitably deteriorate when it is applied to the practical system. A controller takes fully into account the non-linear ties. Emerging intelligent techniques have been developed and extensively used to improve or to replace conventional control technique because these techniques do not require a precise model. One of intelligent technique, fuzzy logic by Zadeh is applied for controller design in many applications. The advantage of fuzzy control methods [5] is the fact that they are not sensitive to the accuracy of the dynamical model. With the development of T2 FLSs and their ability to handle uncertainty, utilizing type-2 FLCs (IT2 FLCs) has attracted a lot of interest in recent years. The concept of type-2 fuzzy sets was first introduced by Zadeh as an extension of the concept of well-known ordinary fuzzy sets, type-1 fuzzy sets. A type-2 fuzzy set [6] is characterized by a fuzzy membership function i.e. the membership grade for each element is also a fuzzy set in [0, 1], unlike a type-1 fuzzy set, where the membership grade is a crisp number in [0, 1]. The membership functions of type-2 fuzzy sets are three dimensional and include a footprint of Uncertainty (FOU), which is the new third dimension of type-2 fuzzy sets. The footprint of uncertainty provides an additional degree of freedom to handle uncertainties. In this paper, the application of type-2 fuzzy logic concepts to the position control of a simple dc motor is illustrated. The device model is straightforward and the physical implications of position control are readily perceived.

GENETIC ALGORITHM (GA)

GA was first proposed in 1975 [7]. GAs are theoretically provide a robust search in complex spaces [8]. Figure 1 contains the flow chart of a basic GA. First, a chromosome population is randomly generated. The fitness of all individuals with respect to the optimization task is then evaluated by a scalar objective function (fitness function).

Genetic operators such as crossover and mutation are applied to the parents in order to produce a new generation of candidate solutions. As a result of this evolutionary cycle of selection, crossover and mutation, more and more suitable solutions to the optimization problem emerge within the population. Increasingly, GA is used to facilitate FLSs design [9]. However, most of the works discuss type-1 FLC design. This paper focuses on genetic algorithm of type-2 FLCs. There are two very different approaches for selecting the parameters of a type-2 FLS [4]. Type-2 FLCs designed via the partially dependent approach are able to outperform the corresponding type-1 FLCs [9], The type-2 FLC has a larger number of degrees of freedom because the fuzzy set is more complex. The additional mathematical dimension provided by the type-2 fuzzy set enables a type-2 FLS to produce more complex input-output map without the need to increase the resolution.

Citation: M. Al-Faiz, M. Saleh and A. Oglah. "Type-2 Fuzzy Logic Controllers Based Genetic Algorithm for the Position Control of DC Motor," Intelligent Control and Automation, Vol. 4 No. 1, 2013, pp. 108-113. DOI:10.4236/ica.2013.41015.

To address this issue, a comparative study involving type-2 and type-1 FLCs with similar number of degrees of freedom is performed. The totally independent approach is adopted so that the type-2 FLC evolved using GA has maximum design flexibility.

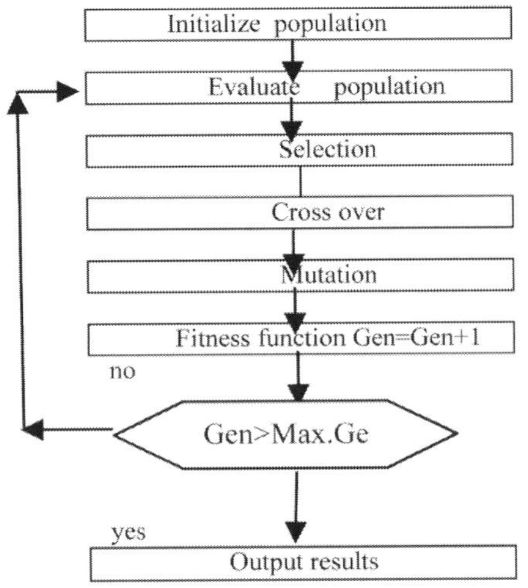

Figure 1. The flow chart of a basic.

INTERVAL TYPE-2

The structure of a type-2 FLS is shown in Figure 2. It is similar to its type-1 counterpart, the major difference being that at least one of the fuzzy sets in the rule base is type-2. Hence, the output of the inference engine is type- 2 sets and a type-reducer is needed to convert them into type-1 sets before defuzzification can be carried out.

Inference

An interval type-2 FLS is employed [4]. "Interval" means that the input/output domains are characterized by interval type-2 sets [5], whereby the membership grades of all elements in the FOU (secondary membership grades) area nifty. The inference engine then matches the fuzzy rules in the rule base. To compute unions and intersections of type-2 sets, compositions of type-2 relations are needed. Just as the sup-star composition is the backbone computation for a type-1 FLC, the extended sup-star composition is the backbone for a type-2 FLC [4].

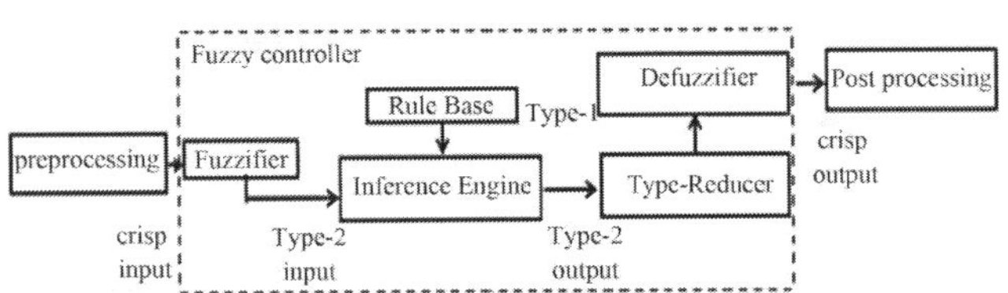

Figure 2. Structure of a type-2 FLS.

Type-Reduction and Defuzzification

The output of the inference engine is a type-2 fuzzy et, it must be type-reduced before the defuzzifier can be used to generate a crisp output. This is the main structural difference between type-1 and type-2 FLCs. The most commonly used type-reduction method is the center-of-sets type-reducer, which may be expressed as [4].

The footprint of uncertainty (FOU) of the membership function (MSF) in the IT2FLS is the area which limited by two MSF, the overhead limitation is the upper membership function UMSF and the down limitation is the lower membership function (LMSF), as shown in Figure 3.

DC MOTOR MODEL

In this study, proposed approach has simulated a DC shunt motor as is shown in Figure 4. The characteristics equations of the DC motor can be defined by the following equations [10];

$$v = R_a i + L_a \frac{di}{dt} + e_a \tag{1}$$

$$T_e = k_m i$$

$$T_e = J\frac{d\omega}{dt} + B\omega \tag{2}$$

$$T_e = J\frac{d^2\theta}{dt^2} + B\frac{d\theta}{dt} \tag{3}$$

$$e_a = k_e\omega = k_e\frac{d\theta}{dt} \tag{4}$$

The transfer functions are defined for speed and position control of DC motor, respectively.

$$\frac{\theta(s)}{v(s)} = \frac{k_e}{s\left(JL_a s^2 + \left(R_a J + BL_a\right)s + \left(k_e^2 + R_a B\right)\right)} \tag{5}$$

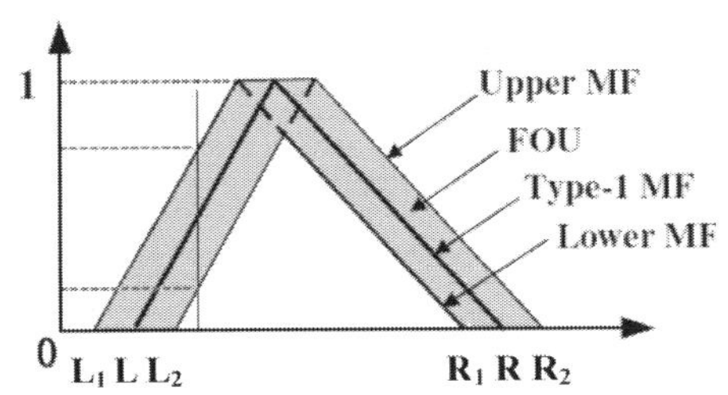

Figure 3. Membership function of FLCT2.

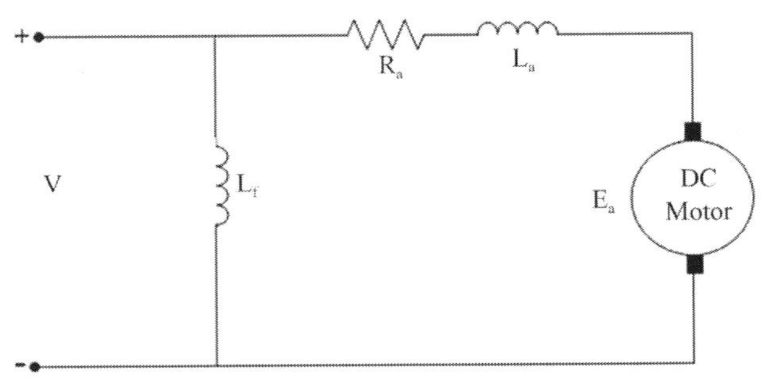

Figure 4. Diagram of DC shunt motor.

where R_a is armature resistance, L_a is armature inductance i is armature current, V is armature voltage, e_a is back emf voltage, K_e is back emf constant, K_m is torque constant, T_m is torque developed by the motor, ω is angular speed of shaft, θ is angular displacement of shaft, J is moment of inertia of motor and load, B is frictional constant of the motor and load. Unlike conventional control, which is based on mathematical model of a plant, a FLC usually embeds the intuition and experience of a human operator and sometimes those of designers and researchers. While controlling a plant, a skilled human operator manipulates the process input (i.e. controller output) based on with a view of minimizing the error within shortest possible time. The controlled variable of fuzzy controller is u(t). Once the fuzzy controller inputs and outputs are chosen, one must think about the membership functions (MSFs) for these input and output variables. In this paper, all membership functions for the conventional fuzzy controller inputs (e and Δe) and the controller output are defined on the common normalized domain [−1, 1]. We use symmetric triangles (except the two MFs at the extreme ends) with equal base and overlap with neighboring MFs. This is the most natural and unbiased choice for MFs. The actual control input voltage for the main fuzzy controller (In the case of PI-type FLC) can be written as

$$u(k)=u(k-1)+\Delta u(k) \tag{6}$$

where k is the sampling instant, is the crisp at k sampling instant and is the incremental change in controller output. showed that an IT2 fuzzy-PI (or the corresponding PD) controller is equivalent to a nonlinear PI (or PD) controller with variable gains and control offset [11].

SIMULATION

In this section. The assumed parameters of the electric DC motor represented in the following transfer function.

$$\frac{\theta(s)}{V(s)}=\frac{5.15725}{s\left(0.000675s^2+0.2527s+27.1\right)} \tag{7}$$

Experiment Analysis of Fuzzy Type-1 & Fuzzy Type-2 without GA

Simulation experiments under different operation status is carried out based on the fore established model and performance comparison with IT2FLC and conventional type-1 fuzzy controller is made. The two curves in Figure 5 are the simulation curves of the rated running state for DC motor respectively under the control of conventional type-1fuzzy controller and the type-2 fuzzy controller. For a clear comparison between the conventional fuzzy controller and the clustering based fuzzy controller, several performance measures such as peak overshoot (%OS), settling time, rise time, integral absolute error (IAE) and integral-of-time-multiplied-absolute error (ITAE) are computed as shown in Table 1. Using fuzzy type-1 and type-2 controller, the rise time and settling time not good whereas for other measures, both the controllers give approximately the same performance.

Fuzzy Type-1 & Fuzzy Type-2 with GA

In this section, DC motor is used in simulation. The genetic algorithm is very useful in optimization technique.

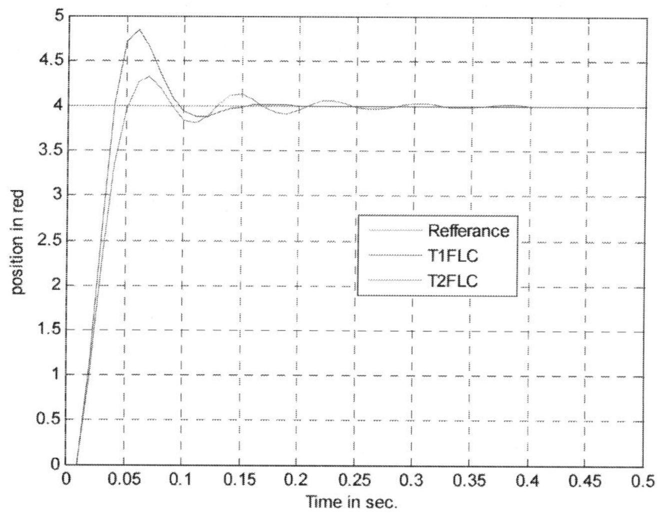

Figure 5. Response of T1FLC & T2FLC OF DC motor.

Table 1. Numerical result of experiment on DC motor without disturbance.

Controller	MP%	*ts* (sec)	*tr* (sec)	ITAE	IAE
T1FLC	21.25	0.165	0.045	22.7536	15.2709
T2FLC	8.00	0.410	0.055	23.2048	15.5737

it is apply in this paper of fuzzy type two logic controller in multi position of simulation process, to measured the optimal parameters of fuzzy such as scaling factor input and output data, start-end point of triangular memberships function (bottom triangle) and center of memberships function (top triangle). The number of variable parameters are thirty two parameters. Figure 6 compares the simulation results of the two controllers when the external disturbance is zero. The performance of the two controllers is listed in Table 2. Observe that GA type-1 FLC has a noticeable overshoot and a slower convergence to the set point in comparison to GAIT2FLC. Clearly, the GAIT2FLC has zero overshoot. It is easy to see that GAIT2 FLC has a faster settling time than GA type-1 FLC. However, under disturbance, the performance of the scheme shows improved results. Figure 7 shows the zoom of response of the system with a 30% disturbance applied at t = 0.36 sec. Table 3 shows the values of peak overshoot, settling time, IAE and ITAE computed under this condition. At the time t = 0.36 sec, the external torque is decreased by a step of 25% Figure 8. The system again reaches the steady state after transient period. Table 4 shows the values of peak overshoot, settling time, IAE and ITAE computed under this condition. The illustrated figures verify that a significant improvement has been achieved using the proposed genetic algorithm (GA) type-2 fuzzy controller.

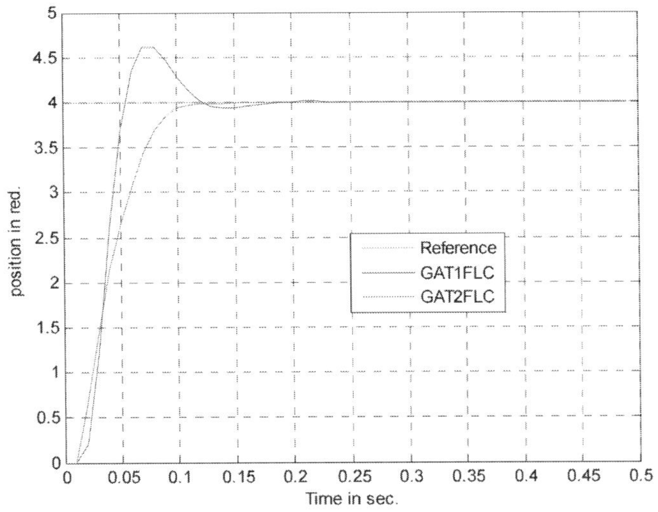

Figure 6. Response of T1FLC & T2FLC OF DC motor without disturbance.

Table 2. Numerical result of genetic simulation on DC motor disturbance.

Controller	MP%	*ts* (sec)	*tr* (sec)	ITAE	IAE
GAT1FLC	21.665	0.2400	0.0550	111.2549	20.2650
GAT2FLC	00.00	0.2150	0.0950	102.1742	18.6110

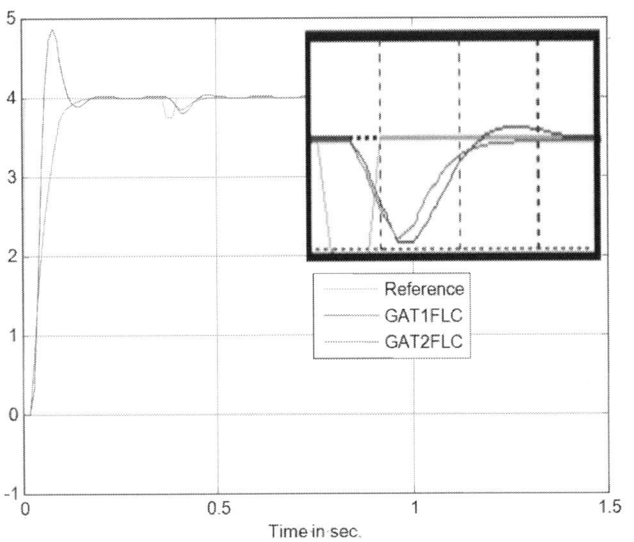

Figure 7. Response of GAT1FLC & GAT2FLC on DC motor with disturbance.

Table 3. Numerical result of simulation on DC motor with disturbance.

Controller	MP%	*ts* (sec)	ITAE	IAE
GAT1FLC	0.725	0.513	32.7145	21.9560
GAT2FLC	00.00	0.487	30.4870	20.4616

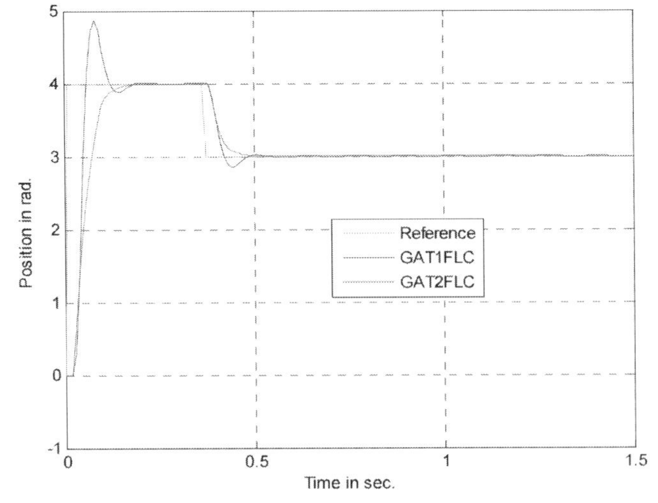

Figure 8. Response of GAT1FLC & GAT2 on DC motor sudden decrease in position.

Table 4. Numerical result of simulation on DC motor with sudden decrease in position.

Controller	ITAE	IAE
GAT1FLC	36.8605	24.7386
GAT2FLC	34.1573	22.9244

Initially the motor is operated at the steady state. At the time t = 0.36 sec, an increased step of 25% of initial set point. As shown in Figure 9 the motor position tracks the new set point after a transient period. Obviously, the external torque is assumed constant. Table 5 shows the values of IATE and IAE under these conditions. Comparisons with the conventional fuzzy controller indicate the improvement achieved.

Fuzzy Type-1 & Fuzzy Type-2 with Random Noise GA

In this section, we show in Figure 10 the simulation results for control of DC motor using both the GA type-2 fuzzy controller and the conventional fuzzy controller when used random noise reference signal.

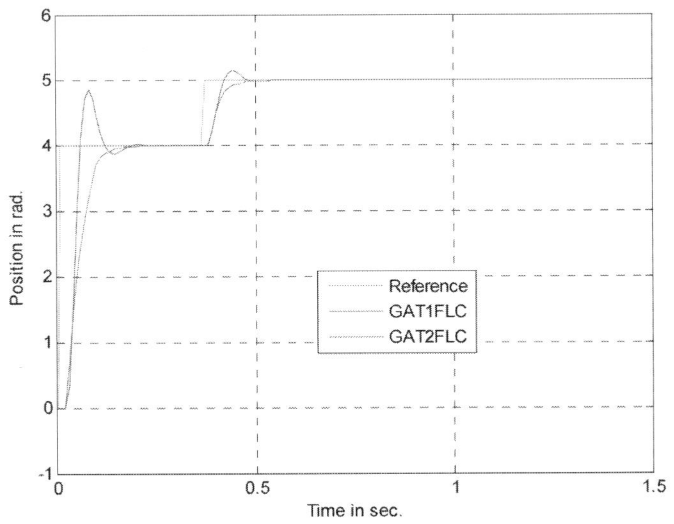

Figure 9. Responce of GAT1FLC & GAT2FLC on DC motor sudden increased in position.

Table 5. Numerical result of simulation on DC motor with sudden increased in position.

Controller	MP%	*ts* (sec)	*tr* (sec)	ITAE	IAE
GAT1FLC	2.900	0.5020	0.0600	36.8607	24.7387
GAT2FLC	00.00	0.5450	0.0700	34.1576	22.9246

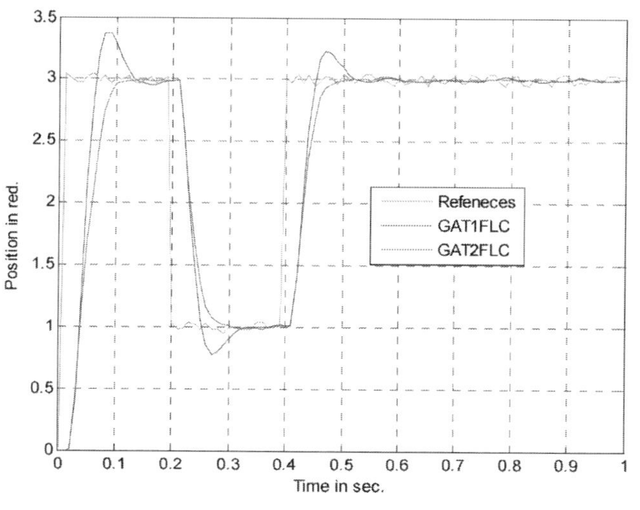

Figure 10. Response of GAT1FLC & GAT2FLC on DC motor with random noise.

CONCLUSIONS

In this paper, The GAFLCT2 has been proposed for position control of DC motor. Performance of the proposed GAFLST2 was also compared with corresponding conventional GAFLCs with respect to several indices such as rise time, settling time, maximum peak overshoot (MP%), integral of absolute error (IAE). and integral time of absolute error (ITAE).

The simulated results show that, using a type-2 FLC in real world applications can be a good option since this type of system is a more suitable system to manage high levels of uncertainty, as we can see in the results shown in Tables 2-5. Simulation results indicate that the performance of the GA FLCT2 will better. That is mean the system will sense for change the value of IAE and ITAE. The results demonstrate that a type-2 FLC can outperform type-1 FLCs that have more robustness design parameters. The main advantage of the type-2 FLC appears to be its ability to eliminate persistent oscillations, especially when unmodelled dynamics were introduced. This ability to handle modeling error is particularly useful when FLCs are tuned offline using GA and a model as the impact of unmodelled dynamics is reduced. The significance of the work is focused to manage the uncertainty of the system. The nonlinear of the systems are big problem therefore the one of successful methods to eliminate or reduce nonlinearity system by using fuzzy type two. It is a good option for real time applications that limited time is needed such as robot system.

REFERENCES

1. L. A. Zadeh, "Fuzzy Sets," Information and Control, Vol. 8, No. 3, 1965, pp. 338-353.doi:10.1016/S0019-9958(65)90241-X
2. J. John and R. Langari, "Fuzzy Logic Intelligence, Control and Information," Prentic Hall, Englewood Cliffs, 1998.
3. P. King and E. Mamdani, "The Application of Fuzzy Control to Industrial Process," Automatica, Vol. 13, No. 3, 1977, pp. 235-242. doi:10.1016/0005-1098(77)90050-4
4. J. M. Mendel, "Rule-Based Fuzzy Logic Systems: Introduction and New Directions," Prentice-Hall, Englewood Cliffs, 2001.
5. Q. Liang and J. M. Mendel, "Interval Type-2 Fuzzy Logic Systems: Theory and Design," IEEE Transactions on Fuzzy Systems, Vol. 8, No. 5, 2000, pp. 535-550. doi:10.1109/91.873577
6. Q. Liang and J. M. Mendel, "MPEG VBR Video Traffic Modelling and Classification Using Fuzzy Technique," IEEE Transactions on Fuzzy Systems, Vol. 9, No. 1, 2001, pp. 183-193.doi:10.1109/91.917124
7. J. H. Holland, "Adaptation in Natural and Artificial Systems," University of Michigan Press, Ann Arbor, 1975.
8. M. Sakawa, "Genetic Algorithms and Fuzzy M Ultiobjective Optimization," Kluwer Academic, Boston, 1975.
9. A. Homaifar and E. McCormick, "Simultaneous Design of Membership Functions and Rule Sets for Fuzzy Controllers Using Genetic Algorithms," IEEE Transactions on Fuzzy Systems, Vol. 3, No. 2, 1995, pp. 129-138. doi:10.1109/91.388168
10. N. Tutkun and D. Maden, "Determination of the PI Controller Parameters for a DC Shunt Motor Using Genetic Algorithm Method," International Conference of Electronics and Computer Engineering, Bursa, 2-5 December 2010, pp. 330-333.
11. X. Du and H. Ying, "Derivation and Analysis of the Structures of the Interval Type-2 Fuzzy-PI and PD Controllers," IEEE Transactions on Fuzzy Systems, Vol. 18, No. 4, 2010, pp. 802-814. doi:10.1109/TFUZZ.2010.2049022

CHAPTER 5

DYNAMICAL SIMULATION OF A NONLINEAR STEPPER MOTOR SYSTEM

Radosław Kępiński, Jan Awrejcewicz, and Donat Lewandowski

Keywords: Stepper motor, Nonlinear dynamics, Modeling, Dynamical simulation

INTRODUCTION

Nowadays, stepper motors are widely used in both industry and consumer devices, such as scanners, ink-jet printers or digital cameras. Their properties are advantageous in many mechatronic devices, especially where there is a need of a cost effective positioning system [1]. Such synchronous electric motors are characterized by a good performance in open loop configurations, allowing for simplification of the electrical design.

Naturally, there are some drawbacks of the open loop configurations. Specifically, positioning errors can occur, as such systems lack position feedback and therefore it cannot be verified if the rotor achieved desired position. Those errors are usually a result of the phenomena specific to stepper and synchronous motors such as stalling (i.e. losing synchronous speed) and loss of step. These events can occur as a consequence of a sudden peak of load torque or exceeding allowed maximum velocity.

To combat these undesired phenomena, one can chose a closed loop approach [2–4]. For instance, the motor driver can wait for confirmation signal from the optical encoder after issuing every step command. The realistic system described in this paper, however, uses a simpler method. Instead of checking the position continuously, it is only verified after each positioning sequence, which is more than enough to ensure sufficient reliability in this particular case.

The goal of the following research is to create a mathematical model that could be used to evaluate motor's performance in a variety of scenarios. In particular it has to allow investigations of the system behavior under varying load conditions and using different control methods.

PHYSICAL SYSTEM

In the following considerations we analyze a physical system being a part of an industrial device dedicated to measuring and cutting PVC bead strings (e.g. used in window and door blinds). The beads are being fed to the device from an injection molding machine at a variable speed. This requires cutting device speed to be appropriately adjusted. More importantly, the strings must be kept tightened to allow for a smooth cut. Because of that, the device has been fitted with a rotational tensioning arm with an inclination sensor. Tensioner angular position is fed back to the PLC allowing to maintain constant position of the arm and consequently—constant tension.

The drive system consists of a low power hybrid stepper motor powered by an appropriate driver. Incremental optical encoder has been installed on motor's shaft to ensure correct position after each positioning sequence (i.e. before cutting the string). Motor angular velocity must be adjusted according to the string feeding rate, therefore feedback signal from the inclinometer is run through a PID controller and converted to a pulse signal which drives the stepper motor. The whole machine, including human–machine interface, PID and motor speed is controlled by the PLC. The outline of the mentioned closed loop system is presented in Fig. 1.

Citation: Radosław Kępiński, Jan Awrejcewicz, and Donat Lewandowski. Dynamical simulation of a nonlinear stepper motor system, International Journal of Dynamics and Control, March 2015; Volume 3, Issue 1, pp 31–35. DOI:10.1007/s40435-014-0093-6.

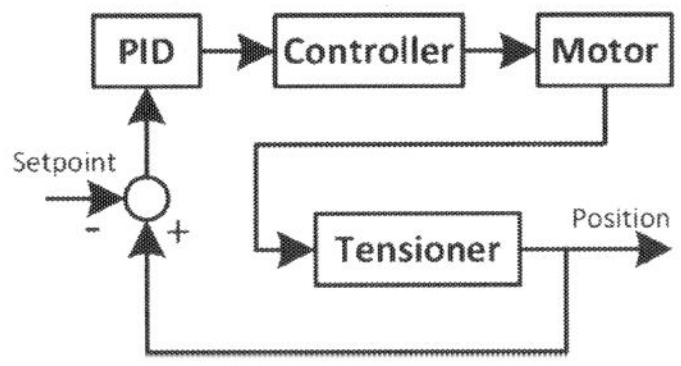

Figure 1. Control system with feedback.

Position setpoint is provided by the machine operator, who chooses the optimal string tension experimentally. The deviation from the setpoint is converted to a clock signal applied to stepper motor controller. Then it is converted to an analogue voltage signal powering two phases of a hybrid stepper motor (V_a and V_b). Inclinometer provides a digital feedback data, transferred back to the PLC via a RS-232 interface.

MATHEMATICAL MODEL

Stepper Motor

The equations governing the dynamics of a hybrid stepper motor can be found in a number of publications [1, 5–8]. They have been cast into the following system of four non-linear ordinary differential equations:

$$\begin{aligned}
\frac{di_A}{dt} &= \frac{1}{L}\left[V_A - Ri_A + K_m\omega\sin(N_r\Theta)\right],\\
\frac{di_B}{dt} &= \frac{1}{L}\left[V_B - Ri_B + K_m\omega\cos(N_r\Theta)\right],\\
\frac{d\omega}{dt} &= \frac{1}{J}\left[-K_m i_A\sin(N_r\Theta) + K_m i_B\cos(N_r\Theta) - B\omega - \tau_l\right],\\
\frac{d\Theta}{dt} &= \omega,
\end{aligned} \tag{1}$$

where: i_a, i_b, V_a, V_b are currents and voltages in phases A and B; R, L are winding's resistance and inductance, respectively; K_m is the motor torque constant, B is viscous friction coefficient, N_r is the number of rotor teeth, JJ is rotor's moment of inertia, τ_l is the load torque, ω is the rotor speed and Θ is its angular position.

First two of the above equations determine the current in motor's windings (phase A and B), while the two latter correspond to Newton's 2nd Law of motion.

Controller

Stepper motor controller is compelled to provide appropriate voltage on motor's connectors and change them as required by the demanded motion sequence. The most basic circuit that can be used as a controller is a dual H-bridge. It consists of four transistors that are switching power sequentially, allowing the rotor to move in the desired direction in consecutive steps. While this approach has a big advantage of simplicity and low cost, it doesn't allow for a very good drive performance. Because of that, for more demanding tasks, dedicated integrated circuits are commonly used.

Those more advanced circuits allow micro-stepping, which by applying the voltages to both phases simultaneously, enable the rotor to take positions between the full steps. The ratio of V_a to V_b constrains the exact position of the rotor. This method reduces the vibrations excited during motion and makes drive's operation more smooth. Those commercially available controllers also use the so called chopper circuits. They use supply voltage that is much higher than motor's nominal voltage and turn it off when the current in phase goes slightly above the nominal value. This method allows the windings to be energized much more rapidly (as for each step current rises much faster), which results in much better performance of the motor, namely higher maximum speed and higher torque.

To recreate above considerations in the computer model, voltages substituted in Eqs. (1) are obtained by the following rule:

$$V = \begin{cases} V_{cc} & \text{if } i < I_m \\ 0 & \text{if } i \geq I_m. \end{cases} \tag{2}$$

In a realistic device, the frequency of switching the voltage is limited by the circuit parameters. Waveform of the measured phase voltage is shown in Fig. 2. It is worth noting that in this particular case, the supply voltage was being switched at a rate of 42 kHz. The above frequency was taken into consideration when designing the computational model.

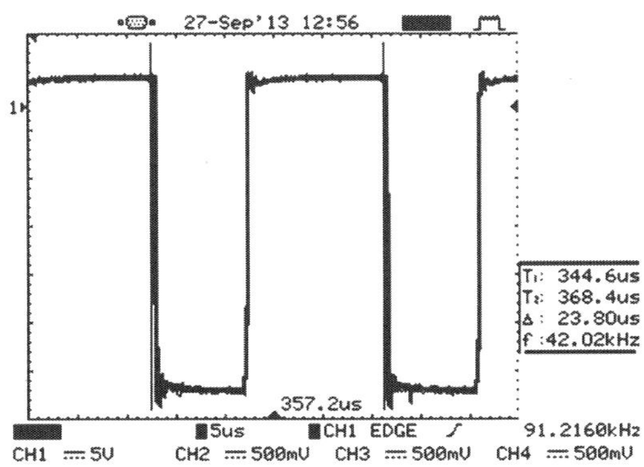

Figure 2. Voltage measurement on phase A using a digital oscilloscope.

If the system is controlled with micro-stepping approach, maximum allowed current I_m for each subsequent micro-step is determined by motor's nominal current and the appropriate current ratio. It can be determined using following two equations:

$$I_{ma} = I_n \sin\left(\frac{n}{2d}\pi\right) \quad \text{for } n \in \{1, 2, \ldots, 4d\}, \tag{3}$$

$$I_{mb} = I_n \cos\left(\frac{n}{2d}\pi\right) \quad \text{for } n \in \{1, 2, \ldots, 4d\}, \tag{4}$$

where I_n is the nominal current, n is the present micro-step and d denotes the division factor (e.g. 2 for half-stepping, 4 for quarter-stepping, etc.). Please note that there are several other methods of determining desired current while micro-stepping, that may be more suitable to a particular motor controller model.

Tensioning Mechanism

The tensioning mechanism is a subject of control in the system. It has been modeled as a single DOF pendulum. The position of the pendulum is at all times restricted by the length of the string currently present in the device (see Fig. 3). Length a, depends on the one hand on the feeding speed and on the other on motor's speed. Using geometric relations function a(φ) has been derived. The inverse function has been approximated by the following third order polynomial:

$$\varphi(a) = 1072{,}9a^3 - 3864{,}7a^2 + 4812{,}7a - 2002, \quad (5)$$

where: a is string's length and φ is arm's position. The above equation is very close to actual relation in angle range the device operates in (between 45° and 90°).

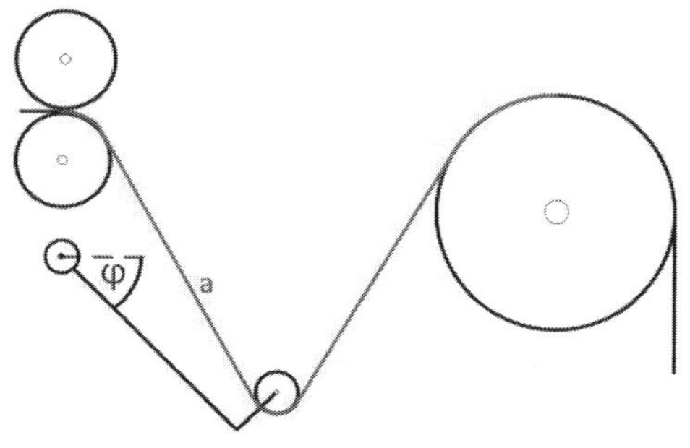

Figure 3. Tensioning mechanism.

COMPUTER SIMULATION

The above mathematical constraints have been applied in a computer model created using *scicos* which is a part of numerical computation software *scilab*. All of the mentioned subsystems were recreated and connected together, as seen in the diagram (Fig. 4). Simulation parameters, namely motor's specification, according to the data sheet are given in Table 1.

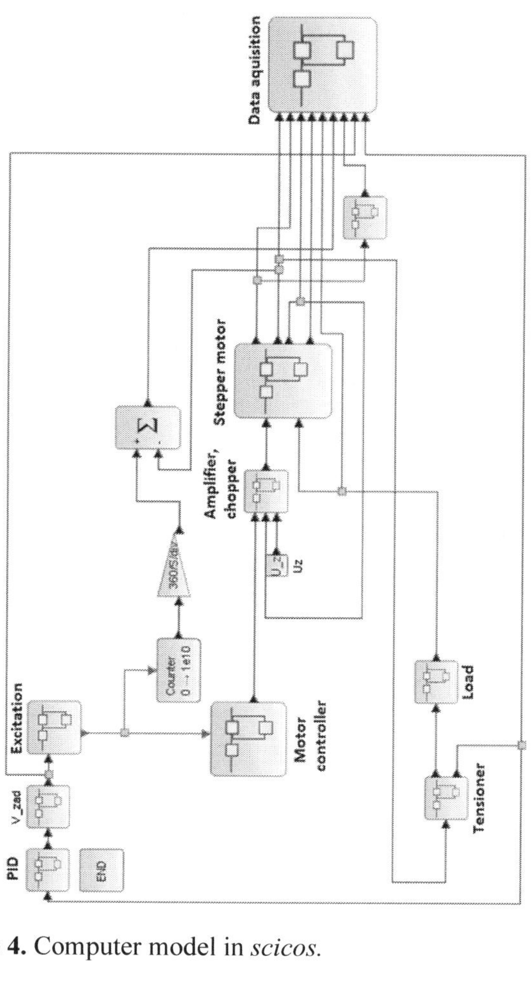

Figure 4. Computer model in *scicos*.

Table 1. Stepper motor properties

Parameter	Value
R	5Ω
L	8.6×10^{-3} H
B	8×10^{-4}Nms/rad
K_m	0.55Nm/A
I_n	1A
J	11×10^{-6}kgm^2
N_r	50

Computer model allows applying different kinds of loads to the motor. In particular, load from tensioning arm is approximated. Assuming that at angle φ=π/2, the tensioner is in vertical position and at φ=0 in horizontal position respectively, the simulated load on the motor is assumed to be:

$$\tau_l = \tau_{max}\cos(\varphi). \quad (6)$$

Apart from the above, the external load can be set to zero or given as step function with a constant load starting from time t_0. Motor's excitation frequency (CLK signal), can be set as a constant value, a polynomial function or a value determined by a PID controller. In the latter case, all of the PID term's constants can be set, which allows tuning the device using the computer simulation instead of testing it on site.

RESULTS

The following plots were made using the computer model described above. Among the objects of interest are phase currents of the stepper motor, as well as rotor's mechanical position. To evaluate the behavior of the model, different load and excitation conditions were tested.

Additionally, to evaluate the simulation results, some simple measurements on the physical object were attempted (compare [9]). In particular, phase current was measured. This was achieved by adding a serial resistor into the phase circuitry. The resistance of it is several orders of magnitude lower than the winding's resistance, so that motor performance is not affected by its presence. By connecting an oscilloscope to both ends of the resistor, one can easily measure voltage drop on the shunt resistor, which is an indirect way of measuring phase current in the circuit, as those two physical quantities are proportional. The downside of this simple method, however, is that the recorded waveforms contain a lot of noise and are not very accurate. This is because of the fact that the measured voltage drop is very low (less than 100 mV).

Investigations shown in Fig. 5 confirm that mathematical model's simulation results are compliant with the physical object. However subtle differences can be observed, such as the amplitude of the phase current. That is likely to be associated with the fact that the amplifier and chopper are not yet accurately modeled.

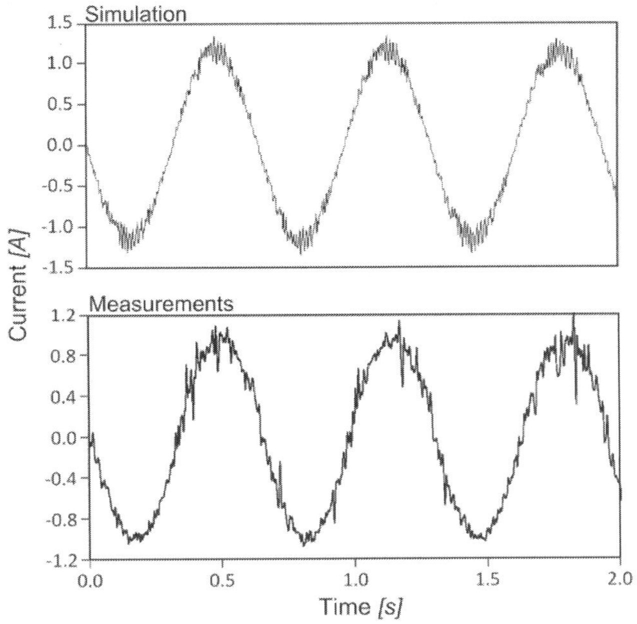

Figure 5. Current in phase A during a test run with excitation frequency f = 100 Hz and micro stepping 1/16.

Figure 6 shows current oscillations when the motor is advancing to the next step. As micro-stepping is used in the considered scenario, those oscillation's amplitude is proportional to maximum allowed current in a given micro step, as can be seen in Eqs. (3) and (4).

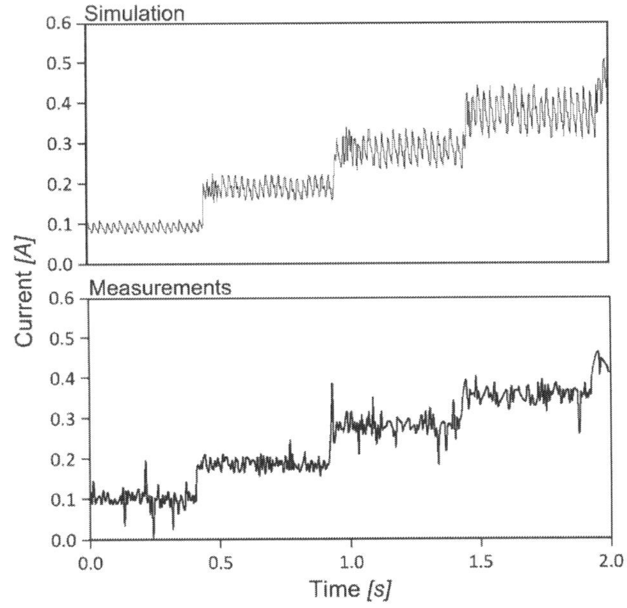

Figure 6. Current in phase A during a test run with excitation frequency f = 4 Hz and micro stepping 1/16.

Discussed computational model is able to recreate some of the phenomena mentioned in Sect. 1. Figure 7 shows the rotor travel while accelerating. The profile of the motion is such that the excitation frequency is linearly increasing for 5 s from 0 Hz to 20 kHz. Instead of a constant acceleration, we observe loss of steps after about 3.5 s with motor never actually exceeding 10 kHz. This, of course, qualitatively fits with real device observations, even though small deviations from an actual maximum speed can be observed.

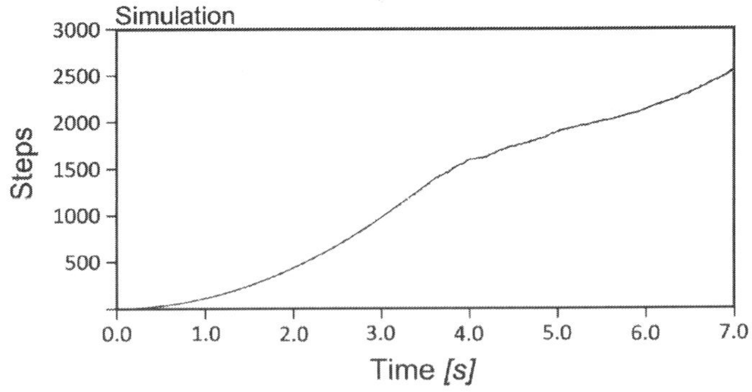

Figure 7. Motor exceeding its maximum speed.

Although simulation results allow to reconstruct motor's behavior, because of the noise, measurement's accuracy is limited. Operation of the system as a whole is presented in Fig. 8. In this particular example, excitation frequency is regulated by a PID controller with non-zero "P" and "I" terms. Observe that it accurately resembles behavior of the actual physical device.

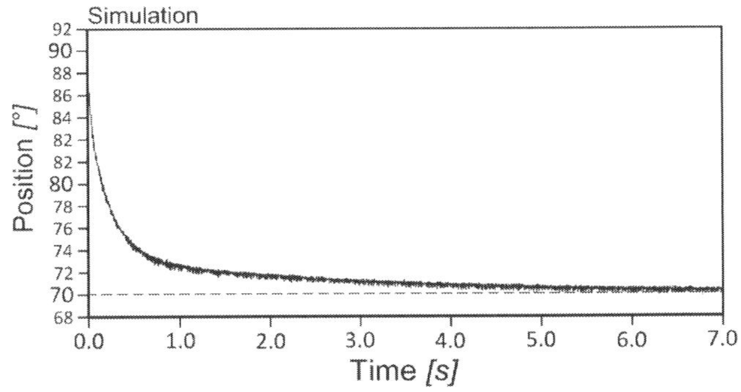

Figure 8. Arm position during PID control (P = 15, I = 5, D = 0).

CONCLUDING REMARKS

The qualitative results obtained during simulations fit well with the expected behavior of an electric stepper motor. In particular, the proposed solutions to simulating real stepper motor controllers, allow for realistic simulation conditions, resembling commercially available devices. This can lead to a better understanding of pros and cons of certain control methods as well as allowing to a broader usage of electric stepper motors in environments, where their features are desirable, offer more performance or lower cost than their DC or asynchronous counterparts. The mathematical model of a stepper motor system allows for investigating and avoiding some of the phenomena that can hinder system's performance. Among those are electromechanical resonances that result in increased vibrations, stalling, losing steps, noisy operation and decreased performance and reliability.

It is worth noting that the proposed computational model is relatively complex. That means that the simulations with integration times over 5 s take a significant time to complete. Isolating most resource heavy parts and optimizing them may be crucial to continue the above analysis. In further research, more physical object measurements will be made, which will open possibilities for fine-tuning the computer model. In particular, precise phase current measurements need to be done to verify the simulation results. This will require building a circuit that allows amplifying the measured signal as well as filtering out the background noise and interferences. Moreover, physical properties of the motor and the whole system should be verified as they can deviate slightly from the values provided in the data sheets.

ACKNOWLEDGMENTS

Above paper was presented during 12th Conference on Dynamical Systems—Theory and Applications held in Lodz, Poland on 2–5 December 2013.

REFERENCES

1. Acarnley P (2002) Stepping motors: a guide to theory and practice. The Institution of Electrical Engineers, London
2. Ferris TA (1981) A low cost, high performance incremental motion control system using a closed loop stepping motor. In: Proceedings of the 10th annual symposium on incremental motion control systems and devices, Urbana
3. Adams K, Van Reenen M (1995) A low-cost stepper motor positioning system with minor closed-loop control. Int J Adv Manuf Technol 10:191–197
4. Banihani S, Al-Widyan K, Al-Jarrah A, Ababneh M (2013) A genetic algorithm based lookup table approach for optimal stepping sequence of open-loop stepper motor systems. J Control Theory Appl 11:35–41
5. Bodson M, Chiasson J, Novotnak R, Rekowski R (1992) High performance nonlinear feedback control of a permanent magnet stepper motor. In: Proceedings of 1st IEEE conference on control systems technology, vol 1, pp 510–515. doi:10.1109/CCA.1992.269821
6. Kenjo T, Sugawara A (1995) Stepping motors and their microprocessor control. Oxford University Press Inc., New York
7. Kuo B, Tal J (1979) Incremental motion control: step motors and control systems. SRL Publishing Company, Champaign
8. Lyshevski SE (2000) Electromechanical systems, electric machines and applied mechatronics. CRC Press, Boca Raton
9. Balakrishnan K, Umamaheswari B, Latha K (2013) Identification of resonance in hybrid stepper motor through measured current dynamics in online for accurate position estimation and control. IEEE Trans Ind Inf 9:1056–1063

CHAPTER 6

DI-BSCCO SUPERCONDUCTING ELECTRIC MOTOR

Chin Kat Meng, Nasri A. Hamid, and Mohd. Yusri Abd. Rahman

Keywords: Superconductor, Magnetic flux, Finite Element Analysis, Electric Motor

INTRODUCTION

Since the discovery of superconductivity in 1911 by Kamerlingh Onnes, the dream of creating highly efficient electric machine is getting closer to a reality. Superconducting materials are able to carry current with zero resistance and reduce power loss during the process of transmitting electricity [1]. This distinctive property of superconductor has seized the attention of many re-searches and a lot of studies have been carried out to produce superconducting wire with high stability. Finally, long length superconducting wire has been successfully produced with a very reliable performance [2,3]. Superconducting devices such as magnetic resonance imaging (MRI) and superconducting quantum interference devices (SQUID) have been used in medicine and industry extensively. Several others electrical power devices such as motor, generator, fault current limiter and magnetic storage system have been developed [4-6]. Superconductor required cooling process below the critical temperature, T_c which is also called as transition temperature to become zero resistance. The conventional low temperature superconductor operates at below 20 K, where cooling by liquid helium is required. In 1986, researchers discovered high temperature superconductor where the critical temperature is higher than 77 K. Therefore, the use of cheaper refrigerant such as liquid nitrogen is sufficient to cool the superconducting material and become superconductivity [7].

The distinctive properties of superconductor enable the development of high efficiency energy devices. It is believed that superconducting motor is able to reduce in size and minimize energy loss compared to the conventional copper-base motor [7,8]. In this work, a low power superconducting electric motor is fabricated and its performance was compared with a conventional electric motor. All the specifications for both types of motor are similar except for the coils. The intended applications of the motor are for driving industry devices such as pump and fan that can operate continuously at a constant speed. The features and performance of both types of motor were investigated. The simulation of the flux plot was also presented.

Citation: Meng, C., Hamid, N. and Rahman, M. (2011) DI-BSCCO superconducting electric motor. Natural Science, 3, 36-41. DOI: 10.4236/ns.2011.31004.

EXPERIMENTAL

Motor Design Description

The new architecture for designing of electric motors was first introduced about a decade ago [9]. In our work, similar approach was adopted at the designing stage of the motor. Figure 1 shows the model of motor prototype that was designed. It has three main structures, namely, coil, stator and rotor. Two different coils were fabricated; the first type was wound with high-temperature super-conductor (HTS) DI-BSCCO wire and the second type was wound with normal cooper wire. The design of the motor was made flexible where the coils can be inter-changed easily. The stator was laminated by a metal sheet with 5 mm thick. In addition, to reduce the effect of eddy current loss during the motor operation, each layer of lamination was insulated with varnish. The fabricated coil was directly mounted to the two-leg of the stator and cooled by liquid nitrogen. In addition, the rotor was installed to the stator with the support of the rotor housing. The stator is designed with a deep "U" shape; where the rotor will be mounted at the opening of the "U" with rotor housing and coils will be directly assembled to one of the stator legs. With this design, various tests can be conducted, each with different number of coil [10]. Table 1 shows the summary of the design of the motor.

Coil Structure

For the superconducting wire, its critical current is strongly influenced by the bending diameter of the wire at 77 K [2]. A frame was designed to protect the super-conducting wire from any external impact or damage. Figure 2 shows the dimensional sketch of the coil layout, where the coil was wound around the frame with a double pancake style. Each coil structure has a total of 100 turns. Figure 3 shows the final product of the superconducting coil. Similar design was used to fabricate the conventional copper coil as shown in Figure 4. The conventional coil structure has exactly the same design and dimensions with the superconducting coil.

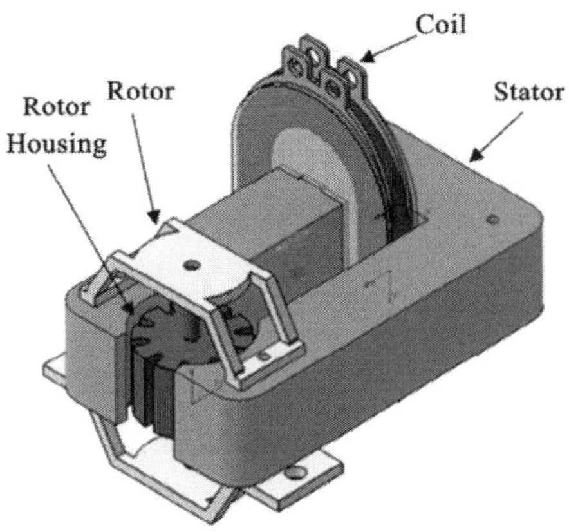

Figure 1. Model of the prototype.

Table 1. Summary of the design of the conventional and superconducting motors.

Motor Type: DC Synchronous Machine	
Details of Field Winding	
Field winding wire type:	Copper
Operating current:	< 2 A
Details of Superconducting Coils Winding Stator	
Stator winding wire type:	HTS
Cooling type:	Immersion
Coolant:	Liquid Nitrogen (LN_2)
Operating Temperature:	77 K
Operating current:	< 65 A
Details of Copper Coils Winding Stator	
Stator winding wire type:	Copper
Operating current:	< 16 A
Stator Lamination Sheet	
Metal sheet type:	Cooled-roll metal sheet
Thickness:	0.5 mm

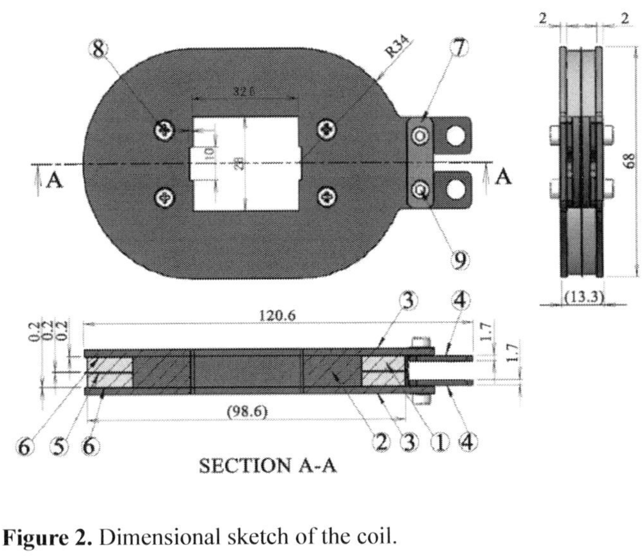

Figure 2. Dimensional sketch of the coil.

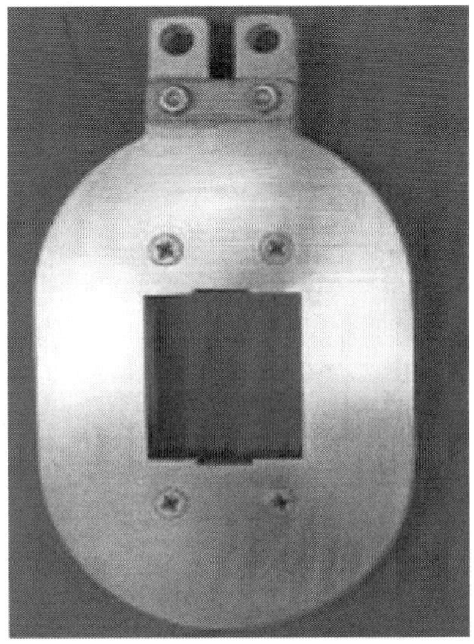

Figure 3. Superconducting coil structure.

Table 2 shows the materials and specifications for the superconducting coil and its components. Selection of the materials for the coil unit is important to ensure the effective distribution of magnetic flux. Ferromagnetic material must be avoided in the coil components to eliminate the suppression of magnetic flux [11]. In this design, brass which is a non-magnetic material was used in the construction of the frame and edge plate.

THEORY

Electromagnetic Theory

Rotating machinery working principles are based on electromagnetic laws. One of the main laws is Ampere's law which is defined as "the line integral of the magnetic field strength taken around any closed path is proportional to the total current flowing across any area, bounded by the path" [12]. Mathematically, the law is expressed by:

$$\oint H \cdot dl = \int J \cdot ds \tag{1}$$

where J is conduction current density, H is magnetic field intensity, dl is the differential length, ds is differential surface enclosed by the closed path of length, and $\oint$ is integral over a closed loop or closed path.

Field Strength of Magnetic Circuit

A magnetic field system comprises of N-turns coil wound on the ferromagnetic core. The coil carries a constant current, i through a uniform cross-sectional area, A_c of the core. When the current flows inside the conductor, magnetic field is generated inside the core and a magnetic flux is confined to the ferromagnetic core. In the actual practice, a small portion of the flux will leak to the environment but this can be neglected due to its small percentage compared to the total flux. Therefore, the current through the enclosed path is expressed by:

$$\int J \cdot ds = Ni \tag{2}$$

where N is number of turn wound on the core and i is current inside conductor. From (1) and (2), the following relationship is obtained and is expressed by:

$$\oint H \cdot dl = Ni \quad (3)$$

Figure 5 shows the equivalent magnetic circuit of the motor. Subscripts *r*, *a*, *y*, and *c indicate "rotor", "air-gap", "york" and "core", respectively*. ϕ represents magnetic flux. Symbols ϕ_c and ϕ_r denote magnetic fluxes that are generated by the coil and the rotor, respectively. "*R*" indicates the reluctance inside the circuit. In this case, circuital law can be applied to this flux path, and thus from (3), the total current in the circuit shown in Figure 5 is expressed as:

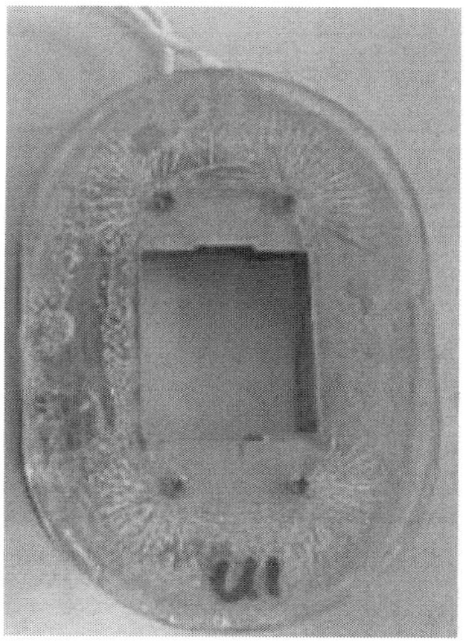

Figure 4. Conventional copper coil structure.

Table 2. Specification of the superconducting coil and components.

No.	Item	Material	Specification
1	HTS coil	Type H	Double Pancake 100 turns
2	Frame	Brass	-
3	Edge Plate	Brass	-
4	Terminal	Copper	-
5	Separator	GFRP	Thickness 0.2 mm
6	Edge Separator	GFRP	Thickness 0.2 mm
7	Insulation Plate	GFRP	Thickness 0.2 mm
8	Screw Blot	-	M3
9	Screw	-	M3

*GFRP = Glass Fiber Reinforced Polymer

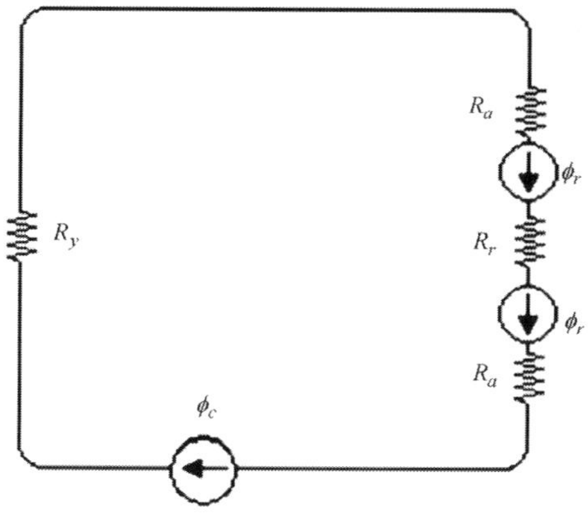

Figure 5. Equivalent magnetic circuit of the system.

$$\sum Ni = HL \quad (4)$$

or

$$N_c i_c + 2N_r i_r = H_y \cdot L_y + 2H_a \cdot L_a + H_r \cdot L_r \quad (5)$$

RESULTS AND DISCUSSIONS

Finite Element Analysis

A finite element analysis was applied to investigate the behavior of the magnetic flux inside the motor. The analysis is very useful to verify an ideal magnetic circuit and to detect any error in the electrical connection, especially for the DC machine. Another advantage of applying finite element analysis before the actual testing is to avoid any short-circuit current between the motor components. Figure 6 shows the model of the motor which was constructed from the finite element analysis. The model was built with a full scale dimension.

Figure 7 shows the magnetic flux path in the circuit. From the flux plot, it indicates that the flux loss was in the form of flux leakage and fringe effect. The simulated result shown in Figure 8 indicates that the flux density was inside the body of the motor. In the figure, it is obvious that a high flux density is in the area around the air-gap and the rotor. In contrast, the other parts of the motor do not show high flux density, indicating a low distribution of flux. The high flux density in the rotor and air-gap area is needed to establish a high force that enables the rotor to rotate.

Motor Assembly and Testing

The lamination of metal sheet on the motor formed the stator. The superconducting coil was mounted on one leg of a stator. A rotor was then mounted on the rotor housing and the housing was attached to the stator. Thus, a complete superconducting electric motor was assembled. The procedure was repeated to assemble a copper electric motor. Photograph of the completed assembly of superconducting coil motor and copper coil motor are shown in Figure 9 and Figure 10, respectively.

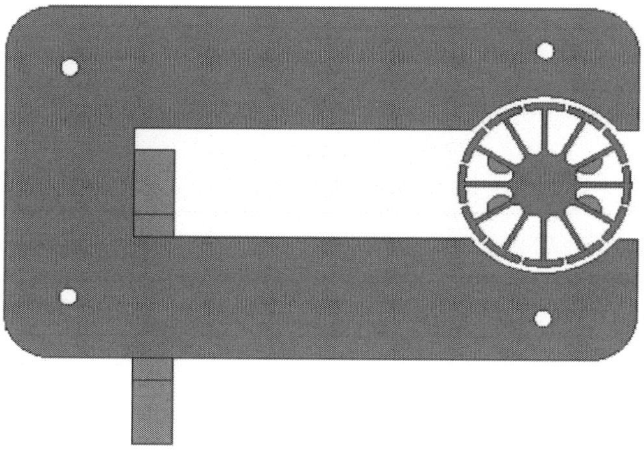

Figure 6. Model generated from the finite element analysis.

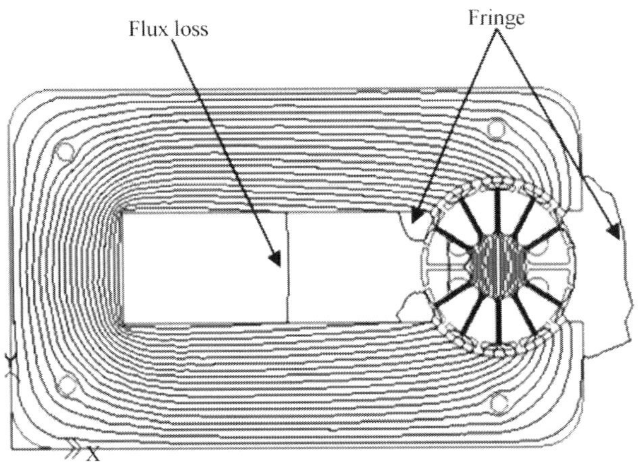

Figure 7. Simulation of the magnetic flux path.

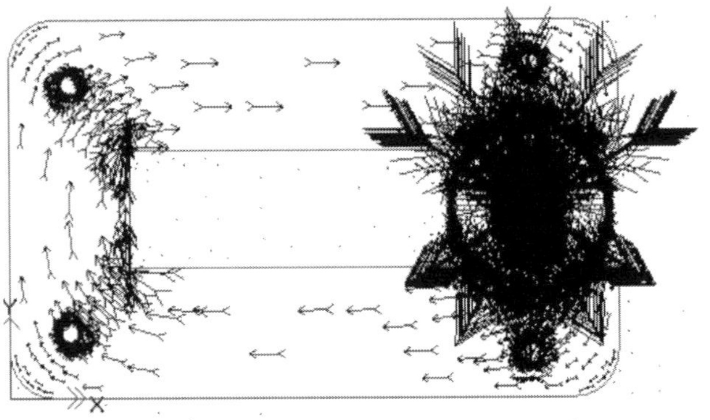

Figure 8. Simulation of the magnetic field density.

Figure 9. Superconducting coil motor assembly.

Figure 10. Conventional copper coil motor assembly.

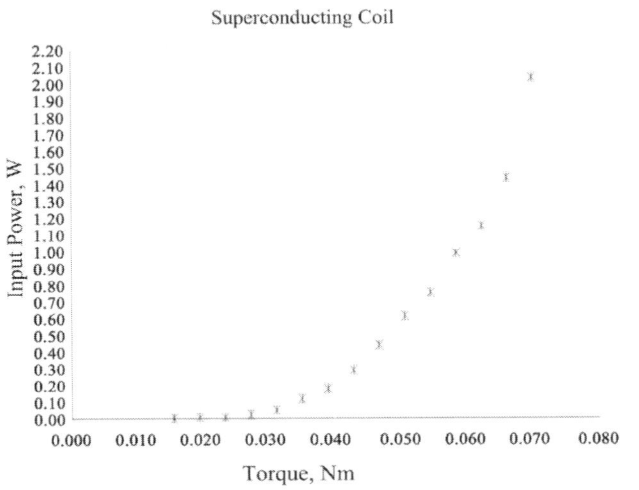

Figure 11. The performance of motor with superconducting coil.

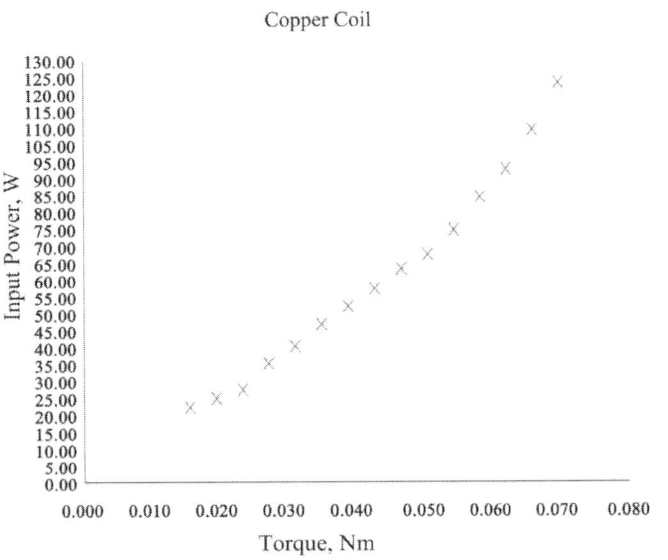

Figure 12. The performance of motor with copper coil.

Motor torque measurement is used to gauge the performance of the fabricated motor. Motor torque is one of the motor ratings that is normally used to rate the motor rotary force produced on its output shaft [13]. In the measurement of motor torque, lever length is referred to the radius of the shaft, and the force applied to the lever is generated by the weight attached to the motor shaft. DC current is used to provide the input current to both the field coil and the rotor. In testing both types of motors, a constant current was supplied to the rotor, while the current applied to the coil was varied at a fixed interval with a load mounted on a shaft.

The data was taken once the rotor started to rotate. In testing the motor with the superconducting coil, a rapid cooling of the coil affected its performance. As such, the coil was cooled gradually by pumping a small amount of liquid nitrogen into the reservoir at a time until equilibrium temperature was reached. Testing conventional copper coil was repeated with the same conditions but without any cooling process. From the testing results, the input power against the horizontal torque generated by the shaft for each type of motor were plotted and are shown in Figure 11 and Figure 12. The figures show that the input power to establish a certain level of torque for the superconducting coil motor is much lower than the conventional copper coils motor. This shows that for the same volume of conductor used in the construction of the motor, the motor with superconducting coil per-formed better performance in term of higher efficiency than the motor with copper coil. One of the possible factors in the enhancement of the efficiency of the motor system is due to lower hysteresis loss inside the super-conductor coils [14].

CONCLUSIONS

The design and feature of the motor components have been presented in this paper. The results of finite element analysis on the motors design have also been presented. Successful preliminary testing of the prototypes of the bench-top motor proved that the performance and energy efficiency of motor with superconductor coils is superior. In addition, the superconducting motor is lighter and compact with almost zero flux and noise leakages. The encouraging results indicate the potential of superconductor applications in machine devices and the possibility of commercialization in the near future.

ACKNOWLEDGEMENTS

This project is supported by the Ministry of Science, Technology and Innovation (MOSTI), Malaysia under the eScience Grant project no. 03-02-03-SF0132.

REFERENCES

1. Schiferi, R., Flory, A., Livoti, W.C. and Umans, S. (2006) High temperature superconducting synchronous motors, economic issues for industrial applications. *Petroleum and Chemical Industry Conference, Industry Applications Society 53rd Annual Meeting*, Philadelphia, Pennsylvania, 2006, 1-9.
2. Okazaki, T., Hayashi, K. and Sato K. (2006) Industrial applications of HTSC coils using next generation BSCCO wire. *SEI Technical Review*, 61, 24-28.
3. Ayai N., Kato, Fujikami, J., Fujino, K., Kobayashi, S., Ueno, E., Yamazaki, K., Kikuchi, M., Ohkura, K., Hayashi, K., Sato, K. and Hata, R. (2006) DI-BSCCO wires by controlled over pressure sintering. *Journal of Physics: Conference Series*, 43, 47-50.
4. Chu, C.W. (2000) High temperature superconducting materials: present status, future challenges, and one re-cent example – the superconducting ferromagnet. *Physica C*, 341-348, 25-30.
5. Noe, M. and Oswald, B.R. (1999) Technical and economical benefits of superconducting fault current limiter in power systems. *IEEE Transactions on Applied Superconductivity*, 9, 1347-1350. doi:10.1109/77.783552
6. Friedman, A., Shaked, N., Perel, E., Sinvani, M., Wolfus, Y. and Yeshurun, Y. (1999) Superconducting magnetic energy storage device operating at liquid nitrogen temperatures. *Cryogenics*, 39, 53-58. doi:10.1016/S0011-2275(98)00126-X
7. McCulloch, M.D. and Dew-Hughes D. (1998) Brushless AC machines with high temperature superconducting rotors. *Materials Science and Engineering B*, 53, 211-215. doi:10.1016/S0921-5107(97)00329-2
8. Iwakuma, M., Tomioka, A., Konno, M., Hase, Y., Satou, T., Iijima, Y., Saitoh, T., Yamada, Y., Izumi, T. and Shiohara, Y. (2007) Development of a 15 kW motor with a fixed YBCO superconducting field winding. *IEEE Transactions on Applied Superconductivity*, 17, 1607-1610. doi:10.1109/TASC.2007.898480
9. Jabbar, M.A. and Yeo, S.W. (2000) A New Architecture of Design Software for Electric Motors. *Proceedings of Electrical and Computer Canadian Conference*, 1, 360-364.
10. Edmonds, J.S., Sharma, D.K., Jordan, H.E., Edick, J.D. and Schiferi, R.F. (1992) Application of high temperature superconductivity to electric motor design. *IEEE Transactions on Energy Conversion*, 7, 322-329. doi:10.1109/60.136228
11. Fitzgerald, A.E., Kingslev, J. and Umans, S.D. (1990) Electric Machinery. McGraw Hill, New York.
12. Hsu, J.S. (2001) Flux guides for permanent magnet machines. *IEEE Transactions on Energy Conversion*, 16, 186-191. doi:10.1109/60.921471
13. Song, M., Yoon, Y.S., Jang, W.K., Ko, T., Hong, G.W. and Jang, I.B. (1999) The design, manufacture and characteristic experiment of a small-scaled high-T_c superconducting synchronous motor. *IEEE Transactions on Applied Superconductivity*, 9, 1241-1244. doi:10.1109/60.921471
14. Kirtley, J.L. (1999) Applications of superconductors in electric utility systems. IEEE Power Engineering Society Summer Meeting, 2, 1181-1186.

CHAPTER 7

POWER FACTOR IMPROVEMENT OF AC-DC CONVERTER BASED ON SEPARATELY EXCITED DC MOTOR USING PASSIVE FILTER

Shazia Baloch, Mukhtiar Ahmed Mahar, Raza Haider, and Shakeel Ahmed

Keywords: DC Motors, AC-DC Converters, Power Factor, Total Harmonic Distortion, Passive Filters

INTRODUCTION

DC (Direct Current) power supplies are important and extensively used in many electrical and electronics applications such as computers and telecommunication equipment. Because of the friendly technology electrical vehicle is rapidly increasing; thus DC power is becoming the heart of these systems [1] [2]. To achieve this DC recently a large number of different AC-DC converters have been developed and due to their low cost, high efficiency and fast switching actions these converters are used in DC motors on a large scale. DC motor drives have very essential industrial applications and are widely used because of its high performance. Speed control of DC motor is very easy by power electronic converters. And due to its favorable cost, reliability, simplicity and ease of applications DC motors have long been the backbone of industrial application [3]. AC-DC converters are playing a vital role in power electronics. In these power electronic converters, the AC supplies can be converted to DC. The non-linear natures of power electronic devices are of prominent harmonic distortion and low power factor which are of great concern [4]. It is also well known fact that AC-DC supplies suffer from two drawbacks i.e. current distortion and low power factor [5].

According to IEEE standards 519-1992, the harmonic distortions produced by power electronic devices should be under specific value that any individual harmonic should not be exceed more than 3% and total harmonic distortion should not be more than 5% [6].

Citation: Baloch, S., Mahar, M., Haider, R. and Ahmed, S. (2016). Power Factor Improvement of AC-DC Converter Based on Separately Excited DC Motor Using Passive Filter. Open Journal of Energy Efficiency, 5, 71-77. DOI:10.4236/ojee.2016.53007.

Therefore, it is evidence that THD should not exceed more than 8% when load applied is varying and continuously changing and should be 5% when full-load is applied [7].

AC-DC CONVERTERS

A power electronic device which converts main AC voltage to DC voltage is called AC-DC converters. AC-DC converters are used in many applications such as in DC Motor speed control, Power supplies, Converters at the input end of DC transmission line, Portable hand tool drives and Solid state switch applications [8].

Effects of AC-DC Converters

- Voltage drops at the buses.
- Causes blown of fuses on power factor correction capacitors due to flow of high current and high voltage.
- Electromagnetic interference with the communication circuits.
- Heating of transformers, generators.
- Increase losses in the equipment connected to the utility such as transformers and motors.

HARMONICS

The harmonics are defined as "a wave whose frequency is an integer multiple of the fundamental frequency" it can be odd harmonics or even harmonics. As power system is designed to operate with pure sinusoidal voltage and current with a constant frequency which is also known as fundamental frequency but by connecting the non-linear power electronics loads like diode and thyristors to the system causes excessive harmonics current which causes distortion of voltage and current. This in turns results power quality problems [9].

Main Problem of Harmonics

- Additional losses in supply generators and transformers.
- Neutral conductor overloading.
- Interference to communication.
- Over loading of power factor correcting capacitors supply voltage distortion.
- Error in energy meters.
- Malfunction of sensitive equipment.
- Reduced power factor to low values.

SINGLE TUNED PASSIVE FILTERS

By using passive elements like inductor, capacitor and resistor passive filters are designed so Passive filter consists of a certain combination of inductor capacitor and resistor as shown in Figure 1, which used to reduce the harmonic current [9] [10].

Many topologies for harmonic mitigation and active filter are used to control the harmonic. But power more than 1 MW passive filters are the best choice because some of the advantages such as low cost, simple design and high reliability. A single installation of passive filter can service many purposes i.e. reactive power compensation, power factor improvement, total harmonic distortion to acceptable limits. A well designed passive filter can be implemented in large sizes of MVAr ratings and provide almost maintenance free service [11].

RESEARCH OBJECTIVES

The objective of this research work is to develop an experimental model of single phase AC-DC converters based separately excited dc motor to analyze the harmonics caused by nonlinear nature of power electronics AC-DC converters and using single tuned passive filter to reduce harmonics according to IEEE standards 519- 1992 and power factor improvement.

RESEARCH METHODOLOGY

This research work is based on experimental setup and the methodology used for this research work is given as

- Development of experimental setup of dc motor with rectifier.
- Power factor and harmonic analysis of experimental models.
- Power factor improvement by using passive tuned filter.

EXPERIMENTAL SETUP

Figure 2 shows the experimental setup of rectifier based separately excited dc motor it can be seen from the Figure 2 that the input supply is AC. To convert this AC into DC AC-DC converter is used. Power quality analyzer is used to measure the harmonic distortion and the power factor before and after installation of filter. Single tuned passive filter is used to improve the power factor and reduce the THD caused by AC-DC converter according to IEEE standard 519-1992. Figure 3 shows the developed experimental model of the research work.

HARMONICS AND POWER FACTOR BEFORE AND AFTER INSTALLATION OF FILTER AT-LOAD

Figure 4 and Figure 5 show the wave spectrum of AC-DC converter at-load without and with installation of passive filter. It can be seen from the wave spectrum in Figure 4 that without installation of filter the THD measured is 47.5% which is very high to reduce the high THD passive filter is used and it can be seen from Figure 5 that after installation of passive filter the THD reduced from 45.5% to 1.6% at-load.

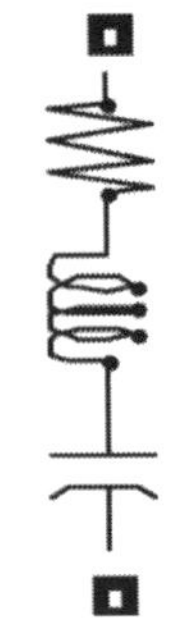

Figure 1. Tuned passive filter.

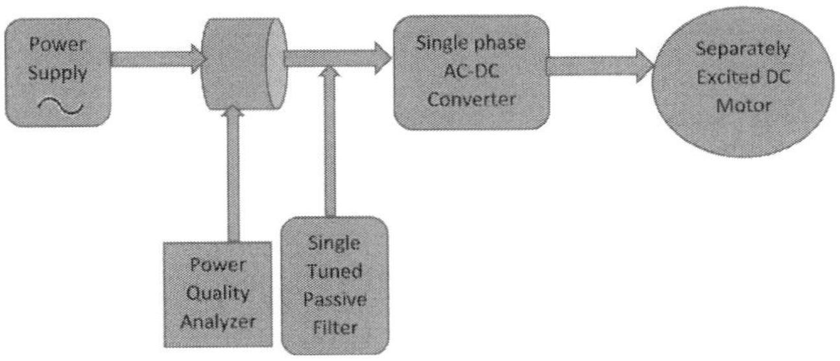

Figure 2. Experimental setup of rectifier based separately excited DC motor.

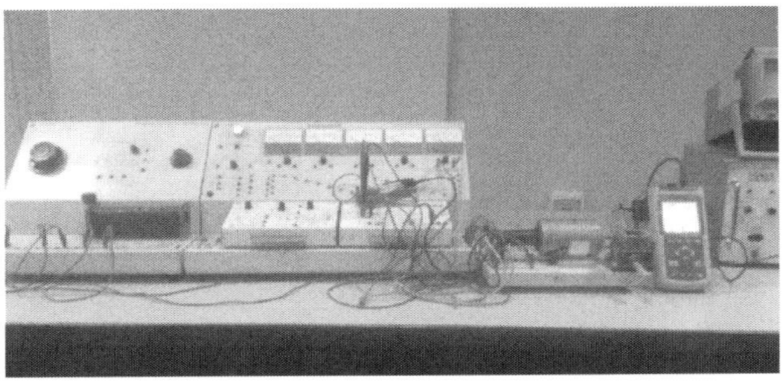

Figure 3. Experimental model of research work.

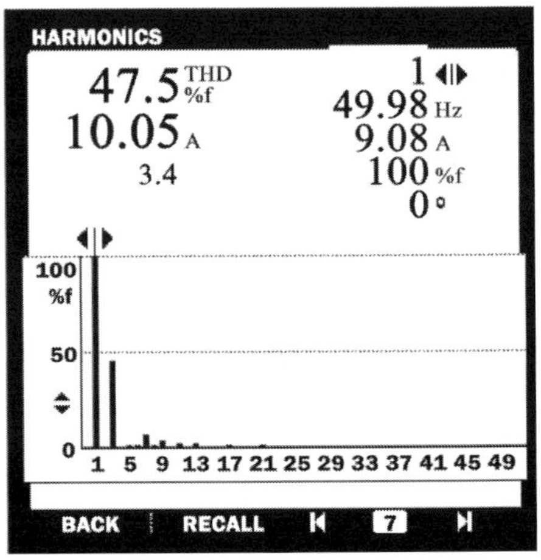

Figure 4. Wave spectrum of THD without filter.

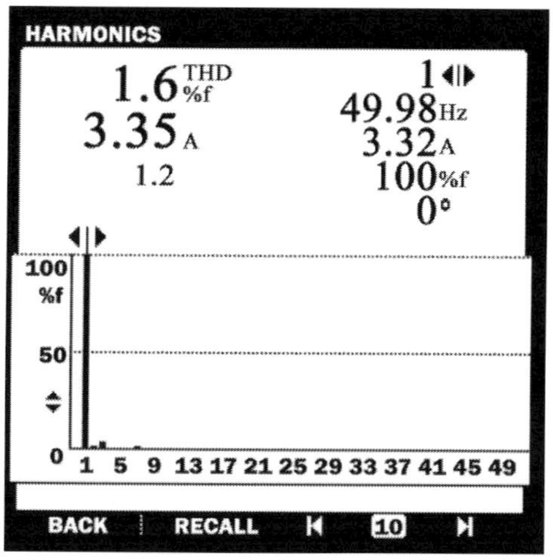

Figure 5. Wave spectrum of THD with filter.

Similar to THD the power factor without filter is 0.67 as shown in Figure 6 which is very low butFigure 7 shows that after installing filter the low power factor has been improved from 0.67 to 0.99.

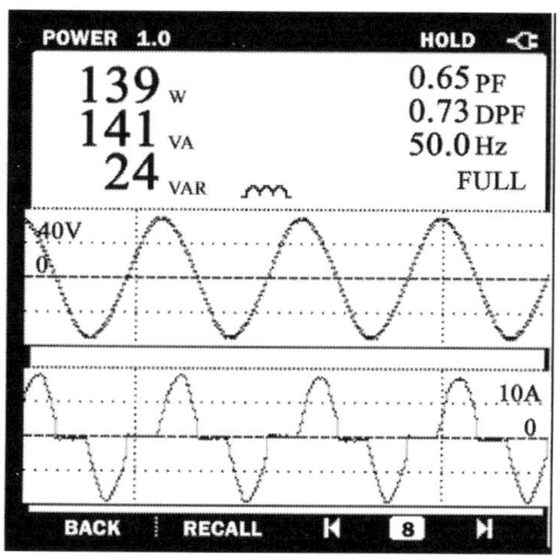

Figure 6. Power factor without filter.

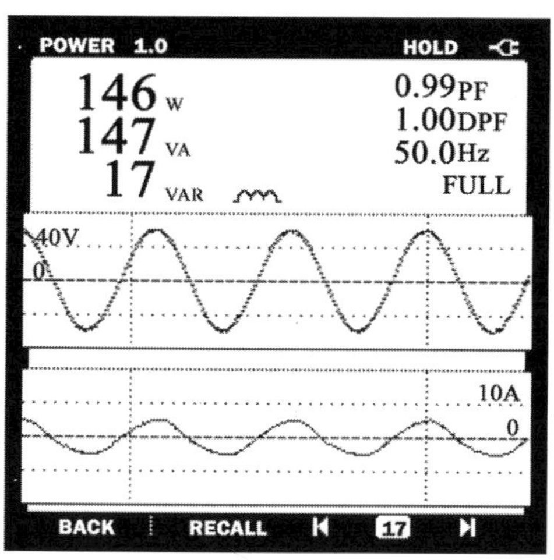

Figure 7. Power factor with filter.

Table 1. Power factor and total harmonic distortion with and without filter at-load.

S.No	Conditions	Power Factor (PF)		Total Harmonic Distortion (THDi) %	
		Without Filter	With Filter	Without Filter	With Filter
1.	At load	0.65	0.99	47.5	1.6

Table 1 shows a comparison of power factor and THD with and without filter at-load. It can be seen from the Table 1 that without filter the power factor at-load is very low i.e. 0.65. To improve this power factor single tuned passive filter is used. After installation of filter the power factor has been improved up to 0.99. Similarly without filter the total harmonic distortion was very high i.e. 47.5% but after installation filter it has been reduced to 1.6%.

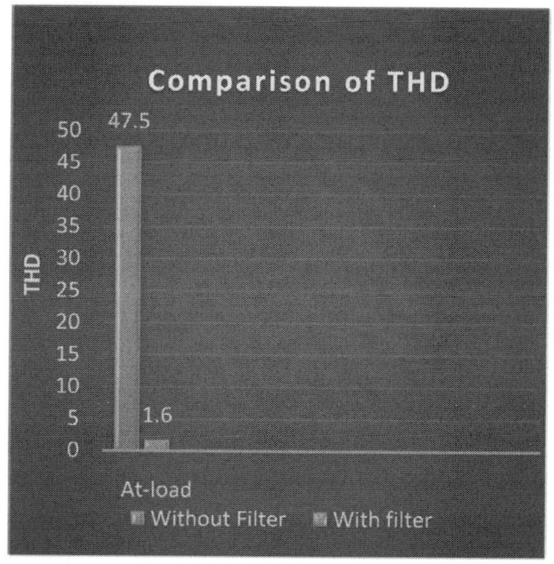

Figure 8. Graphical representation for the comparison of THD with and without filter.

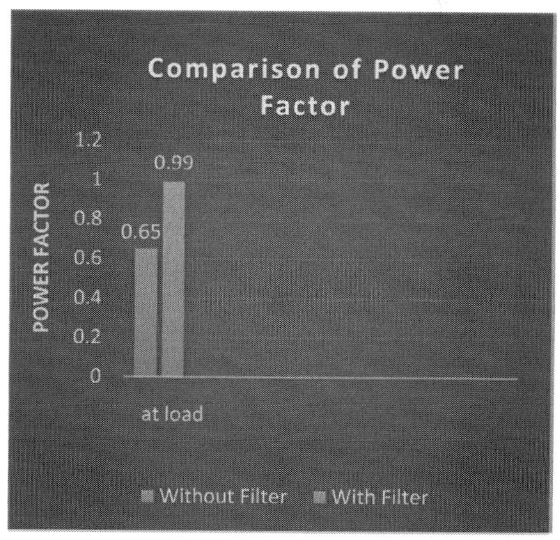

Figure 9. Graphical representation for the comparison of power factor with and without filter.

GRAPHICAL REPRESENTATION OF PF AND THD WITH AND WITHOUT FILTER

Figure 8 and Figure 9 show the comparison of power factor and THD by graphical representation. It can be seen from the comparison that without installation of filter the power factor is very low and total harmonic distortion is very high but after installation tuned passive filter the power factor improved to unity and total harmonic distortion reduced according to IEEE standard 519-1992.

CONCLUSION

In this research work power factor and total harmonic distortion of separately excited DC motor controlled with AC-DC converter has been analyzed at-load. The tuned passive filter is also designed in this research work. From the experimental results, it is concluded that the total harmonic distortion (at-load) has reduced within the range of IEEE Standards 519-1992. It has been seen that without filter the power factor of DC motor was very low i.e. 0.65 at-load and total harmonic distortion was very high 47.5%. After development of single tuned filter, it is clear from experimental results that THD reduced within the range of IEEE standards 519-1992 and the power factor is improved near to unity at-load.

FUTURE WORK/SUGGESTIONS

As this research work was done on single phase converter using single tuned passive filter the same work can be extended on three phase AC-DC converter based DC drives. This research work can be extended to implement passive filters in field control methods of DC drives. Instead of single tuned passive filter we can also use parallel input resonant filter, series input resonant filter.

REFERENCES

1. Ismail Hossain, Md. and Alam, M.J. (2014) Cuk Topology Based Power Factor Correction and Output Voltage Regulation of Ac-Dc Converter. 2014 International Conference on Electrical Engineering and Information & Communication Technology (ICEEICT), 10-12 April 2014, 1-6. http://dx.doi.org/10.1109/ICEEICT.2014.6919126
2. Gupta, R. and Ruchika, R. (2013) Thyristor Based DC Motor Control with Improved PF & THD. International Journal on Electrical Engineering and Informatics, 5, 519-536. http://dx.doi.org/10.15676/ijeei.2013.5.4.10
3. P.karpagavalli, P. and Jeyakumar, A.E. (2014) Power Factor Improvement in Single Phase dc Motor Drive Using Double Sided Pwm Switching. Journal of Theoretical and Applied Information Technology, 69, 232.
4. Lenine, D., Sai Babu, Ch. and Surya Kumari, J. (2010) A Static Improvement of Predictive Control for Single Phase Voltage Fed Power Factor Correction Converters. 2010 International Conference on Advances in Recent Technologies in Communication and Computing (ARTCom), Kottayam, 16-17 October 2010, 116-120. http://dx.doi.org/10.1109/ARTCom.2010.16
5. Ahmed, M.M.R., Putrus, G.A. and Shokralla, S.S. (2005) A New Approach for Speed Control of A.D.C. Motor Fed from AC to DC Sepic Converter. 18th International Conference and Exhibition on Electricity Distribution, Turin, 6-9 June 2005, 1-5. http://dx.doi.org/10.1049/cp:20051219
6. Pyakuryal, S. and Matin, M. (2013) Harmonic Analysis for a 6-Pulse Rectifier. IOSR Journal of Engineering, 3, 57-60.http://dx.doi.org/10.9790/3021-03325760
7. Memon, Z.A., Uqaili, M.A. and Soomro, M.A. (2011) Experimental Analysis of Harmonic Mitigation Effects on Three Phase Six Pulse Converter by Using Shunt Passive Filter. Mehran University Research Journal of Engineering & Technology, 30, 653-656.
8. Rashid, M. (2011) Power Electronics: Circuits, Devices, and Applications. 3rd Edition, Prentice Hall, New Jersey.
9. Soomro, D.M. and Almelian, M.M. (2015) Optimal Design of a Single Tuned Passive Filter to Mitigate Harmonics in Power Frequency. ARPN Journal of Engineering and Applied Sciences, 10, 9009-9014.
10. Zobaa, A.F., Abdel Aziz, M.M. and Abdel Aleem, S.H.E. (2010) Comparison of Shunt-Passive and Series-Passive Filters for DC Drive Loads. Electric Power Components and Systems, 38, 275-291. http://dx.doi.org/10.1080/15325000903273262
11. Das, J.C. (2003) Passive Filters-Potentialities and Limitations. 2003 Conference Record of the 2003 Annual Pulp and Paper Industry Technical Conference, Charleston, 16-20 June 2003, 187-197. http://dx.doi.org/10.1109/PAPCON.2003.1216916

CHAPTER 8

INDUCTIVE-MOTORS LOAD PARTITION BASED ON COHERENCE

Chengjun Xia, Yun Zhou, Kun Men, and Yinggeng Xie

Keywords: Load Model, Inductive-Motor Load, PSO, FCM, Clustering Validity, Dynamic Equivalent

INTRODUCTION

To simplify the analysis of the large scale interconnected power system, a series of power system equivalent methods had been proposed in recent years. Numerous investigations had shown that the load model has an important and complex role on the dynamic simulation and response of power system after fault. Because most power system loads involve many inductive motors, it makes sense to develop an effective identification theory for industrial loads which comprised of different induction motors. In order to obtain more accurate results in transient analysis and simulation, it is necessary to identify inductive motors based on coherence [1]-[3].

For actual power system, there is no coherence load in the strict sense because of the uncertainty of load model. The similar dynamic characteristics exist in coherence load due to its ambiguity. In [4], clustering method is used to partite motor load, and criteria of coherence were proposed in dynamic control and analysis in power system. In [5], dynamic performance of power system load representation is analyzed and the importance of load modeling is emphasized. In [6], the identifiability of the equivalent motor parameters is investigated in the composite load model, case studies verify the validity of the proposed method.

This paper proposes a method for the identification of inductive motors load partition based on coherence. In order to partite motors load rapidly and accurately, fuzzy c-means clustering along with particle swarm optimization (PSO-FCM) algorithm is applied to identify coherent motors base on its physical essence of fuzziness. Then fuzzy partition entropy function is used as a measurement index to measure the performance of algorithm and evaluate the effectiveness of fuzzy c-mean clustering algorithm based on particle swarm optimization. The computational results show that control strategy works effectively and can be applicable to the identification problem.

Citation: Xia, C., Zhou, Y., Men, K. and Xie, Y. (2014). Inductive-Motors Load Partition Based on Coherence. Journal of Power and Energy Engineering, 2, 162-169. DOI: 10.4236/jpee.2014.29023.

LOAD MODELS IN POWER SYSTEM

Static Load Models [7]

The load models are traditionally classified into two broad categories: static models and dynamic models. A static load model expresses the characteristics of the load at any instant of time as algebraic functions of the bus voltage magnitude and frequency at that instant, including ZIP model, exponential model and polynomial model. An alternative model which has been widely used to represent the voltage dependency of loads is the ZIP model, as it is composed of constant impedance (Z), constant current (I), and constant power (P) components.

Dynamic Load Models

There are, however, many cases where it is necessary to account for the dynamic of load components. Studied of inter-area oscillations, voltage stability, and long-term stability often require load dynamics to be modeled. Study of system with large concentrations of motors also requires representation of load dynamics.

The dynamic characteristics of induction machine should be taken in account in terms of influence of induction machine on power system. The dynamic equation of an induction machine may be written as:

$$\begin{cases} \frac{dE'_d}{dt} = -\frac{1}{T'_0}[E'_d - (X - X')I_q] - \omega_b(\omega_r - 1)E'_q \\ \frac{dE'_q}{dt} = -\frac{1}{T'_0}[E'_q - (X - X')I_d] - \omega_b(\omega_r - 1)E'_d \end{cases} \quad (1)$$

$$\begin{cases} I_d = \frac{1}{R_s^2 + X'^2}[R_s(V_d - E'_d) + X'(V_q - E'_q)] \\ I_q = \frac{1}{R_s^2 + X'^2}[R_s(V_q - E'_q) - X'(V_d - E'_d)] \end{cases} \quad (2)$$

$$\frac{ds}{st} = \frac{1}{T_J}[T_m - T_e] \quad (3)$$

here, E'_d and E'_q are transient electric potential, ω is rotor angular velocity, $\omega_b = 2\pi f_{base}$, $f_{base} = 50Hz$, $\omega_r = 1 - s$, $\omega = 1 - s_0$; T_J; TJ is motor inertia time constant, S is slip speed of rotor. $T_m = T_0(A\omega^2 + B\omega + C)$ is mechanical torque, $T_e = E'_d i_d + E'_q i_q$ is electromagnetic torque; $X = X_s + X_m$ is rotor open circuit reactance.

Synthesis Load Model

For actual power system, the modeling of loads is complicated because a typical bus represented in stability studies is composed of a large number of devices. The exact composition of load is difficult to estimate. The characteristics of load can be represented by the ZIP models and induction machine. The synthesis load model consists of third-order induction motor in paralleled with ZIP. The schematic structure of this synthesis load model is shown in Figure 1.

In Figure 1, *Rs* is stator equivalence resistance, X_s is stator equivalence leak reactance, R_r is rotor equivalence resistance, X_r is rotor equivalence leak reactance, X_m is value of mutual inductance between stator and rotor windings.

Model order reduction techniques are often used to reduce the model complexity. When analyzing the transients of circuits and electrical machines, the detailed model is often built first to describe the plant as accurately as possible. Then the model order can be reduced according to the specific problems under investigation.

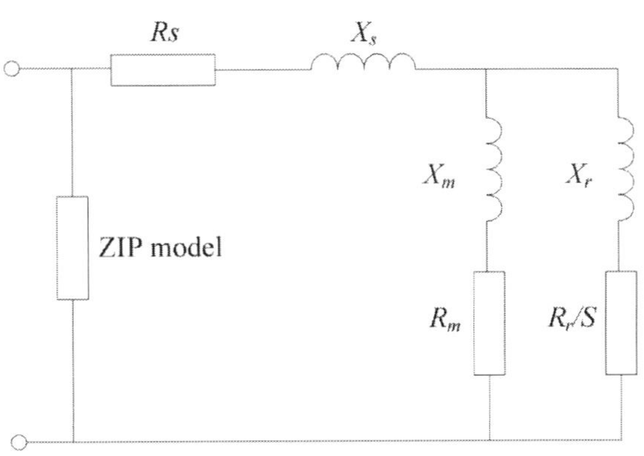

Figure 1. Equivalent circuit of the synthesis load model.

THE CHARACTERISTICS FOR LOAD PARTITION IDENTIFICATION

The dynamic characteristics of equivalence load will be poor in the circumstance because there are amount of discrepant load in equivalence load. To guarantee the accuracy of dynamic simulation of power system, it's necessary to partite load in appropriate way [8]. The essence of partition equivalence load is to observe the dynamic characteristics of aggregation load. For induction machine, the best tool to study the performance is slip. The relationship between slip and rotor speed is $s = (n_0 - n)/n_0$, so s is corresponding to speed completely. If T_m is constant, the formula of Te can be expressed as following:

$$T_e = \frac{U^2 / X_l}{\frac{s}{s_{cr}} + \frac{s_{cr}}{s}} \quad (4)$$

where, $S_{cr} = R_r/X_l$ is critical slip, X_l is sum of rotor leak reactance and stator leak reactance. Therefore, the linearized forms of Equation (4) is

$$\Delta T_e = \zeta \Delta s \quad (5)$$

$$\zeta = \frac{U_N^2}{X_l} \cdot \frac{s_{cr} - s_N^2 / s_{cr}}{\left[s_{cr} + s_N^2 / s_{cr}\right]^2} \quad (6)$$

Generally, the value of S_N is quite small, so compare S_N^2 / S_{cr} to S_{cr}, it can be ignored. Then,

$$\zeta \approx \frac{U_N^2}{X_l} \times \frac{1}{s_{cr}} = \frac{U_N^2}{X_l} \times \frac{X_l}{R_r} = \frac{U_N^2}{R_r} \quad (7)$$

Therefore, the linearized forms of Equation (7) is

$$\Delta \dot{s} = -\frac{U_N^2}{T_J R_r} \Delta s \quad (8)$$

From the given Equation (8), the corresponding solution is

$$\lambda = -\frac{U_N^2}{T_J R_r} \quad (9)$$

So λ is determined by $T_J R_r$. If values of $T_J R_r$ of the induction machine are similar, then they own similar dynamic characteristics. Finally, to evaluate the state of induction machine, ($T_J R_r$, K_L) is chosen to be a characteristic quantity which is used to identify coherence inductive motors in power system. So, coherent inductive motors own similar characteristic quantity.

IMPROVED FUZZY C-MEANS CLUSTERING ALGORITHM

Fuzzy c-Means Clustering

Fuzzy c-means clustering (FCM) approach is widely implemented to tackle practical problems with uncertainty, imprecision and vague peculiarity. Initial categories and parameters are given before the program began to iterate. Satisfying solution for project can be obtained though modification, as well as iteration. Robust results for categories can be achieved based on fuzzy clustering. The formula of FCM can be expressed as following:

$$\min J_m = \sum_{i=1}^{N}\sum_{j=1}^{K} u_{ij}\left\|x_i - c_j\right\|^2 = \sum_{i=1}^{N}\sum_{j=1}^{K} u_{ij}(d_{ik})^2 \quad (10)$$

where, x_i is sample, c_j is a set of class center, $U = (u_{ij})$ is a membership matrix of samples.

To solve Equation (10) is impossible. Formula for soft clusters was given by mathematician Bezdek [9] [10]:

$$u_{ij} = \frac{1}{\sum_{k=1}^{c}\left(\frac{d_{ij}}{d_{ik}}\right)^{\frac{2}{m-1}}} \quad (11)$$

$$c_j = \frac{\sum_{i=1}^{N} u_{ij}^m x_i}{\sum_{i=1}^{N} u_{ij}^m} \quad (12)$$

$$d_{ij} = \left[\sum_{j=1}^{m}(x_{ij} - c_{kj})^2\right]^{\frac{1}{2}} \quad (13)$$

where, m is constant of vague degree, d_{ij} represents the distance between sample and class center.

Particle Swarm Optimization

In PSO, the solution of the objective function can be imaged as a point in d -dimension, just as a bird fly in the sky. The process of finding the solution is based on the analogy of swarm of bird flocking and fish schooling. A swam of particles represent a candidate solution to the optimization problem [11] [12].

Each particle adjusts its position according to its own experience, and the experience of its neighbouring particles. The position and velocity of the ith particle in the d-dimensional search space are represented in (14) and (15).

$$x_{id}^{k+1} = x_{id}^{k} + v_{id}^{k+1} \quad (14)$$

$$v_{id}^{k+1} = \omega v_{id}^{k} + c_1 rand_1(p_{id} - x_{id}) + c_2 rand_2(p_{gd} - x_{id}) \quad (15)$$

where, X_i^k is position of individual i at iteration k, V_i^k is velocity of individual i at iteration k, ω is weight pa rameter, c_1 is cognitive factor, c_2 is social factor, p_{id} is the best position of individual i until iteration k, P_{gd} is the best position of the group until iteration k, $rand_1$, $rand_2$ are random numbers between 0 and 1.

To evaluate the merit rating of the particles, we can set a function to measure which called the fitness function. According to the fitness function, the best particle is selected. And initial positions and velocities are vector. Particles can move to a better position by sensing its own position and the particles nearby. In every step each agent's position is updated by the best fitness value called as p_{ibest}. Also each agent knows the best value in the group called as p_{gbest} among the p_{ibest}. After a specified number of iterations, the agents searching points gradu-ally reach the global optimal point using p_{ibest} and p_{gbest}.

An inertia weight in PSO algorithm was introduced in 1998, in order to provide better control exploration [13]. The inertia weight is used to control the impact of the previous history of velocity, and thus to influence the tradeoff between global and local exploration abilities of the moving agents. In PSO, the balances between the global and local exploration abilities are mainly controlled by inertia weights. Both convergence speed and ac-curacy of solution will be affected. The inertia weight is suggested to decrease linearly from 0.9 to 0.4 during the run. The inertia weight is formulated as in (16).

$$\omega = \left(\frac{T_{\max} - t}{T_{\min}}\right)^{\omega_n}(\omega_{\max} - \omega_{\min}) + \omega_{\min} \quad (16)$$

where ω_n is the attenuation index of inertia weight, $\omega_{\max}$ is the initial value of the inertia weight, $\omega_{\min}$ is the final value of the inertia weight, T_{max} is the maximum iteration number; t is the current iteration number.

Particle Swarm Optimization Fuzzy c-Means Clustering Algorithm

As fuzzy c- means algorithm is easy to convergent to partial optimum, PSO is introduced to program instead of the process based on gradient descent. The performance of the algorithm is fantastic and effective in the process of iteration, and it is insensitive to initial values. Finding the optimal particle center is main task in the process of iteration. Hence, class center c_i is encoded as a particle. The particle can be expressed as following:

$$x_l = \left(c_{1,1},\ldots,c_{1,D},\ldots,c_{i,j},\ldots,c_{K,1},\ldots,c_{K,D}\right) \quad (17)$$

where, c_{ij} is the center i^{th} with clustering method j^{th}.

To evaluate the performance of particle, objective function is recommended. In the paper, J_H is used to be objective function to compare the performance among particles. The objective function can be expressed as following:

$$f(x_l) = J_H \quad (18)$$

Solving Equation (18), we can obtain the fitness for every particle. Particle optimum point can be obtained after comparing the global optimal point and particle optimal point. If $f(x_{id})$ is greater than $f(p_{gd})$, then fitness of global optimal point will be updated, vice versa.

$$p_{id}(t) = \begin{cases} x_{id}, & f(x_{id}(t)) \geq f(p_{id}(t-1)) \\ p_{id}(t-1), & f(x_{id}(t)) < f(p_{id}(t-1)) \end{cases} \quad (19)$$

The iteration structure of particle swarm optimization fuzzy c-means clustering algorithm is presented in Figure 2.

Evaluating the Effectiveness of Clustering Algorithm

To evaluate the effectiveness of fuzzy c-mean clustering algorithm based on particle swarm optimization, fuzzy partition entropy $S(U,K)$ is used as a measurement index to measure the performance of algorithm.

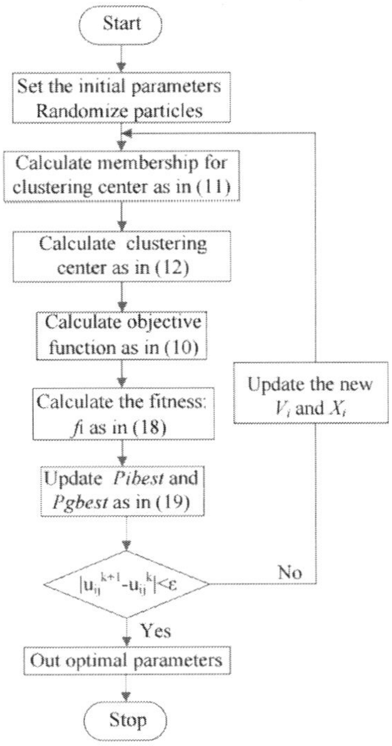

Figure 2. Block diagram of PSO-FCM algorithm.

The small $S(U,K)$ is, the dispersion degree between particle is smaller, that is to say, the performance of clustering algorithm is more perfect. $S(U,K)$ can be expressed as:

$$S(U,K)=\frac{\frac{1}{N}\sum_{i=1}^{N}\sum_{j=1}^{K}u_{ij}^{2}\left\|x_i-c_j\right\|^2}{\min_{i,j}\left\|c_i-c_j\right\|^2} \tag{20}$$

The distance between clusters and the data points assigned to them should be minimized and the distance be-tween clusters should be maximized [14].

SIMULATION AND NUMERICAL ANALYSIS

To verify the effectiveness of the proposed algorithm, source program is developed for identification of inductive-motors load partition based on coherence for simulations and is applied to the standard IEEE-39 bus benchmark system. The parameters used for chaos particle swarm optimization algorithm are as following: number of particles is 30, ω_{max} = 0.9, ω_{min} = 0.2, number of iterations is *itermax* = 100, λ = 1.

Research system and external system are shown in Figure 3. Broken line passing through bus 39, bus 3 and bus 27 is shown in Figure 3. Taking actual research into consideration, research system is one above the broken line. The program is developed to partite the coherence motors in outside system including 13 motors.

It is clearly known that clustering results with the relationship among membership matrix. The results of identification of inductive-motors load partition based on coherence can be written as following.

2 Group: (3, 4, 7, 8, 12, 15, 18);

(16, 20, 21, 23, 24, 27)

3 Group: (3, 4, 7, 8, 12, 15, 18);

(16, 20, 21); (23, 24, 27)

4 Group: (3, 4, 7, 8, 12, 15, 18);

(16, 20, 21); (23, 27); (24)

The value of fuzzy partition entropy $S(U,K)$ is shown in Table 1.

Therefore, it can be inferred that the cluster performance is best in group 2. Optimum membership matrix U can be obtained after calculation and simulation.

$$U=\begin{vmatrix} 0.0084 & 0.0006 & 0.0064 & 0.0033 & 0.0092 & 0.0006 & 0.9841 & 0.0033 & 0.9767 & 0.9609 & 0.9657 & 0.9870 & 0.9657 \\ 0.9916 & 0.9994 & 0.9936 & 0.9967 & 0.9908 & 0.9994 & 0.0159 & 0.9967 & 0.0232 & 0.0391 & 0.0343 & 0.0130 & 0.0343 \end{vmatrix}$$

For this improved PSO- FCM algorithm, the program runs after parameters initialized, and convergence results of iteration are shown in Figure 4.

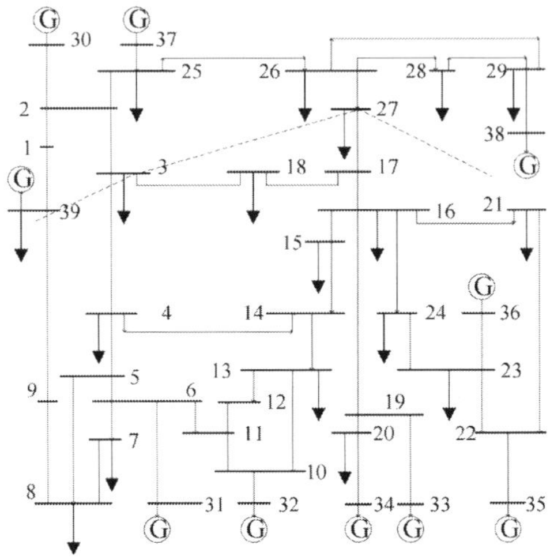

Figure 3. Diagram of IEEE-39 with 10 generators.

The progress of PSO-FCM algorithm can be illustrated easily in Figure 4. After 35 times iteration, the task of iteration is over. And the optimum solution can be obtained. So identification of inductive motors load partition based on coherence can be easily implemented.

To illustrate the stability of the proposed algorithm, source program is developed for identification of inductive motors load partition, and runs for several times, which contain 2 groups of inductive motors. The results are shown in Table 2. It is clearly know that the results are same nearly. The value of fuzzy partition entropy in iteration 3rd, 5th, 6th, 8th, and 10th is 7.7845e−04. The value in iteration 2nd, 7th is 7.7853e−04. The error existed between maximum and minimum is only 0.0034. It is sufficient to illustrate the stability of proposed algorithm.

Considering convergence rate, the PSO-FCM method can always generate precipitous convergence rate to-ward an acceptable solution, which show that the PSO-FCM method has good convergence performance. As shown in Figure 5, the convergence characteristic from 20 to 30 times is different. The program based on PSO-FCM is easier to be convergent with certain reliability. And it can be applied to large-scale power systems.

Table 1. The validity function value.

Group	2	3	4
$S(U,K)$	7.7862e−004	0.0437	0.2261

Table 2. $S(U,K)$ value of ten times simulation (e−004).

Times	1	2	3	4	5
S(U,K)	7.7862	7.7853	7.7845	7.7879	7.7845
Times	6	7	8	9	10
S(U,K)	7.7845	7.7853	7.7845	7.7845	7.7845

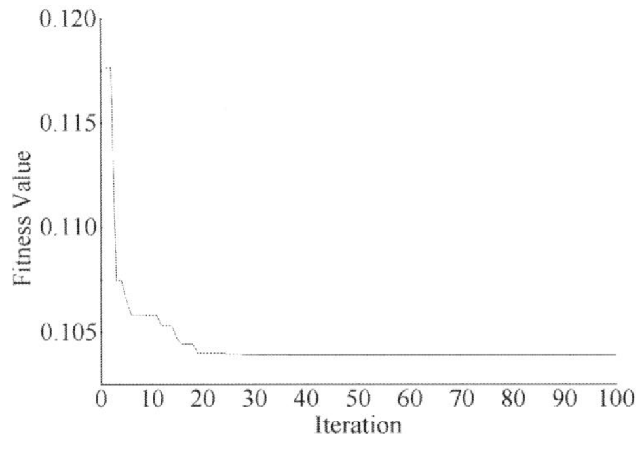

Figure 4. Fitness function evolution curve.

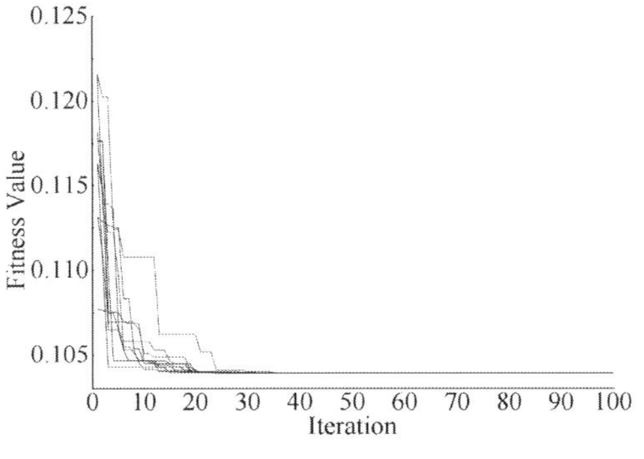

Figure 5. Fitness function evolution curve in ten times.

CONCLUSION

To investigate the identification of inductive motors load partition based on coherence, PSO -FCM algorithm is proposed considering the characteristic of inductive motor load. The proposed algorithm uses PSO to find optimum instead of the process of gradient descent. The coherence motors can be identified according to membership matrix, as well as intensity of coherence motors. For the sake of validating the coherence motor in dynamic studies, a series of simulation tests are performed and compared. And the analysis results of measured simulation curves show that the coherence motors identified by the proposed strategy possess good description ability for load dynamics. What's more, the proposed identification strategy can improve the stability of identification results, by which promotes the practical application of the load model. Case study also shows the effectiveness and feasibility of the proposed identification strategy.

ACKNOWLEDGEMENTS

This work was financially supported by National High Technology Research and Development Program of China (2012AA050209).

REFERENCES

1. Shenghu, L., Zhengkai, L., Xinjie, H. and Shusen, J. (2010) Dynamic Equivalence to Induction Generators and Wind Turbines for Power System Stability Analysi s. International Symposium on Power Electronics for Distributed Generation Systems, Hefei, 16-18 June 2010, 887-892.
2. Baozhen, Z., Yao, Z., Qing, Z. and Zhaobin, D. (2011) Application of Dynamic Equivalence Based on Coherency in South China Power Grid. Power and Energy Engineering Conference, Wuhan, 25-28 March 2011, 1-4.
3. Resende, F.O., Matevosyan, J. and Milanovic, J.V. (2013) Application of Dynamic Equivalence Techniques to Derive Aggregated Models of Active Distribution Network Cells and Microgrids. PowerTech IEEE Grenoble, Grenoble, 16-20 June 2013, 1-6.
4. Haiqiang, Zh., Ping, J., Hui, Y. and Ran, S. (2010) Dynamic Equivalent Method of Interconnected Power Systems with Consideration of Motor Loads. Science China Technological Sciences, 40, 902-908. http://dx.doi.org/10.1007/s11431-010-0110-8
5. He, R., Ma, J. and Hill, D.J. (2006) Composite Load Modeling via Measurement Approach. IEEE Transactions on Power Systems, 21, 663-672. http://dx.doi.org/10.1109/TPWRS.2006.873130
6. Ma, J., Han, D. and He, R. (2007) Research on Identifiability of Equivalent Motor in Composite Load Model. Power Tech IEEE Lausanne, Lausanne, 1-5 July 2007, 1015-1020.
7. Kundur, P. (2001) Power System Stability and Control. 1st Edition, McGraw Hill Inc., New York.
8. Wang, J., Han, M. and Ma, J. (2010) A New Identification Strategy for Improving Convergence Stability of Load Model Parameters. International Conference on Electrical and Control Engineering, Wuhan, 25-27 June, 145-148.
9. Pal, N.R., Pal, K. and Keller, J.M. (2005) A Possibilistic Fuzzy c-Means Clustering Algorithm. IEEE Transactions on Fuzzy Systems, 13, 2005. http://dx.doi.org/10.1109/TFUZZ.2004.840099
10. Chunjuan, Y. (2011) Derivatives of Fuzzy C-Means Method and Their Application Comparisons. International Conference on Computer Science and Service System, Nanjing, 27-29 June 2011, 326-329.
11. Fengrui, Z., Jianshu, C. and Zhenhui, X. (2013) An Improved Particle Swarm Optimization Particle Filtering Algorithm. International Conference on Circuits and Systems, Chengdu, 15-17 November 2013, 173-177.
12. Lin, L., Qi, L., Junyong, L. and Chuan, L. (2008) An Improved Particle Swarm Optimization Algorithm. IEEE International Conference on Granular Computing, Hangzhou, 26-28 August 2008, 486-490.
13. Yi, S., Yunfeng, B. and Mingxin, Y. (2009) A Novel Chaos Particle Swarm Optimization (PSO) and Its Application in Pavement Maintance Decision. IEEE Conference on Industrial Electronics and Applications, Xi'an, 25-27 May 2009, 3521-3526.
14. Lingkui, M. and Chunchun, H. (2007) Cluster Validity Index Based on Measure of Fuzzy Partition. Computer Engineering, 33, 15-17.

CHAPTER 9 TESTING AND ANALYSIS OF INDUCTION MOTOR ELECTRICAL FAULTS

K. Prakasam and S. Ramesh

Keywords: Three Phase Induction Motor, Motor Current Signature Analysis (MCSA), ZSVC, Fault Diagnosis, Current Spectrum Analysis

INTRODUCTION

Induction motors are a popular energy conversion component of many industrial processes and are frequently integrated with commercially available equipment and industrial processes. However, these machines will have a wider chance to fail because of heavy duty cycles, installation, poor working environment, improper excitation and manufacturing factors, etc. With increasing demands for efficiency and reliability, the field of fault diagnosis in induction motors is gaining importance [1] - [3]. There are plentiful methods are presented in existing like based on chemical monitoring, machine pulsation (vibration) monitoring, thermal monitoring, aural secretion monitoring, but all have necessitate of exclusive dedicated sensors or equipments to obtain the specific gesture to process, whereas in this projected current monitoring procedure there is no supplementary sensors required [2]. This is because of the fundamental electrical quantities linked with electromechanical industries such as voltage and currents are voluntarily deliberated by tapping into existing voltage and current transformers at the protection system. In this way, considering the advantages and reduced costs, non-invasive methods have been preferred for practical purposes. During the typical operating stage of the induction motor a combination of faults may occur. If these faults are undetected can at the right time or initially, then it leads to create a possible catastrophic failure [4]. Therefore, a diversity of stipulation monitoring schemes was introduced for an analysis of unexpected surroundings.

Citation: Prakasam, K. and Ramesh, S. (2016). Testing and Analysis of Induction Motor Electrical Faults. Circuits and Systems, 7, 2651-2662. DOI: 10.4236/cs.2016.79229.

The electrical faults and mechanical faults of the induction motors are the two groups of faults which are derived from common internal faults. Based on the fault occurred in the induction motor, the electrical faults have been classified in two ways in industries like stator faults and rotor faults. The stator faults are the primary and a critical one to diagnose, hence the stator faults are only taken into account in this study. The distinctive faults are scheduled as revealed in Figure 1. In an impartial three phase induction motor a stator short circuit fault between turn to turn causes a vast circulating current to pour and afterward engender excessively much heat in the shorted winding turns is measured as an earlier kind of fault in the given three phase induction motor [4] [5]. After that, the inter-turn short circuit is the immediate type of fault inspected in this manuscript.

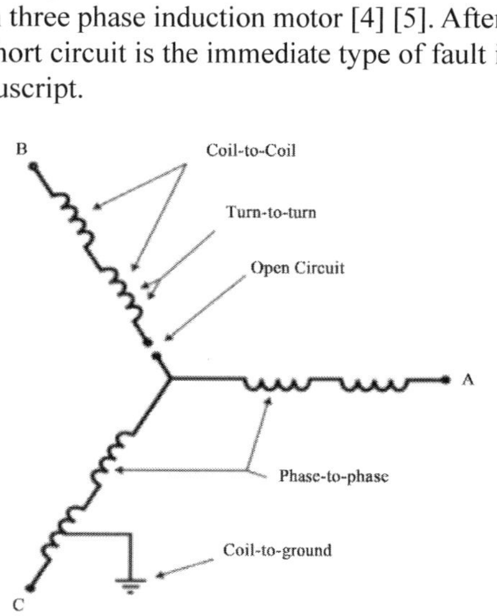

Figure 1. Typical faults in stator winding.

The stator winding inter-turn short circuit is the preface fault point for winding faults such as turn loss in the stator windings of the three phases. The stimulates harmonic frequencies in a three phase induction machine stator current as:

$$f_{\text{stator}} = \left\{ \frac{n}{p}(1-s) \pm k \right\} f_1 \quad (1)$$

where p is the number of pole pairs in the stator side, $n = 1,2,3,\cdots$, and $k = 1,3,5,\cdots$, likewise.

With sinusoidal mmf the stator slot harmonic frequency element for the stator current can be expressed as:

$$f_{\text{stator}} = \left\{ z_2 (1-s) \frac{f_1}{p} \pm f_1 \right\} \quad (2)$$

Z being the rotor slot numbers.

To attain three phase induction motor stator winding mmf time harmonic into account, f_1 is multiplied with harmonic figure of the stator winding mmf in time harmonics and it is indicated by h, ($h = 1,3,5,7,\cdots,$).

The slot harmonic frequency (f_h) for the stator winding current will arrive out as

$$f_h = \left\{ z_2 (1-s) \frac{f_1}{p} \pm h f_1 \right\} \quad (3)$$

Fault indicative practices are gaining more significance in any industry since the necessity to swell consistency and to shrink the leeway of manufacture loss due to machine collapse. By the way of evaluating the signals of a machine operation in usual and slip-up conditions, the discovery of faults such as throng unbalance, misalignment, gear disappointment and budding failures of the components, during the online monitoring method, falling the opportunity of ruinous injuries in the machine stator side and downtime can be determined [6]. A several efforts have been made to design and optimize three phases' fault-tolerant motor drives. It is recognized that FEM recreation is a dominant and precise tool to explore the special effects of open-circuit as well as short-circuit faults. In that inter-turn faults are one of the mainly common and potentially detrimental stator faults. So, the FEM and the ZSVC are specially utilized to find the inter-turn short circuit faults.

Further practice includes machinery vibration analysis, thermal (temperature) analysis, acoustic noise measurement, torque sketch analysis and magnetic meadow analysis. These practices have need for complicated and luxurious sensors, further electrical and mechanical installations of the equipment's, and repeated preservation and are not favored [7].

The utilization of the steady-state spectral elements of the stator winding is one of the most fashionable ways of three phase induction machine condition monitoring. These spectral frequency elements of the stator winding includes electrical quantities like voltage across the points, current flows in the particular area and power drawn and are used to discover inter-turn faults, broken bars (rotor side), eccentricities in the air gap, bearing failures. Right now, countless practices that are supported on steady-state examinations are the MCSA and the Extended Park's Vector Approach (EPVA) [7], as well as an innovative momentary technique that is a mixture of the EPVA and the Discrete Wavelet Transform.

MOTOR CURRENT SIGNATURE ANALYSIS (MCSA)

The method Motor Current Signature Analysis (MCSA) is used to resolve the operating condition of induction motors without suspending the production. The fluctuations in the motor current can be observed by clamping a Hall Effect current sensor. MCSA concentrates its efforts in the spectral analysis of the stator current and has been successfully applied to perceive abnormal levels of the condition of stator windings like open circuit fault, turn-turn, coil-coil, leakage faults, and eccentricity in the air-gap and broken bars in the rotor side among other mechanical troubles [8]. The key purpose of the current spectral analysis is to analyze the stator current in search of current harmonics frankly associated with new-fangled rotating flux components, which are caused by trouble in the motor-flux circulation.

For analyzing or trending dynamic energized scheme the motor current signature analysis is used. The results of MCSA support in the discovery of the following:

- Health of the incoming winding
- Rotor Health
- Eccentricity in the air gap
- Shaft coupling health (with direct, belted and geared systems)
- Issues due to load
- Mechanical Bearing health

The real time graphical results and the comparison of numerical values of simulation and hardware results are stated in table for analyzing the proposed work.

SYSTEM SPECIFICATIONS

The proposed work deals three phase squirrel cage Induction Motor testing with the help of LabVIEW software to predict the typical faults which are artificially introduced into the machine. Figure 2 shows the typical winding arrangements made in the specially wounded induction motor for analyzing the proposed work. A three phase squirrel cage induction motor is chosen for this study and the specifications are listed in Table 1.

The above specified machine is specially wounded and takes the number of outlet terminals to make the faults and testing of the machine as shown in Figure 4.

HARDWARE DESCRIPTION OF FAULT DIAGNOSIS

A personal computer based DAQ system is plugged in for the permutation of the computer with the signal from induction motor. It shows the real time attribute of the investigational induction motor during its operation. Besides giving the graphical outlook to the addict, LabVIEW also gives a superior analysis capability about the measurements on the induction motor in the proposed system. The real time working model of the induction motor is designed with PC and LABVIEW interfaced system is depicted in Figure 3 and the specially wounded Induction motor is depicted in Figure 4.

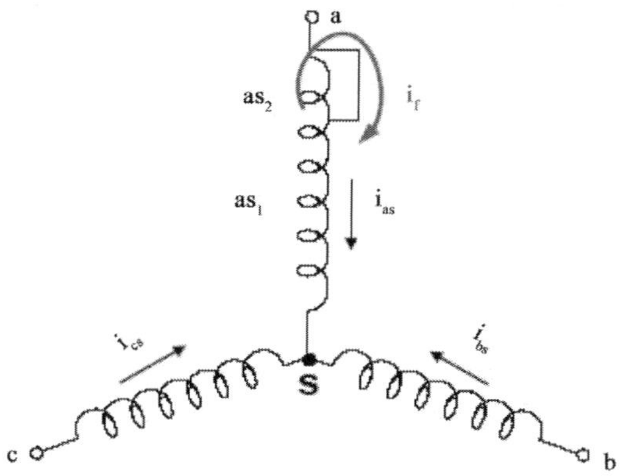

Figure 2. Typical stator winding connection and current flow.

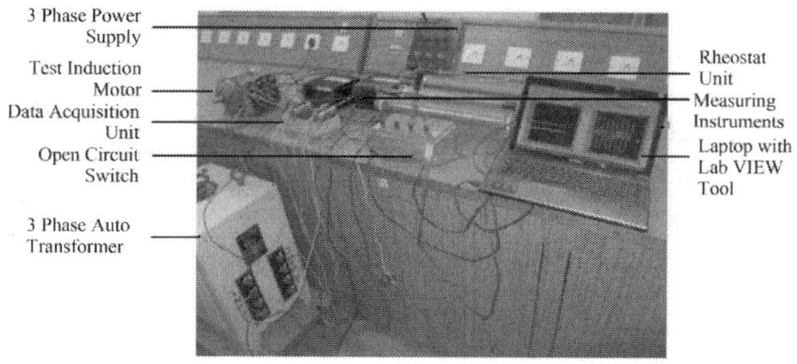

Figure 3. Hardware setup of the proposed system.

Figure 4. Specially wounded induction motor and its terminal box.

Table 1. 0.75 HP 3 phase induction motor characteristics.

Description	Range
Power (hp)	0.75
Voltage (V)	200/415
Current (I)	3.34/1.6
Speed (N)	1390
Phases	3
No. of Poles	4
No. of Coils per Phase	12
Types of Stator Winding	Double Layer, Lap & Y Connected
No. of Stator Slots	36
Make	Spartan Electricals, Bombay

The current signature of the induction motor is acquired from each phase of the power supplied to the Induction Motor. Analysis of the current signature is carried out as a co-simulation in LabVIEW. Graphical waveforms of healthy and faulty motors and comparison of their respective signatures are accomplished with the help of LabVIEW tools for different ratings of Induction Machines. In order to create faults in the motor the different control strategy were used [6] - [9].

In this system, voltage spectrum, current spectrum and ZSVC spectrum were measured. The graphical results for different fault conditions are observed and the deficiency between the normal operating condition and the fault conditions were encountered.

RESULTS AND DISCUSSION

The specially wounded Induction Motor has a nine number of external terminals to create the faults artificially on the outside of the Induction Motor. Based on the terminals connected the output current spectrum of the motor with each of open circuit, short circuit and leakage faults are presented here.

Waveform of Healthy Motor

The electrical signatures of the stator winding is observed on the healthy machines ought to be clear as per the basic concepts. It is not practically possible to observe the clear signatures. Because, due to line fluctuations and inductive load effects of the induction machine, it also has some fluctuations. Under no-load operating condition this wave form is acquired and shown in Figure 5 and Figure 6. The graphical representation shows the three parameters like voltage, current and fault current. There is no fault current for this case so, the representation shows the huge distortion as shown in Figure 7.

Open Circuit Fault

The open circuit faults arise due to aging, high swiftness applications and mechanical anxiety cut off of any of the winding presented in the stator winding. The voltage of the opened phase directly goes to zero and other phases are pretentious by noise and induction effect as shown as fluctuations. The results were depicted with each phase opened and the waveform is shown in Figures 8-10.

Coil-Coil Short Circuit Fault

On the stator winding, turn-to-turn or turn-to-earth incidences cause the leakage current faults. This is the premature fault stage of open circuit and short circuit faults of the stator windings. This result causes the irregular voltage and currents spectrums in all the three phases and build noise, temperature and varies with the rotor velocity. The similar faulty spectrum is observed if leakage occurring at any of the stator winding phases. Winding may experience short circuit, whose pretentious signature due to such leakages as shown in Figure 11 and Figure 12.

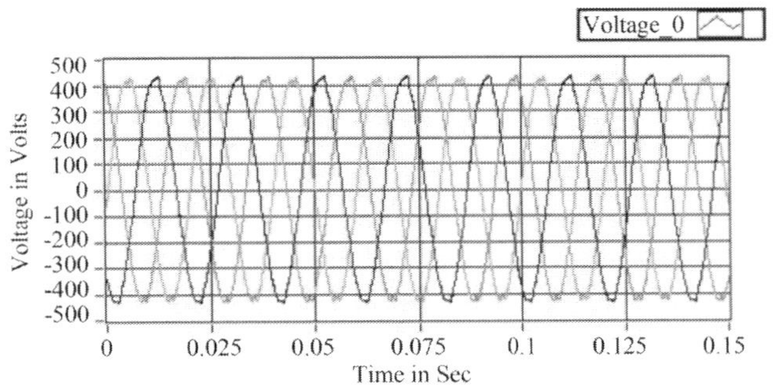

Figure 5. Voltage waveform of healthy induction motor under no-load condition.

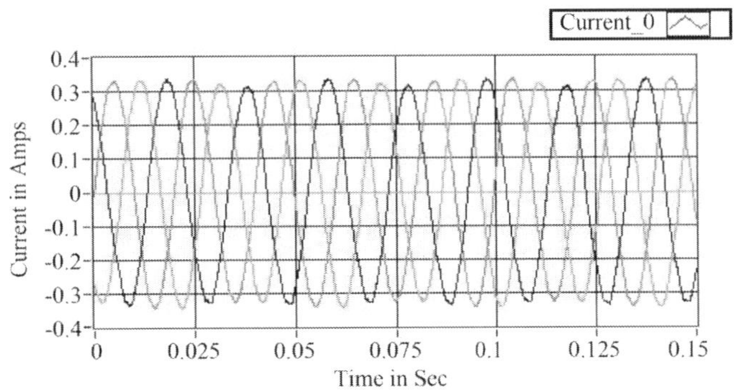

Figure 6. Current waveform of healthy induction motor under no-load condition.

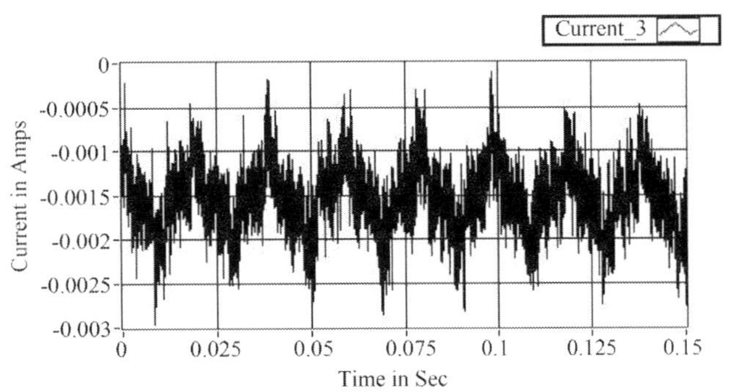

Figure 7. Fault current waveform of healthy induction motor under no-load condition.

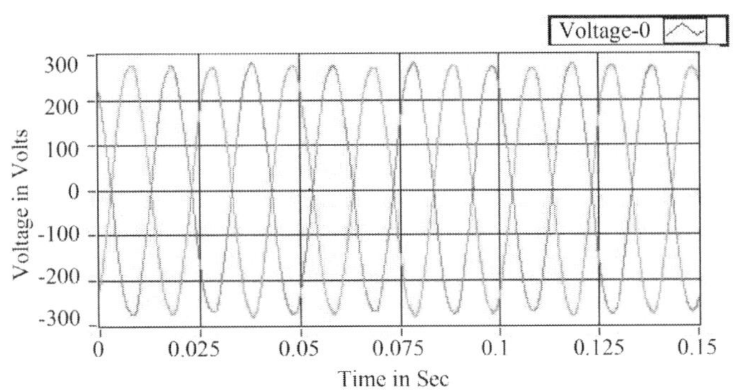

Figure 8. Voltage waveform of the IM with open circuit fault (Phase C opened).

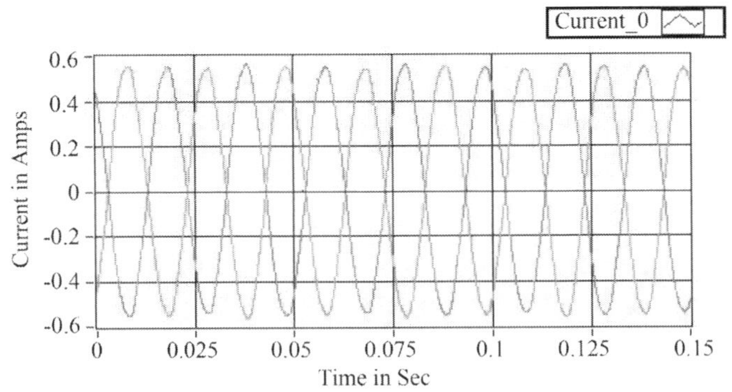

Figure 9. Current waveform of the IM with open circuit fault (Phase C opened).

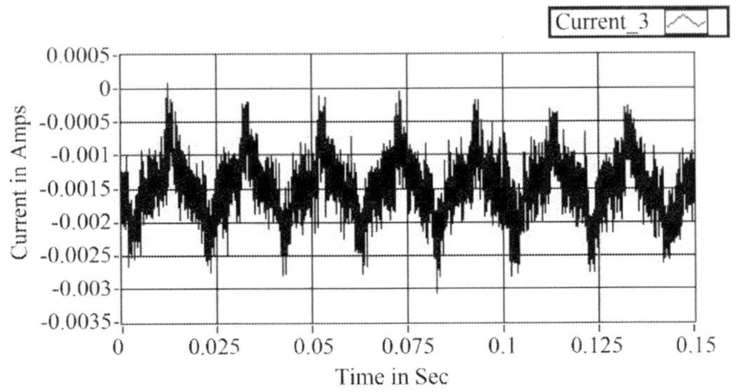

Figure 10. Fault current waveform of the IM with open circuit fault (Phase C opened).

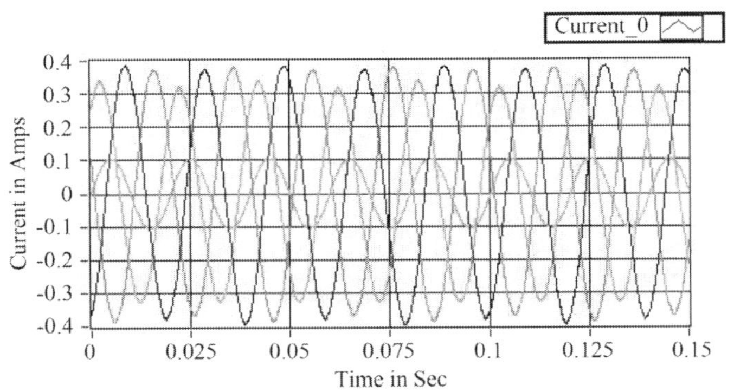

Figure 11. Signature of the faulty IM with leakage fault (Between Phase A & B).

Short Circuit Fault

The two diverse phases in stator windings may short stick between them. Then the induction motor undergoes a short circuit fault and it allows to flow of up normal current and drop down the voltage resulting in a faulty current spectrum. But in our specially wounded induction motor have a separate lead to make short circuit faults. With the help of those terminals the short circuit is created via the couple of three hundred ohm rheostats is connected in series to avoid dead short circuit conditions. By shorting three dissimilar arrangements of the phases of the induction motor stator windings, the defective output current spectrums are acquired as shown in Figure 13 and Figure 14.

Turn-Turn Short Circuit Fault

The shorting of two turns in the same coil of the same phase in stator windings of the induction motor causes turn-turn short circuit fault. During the short circuit fault condition the current flows through the particular area is in up normal value and the voltage will drop down. The results are observed as voltage, current, faulty current output spectrums and ZSVC spectrums are acquired as shown in Figures 15-20.

Table 2 shows the analytical results of the proposed system between the hardware and the simulation models. The result shows the similarity values are obtained in simulation as well as in hardware. In this table the voltage, three phase current and the fault currents were measured to analyze the parameter variations in the stator side. Based on that the results were compared and conclude that the system gives the accurate values. So, the proposed system's accuracy was determined.

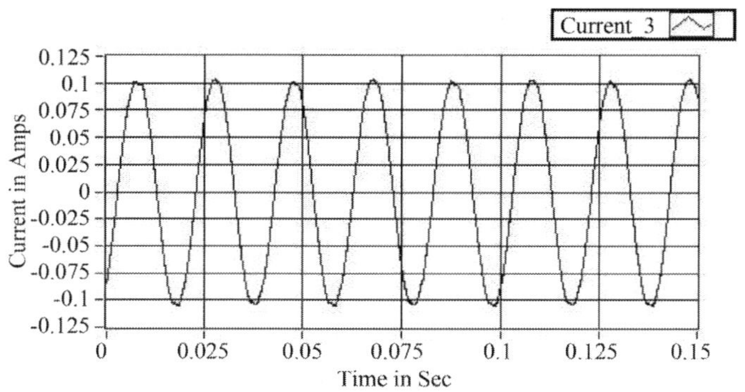

Figure 12. Signature of the faulty IM with leakage fault (Between Phase B & C).

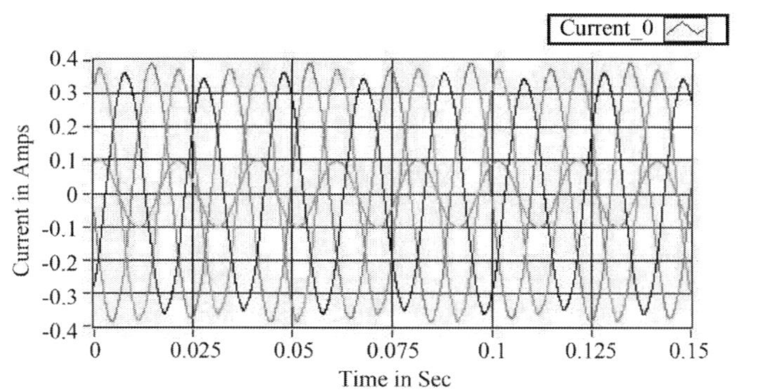

Figure 13. Faulty current spectrum by shorting A & B phases.

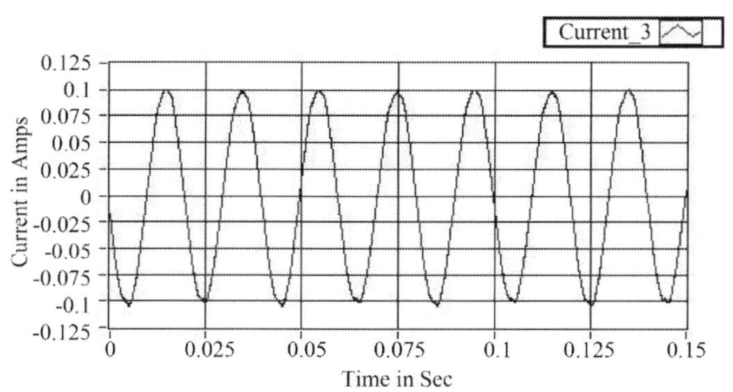

Figure 14. Faulty current spectrum by shorting A & B phases.

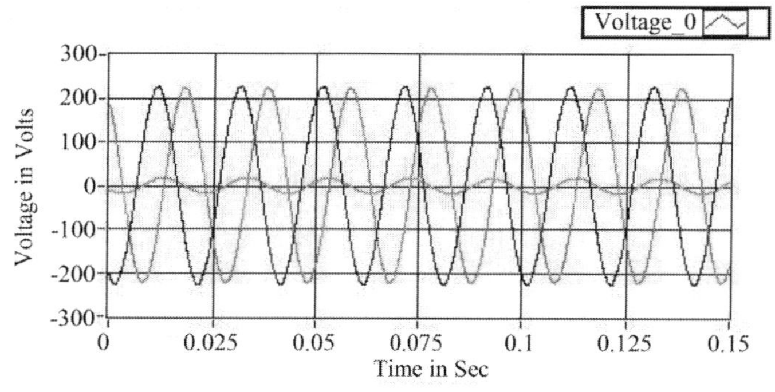

Figure 15. Faulty current spectrum by shorting A & B phases.

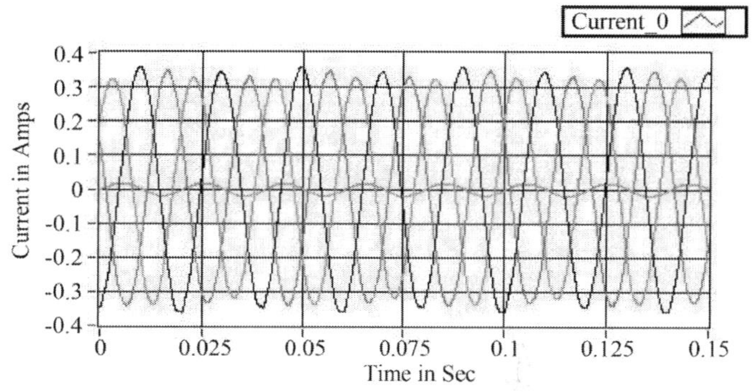

Figure 16. Faulty current spectrum by shorting A & B phases.

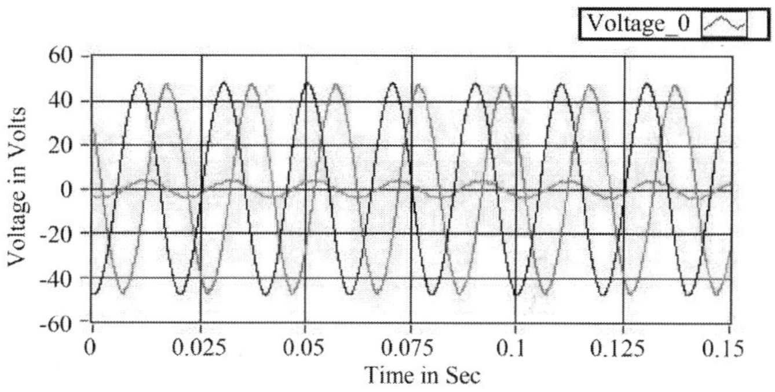

Figure 17. Faulty current Spectrum by shorting A & B phases.

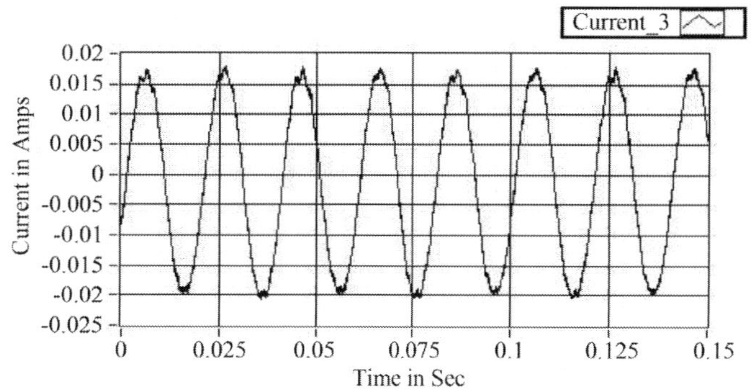

Figure 18. Faulty current spectrum by shorting A & B phases.

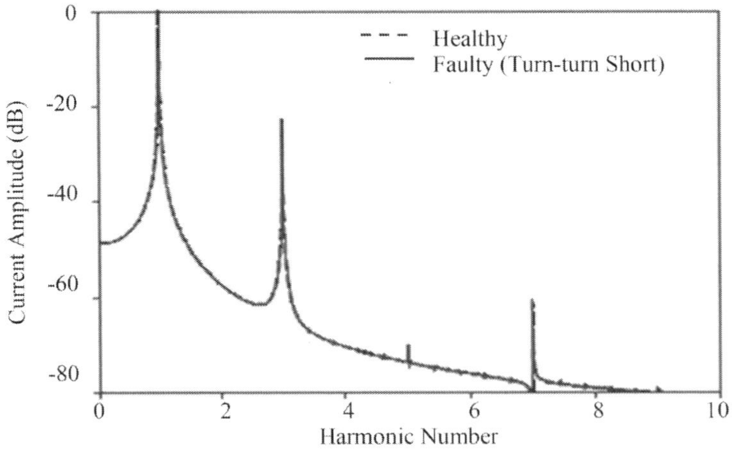

Figure 19. Healthy and faulty current spectra of turn to turn short for one phase.

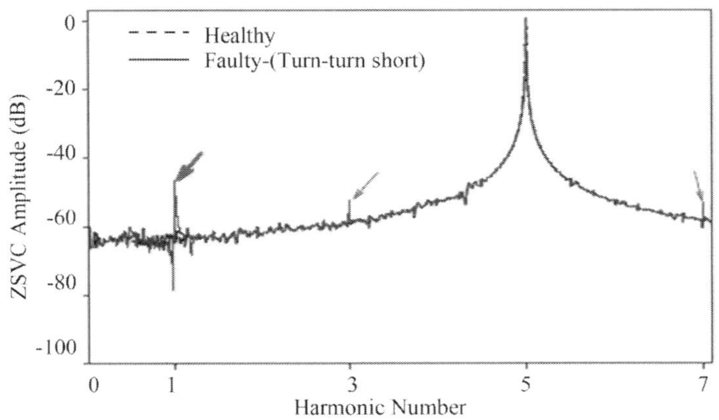

Figure 20. Healthy and faulty ZSVC spectra of turn to turn short for one phase.

Table 2. Hardware and simulation result comparison.

Working Condition	Hardware					Simulation				
	Volt-age	Current			Fault Current	Volt-age	Current			Fault Current
		R	Y	B		R	Y	B		
Normal	415	0.32	0.32	0.32	0.0005	415	0.33	0.33	0.33	0.0005
Open Circuit	415	0.56	0.56	0	0.0005	415	0.54	0.54	0	0.0005
Coil-coil short RY	252	0.37	0.37	0.31	0.1	250	0.35	0.35	0.29	0.09
Phase-phase short RY	180	0.39	0.38	0.36	0.1	176	0.37	0.38	0.35	0.09
Turn-turn short	32	0.34	0.32	0.35	0.0164	31	0.33	0.31	0.34	0.0159

CONCLUSION

Thus the effectiveness of the MCSA method with ZSVC to discover unusual stator faults in the three phase squirrel cage Induction Motor and to discriminate stuck between them has been proved all with its vital rewards. Such as the cleanness of data gaining schemes and the mandatory software, beside with the healthiness of the tools used, which has provided quite acceptable results. Also, the proposed system gives the chance to discover the rotor faults too, like the stator current spectrum analysis. Hence, personal computer aided LabVIEW based on-line condition monitoring system for induction machine was tested and the results were presented in this paper, establishing the efficiency and effectiveness of this method. Also, this projected method has an extensive range of features like on-line MCSA system with artificial intelligence and fault rectification systems, etc. by adding some additional tools in the existing one. The intelligent control tools for the above said hardware systems to improve the performance are in progress and it adds more additional features for this.

ACKNOWLEDGEMENTS

The work for this paper was supported by the Coral Rewinding India (P) Ltd. Erode, Tamilnadu, India.

REFERENCES

1. Yeh, C., Sayed, A. and Povinelli, R. (2007) A Reconfigureurable Motor for Experimental Emulation of Stator Winding Inter-Turn and Broken Bar Faults in Poly Phase Induction Machines. IEEE International Electric Machines & Drives Conference, Antalya, 3-5 May 2007, Vol. 2, 1413-1419. http://dx.doi.org/10.1109/IEMDC.2007.383636 [Paper reference:1]
2. Penman, J., Sedding, H.G., Lloyd, B.A. and Fink, W.T. (1994) Detection and Location of Interturn Short Circuits in the Stator Windings of Operating Motors. IEEE Transactions on Energy Conversion, 9, 652-658. http://dx.doi.org/10.1109/60.368345 [Paper reference:1]
3. Fenger, M., Susnik, M., Laderoute, P. and Thomson, W.T. (2003) Development of a Tool to Detect Faults in Induction Motors via Current Signature Analysis. Proceedings of Cement Industry Technical Conference, 4-9 May 2003, 37-46. http://dx.doi.org/10.1109/citcon.2003.1204707 [Paper reference:1]
4. Neelam, M. and Dahiy, R. (2007) Motor Current Signature Analysis and its Applications in Induction Motor Fault Diagnosis. International journal of systems Applications, Engineering & Development, 2, 29-35. [Paper reference:2]
5. Messaoudi, M. and Sbita, L. (2010) Multiple Faults Diagnosis in Induction Motor Using the MCSA Method. International Journal of Signal and Image Processing, 1, 190-195. [Paper reference:1]
6. Jagadanand, G., Gopi, L., George, S. and Jacob, J. (2012) Inter-Turn Fault Detection in Induction Motor Using Stator Current Wavelet Decomposition. International Journal of Electrical Engineering and Technology (IJEET), 3, 103-122. [Paper reference:2]
7. Prakasam, K. and Ramesh, S. (2015) Investigation of Induction Motor Stator Faults Using Motor Current Signature Analysis and Multisim. Middle-East Journal of Scientific Research, 23, 277-284. [Paper reference:2]
8. Bhowmik, P.S., Pradhan, S. and Prakash, M. (2013) Fault Diagnostic and Monitoring Methods of Induction Motor—A Review. International Journal of Applied Control, Electrical and Electronics Engineering (IJACEEE), 1, 1-18. [Paper reference:1]
9. Bhardwaj, K. and Agarawal, A. (2012) Fault Diagnosis of Three Phase Induction Motor Using Fuzzy Logic Controller and Sequence Analyzer. International Journal of Electrical and Instrumentation Engineering, 2, 112-118. [Paper reference:1]

CHAPTER 10

ENERGY EFFICIENT CONTROL OF THREE-PHASE INDUCTION MOTOR DRIVE

Hussein Sarhan

Keywords: Induction Motor Drive, Efficiency Optimization, Slip Compensation

INTRODUCTION

Induction motors are the most used in industry since they are rugged, inexpensive, and are maintenance free. It is estimated that more than 50% of the world electric energy generated is consumed by electric machines. Improving efficiency in electric drives is important, mainly for economic saving and reduction of environmental pollution [1,2]. Induction motors have a high efficiency at rated speed and torque. However, at light loads, motor efficiency decreases dramatically due to an imbalance between the copper and the core losses. Hence, energy saving can be achieved by proper selection of the flux level in the motor [3,4]. The main induction motor losses are usually split into: stator copper losses, rotor copper losses, core (iron) losses, mechanical and stray losses. To improve the motor efficiency, the flux must be reduced, obtaining a balance between copper and core losses. Many minimum- loss control schemes based on scalar control or vector control of induction motor drives have been reported in literature [4-8]. Induction motor drive can be controlled according to a number of performance functions, such as input power, speed, torque, air gap flux, power factor, stator current, stator voltage, and overall efficiency [9]. Basically, there are three strategies, which are used in efficiency optimization of induction motor drive: Simple state control, model based control, and search control. Search strategy methods have an important advantage compared to other strategies. It is completely insensitive to parameters changes, while effects of the parameters variations caused by temperature and saturation are very expressed in two other strategies [10-12]. In this paper, an efficiency optimization controller of induction motor drive system, based on searching the value of stator voltage that maximizes the efficiency, is developed. Then, the validity of the proposed controller and the performance of the drive system were analyzed by simulation results. To reduce the slip at light loads and low frequencies, a slip compensator has been introduced.

Citation: H. Sarhan. "Energy Efficient Control of Three-Phase Induction Motor Drive," Energy and Power Engineering, Vol. 3 No. 2, 2011, pp. 107-112. DOI: 10.4236/epe.2011.32014.

MODELING AND SIMULATION OF DRIVE SYSTEM

The original drive system studied in this paper consists of IGBT-inverter-based AC to AC converter, three-phase squirrel cage induction motor and *V/f* controller. In order to analyze the system performance, all of these components should be modeled (mathematically de-scribed). The inverter- based AC-to-AC converter is considered to be an ideal system, where the DC voltage at the input of the inverter has no AC component, and the output voltage of the filter at the output of the inverter has no harmonics. For sinusoidal pulse width modulation SPWM, the ratio of the amplitude of the sinusoidal waveform to the amplitude of the triangular waveform is called the modulation index *m*, which can be in the range of 0 to 1 [5]. The stator voltage V_s can be de-fined as:

$$V_s = mV_n \quad (1)$$

where V_n = nominal value of stator voltage.

The frequency of the stator voltage f equals the frequency of the sinusoidal input waveform f_{in}.

$$f = f_{in} \quad (2)$$

Varying the modulation index and the sinusoidal waveform frequency will change the RMS value of the stator voltage and frequency, respectively. Equations (1) and (2) constitute the steady-state model of inverter. The controller with V/f = constant must apply the following function:

$$m = \begin{cases} Kf, & 0 < f < f_n \\ 1, & f \geq f_n \end{cases} \quad (3)$$

where $K = 1/f_n$, and f_n = nominal frequency.

Based on the flux linkages and voltages equations, the electrical and mechanical models of the squirrel cage three-phase induction motor with respect to a synchronously rotating $d-q$ coordinates were adopted for simulation [5,13]. The parameters of modeled and simulated induction motor are given in Table 1.

DESIGN OF EFFICIENCY OPTIMIZATION CONTROLLER

The efficiency of the drive system at any steady state operating point can be calculated as:

$$\eta = \frac{P_m}{P_s} \quad (4)$$

Table 1. Induction motor data.

Parameter	Value
Stator resistance R_1	1.230 Ω
Stator Inductance L_1	0.006 H
Mutual Inductance L_m	1.172 H
Rotor resistance referred to the stator R_2'	1.395 Ω
Rotor Inductance referred to the stator L_2'	0.006 H
Nominal voltage V_n	230/400 V
Nominal output power P_n	4 kW
Nominal power factor $\cos\varphi_n$	0.850
Nominal frequency f_n	50 Hz
Number of poles P	4
Nominal rotational speed n_n	1430 rpm
Moment of inertia J	0.013 kg·m^2

where $P_m = Lw$, mechanical output power and $P_s = \frac{3}{2}\left(v_{sd}i_{sd} + v_{sq}i_{sq}\right)$, total input power supplied to the stator. To design the efficiency optimization controller based on search method, a matrix, as shown in Table 2, consisting of loads, frequencies and stator voltage values (optimal voltage), which maximize the efficiency, was constructed using MATLAB Simulink model of the studied drive system.

The relationship between load torque, frequency and optimal voltage is shown in Figure 1.

Based on the data in Table 2 and Equation (4), the relationship between the maximum efficiency, load torque and frequency is shown in Figure 2.

For the original system with V/f = constant controller, the relationship between efficiency, load torque and frequency is shown in Figure 3. The matrix shown in Table 2 was used to form a 2D lookup module (table) in MATLAB, which serves as the main part of the proposed efficiency optimization controller, shown in Figure 4. The output of the controller is the modulation index m, which correspondences the optimal value of voltage maximizing the efficiency under certain operation condition.

Simulation results showed that the main disadvantage of the proposed efficiency optimization technique was the increase in slip, especially at light loads and low frequencies, as shown in Figure 5.

So, there is a need to compensate (reduce) the slip according to some reference (acceptable) value s_{ref}. This can be achieved by increasing the electromagnetic torque of the motor by increasing the stator voltage in the optimized system.

Table 2. Optimal stator voltage as a function of load torque and frequency.

Voltage, V		Load Torque, N·m				
		5	10	15	20	25
Frequency, Hz	5	32.8	46.4	57.6	65.6	73.6
	10	58.4	86.4	100	116	130.4
	15	83.2	116	143.2	176	190.4
	20	109.6	151.2	185.6	220	246.4
	25	137.6	196.8	236.8	267.2	308.8
	30	160.8	222.4	272.8	324	349.6
	35	189.6	276.8	326.4	369.6	400
	40	220.8	295.2	353.6	400	400
	45	243.2	340	391.2	400	400
	50	272.8	372.8	400	400	400

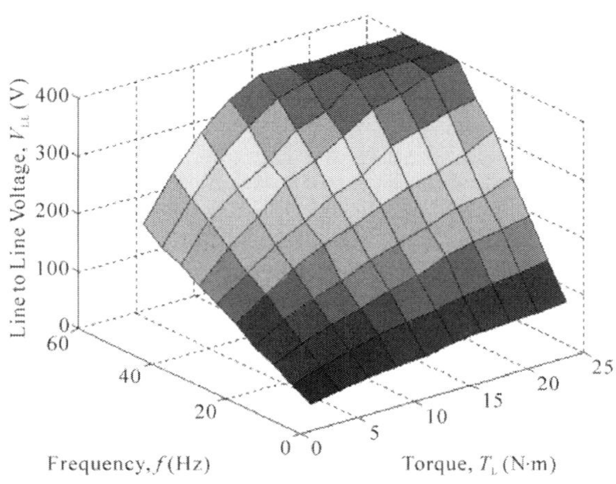

Figure 1. Optimal voltage as a function of load torque and frequency.

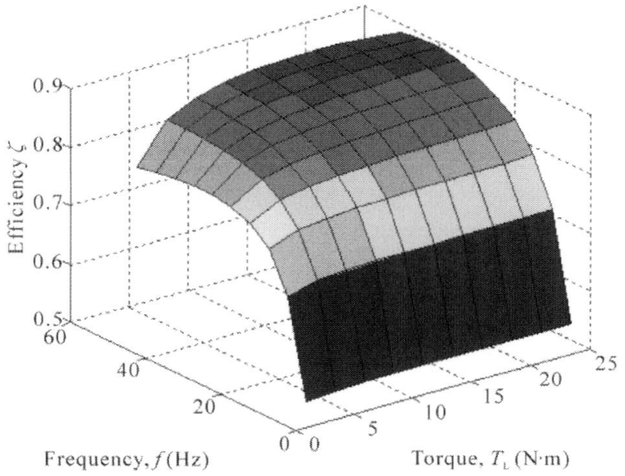

Figure 2. The relationship between maximum efficiency, load torque and frequency.

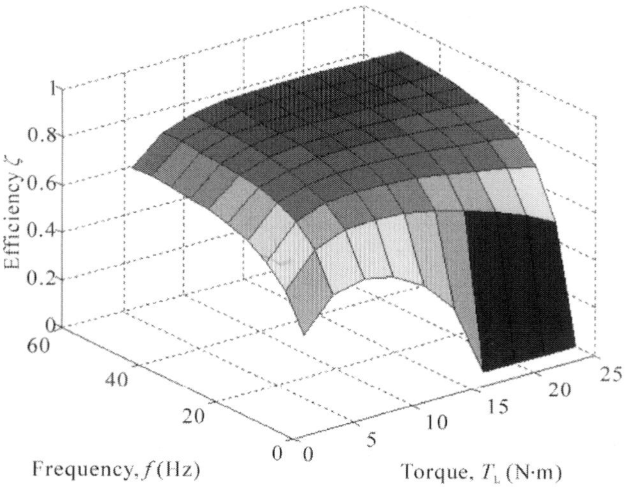

Figure 3. The relationship between efficiency, load torque and frequency for the original system.

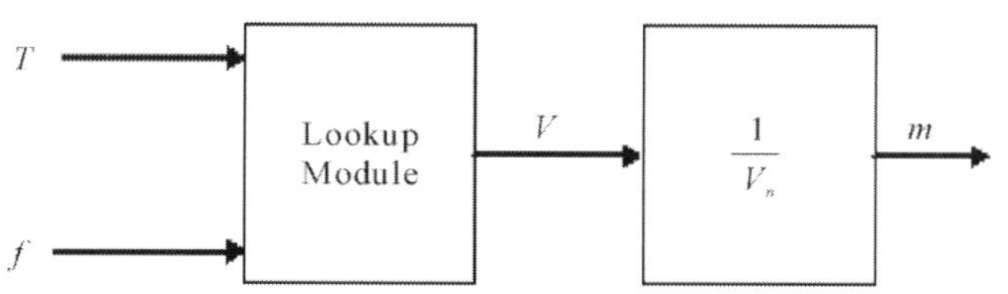

Figure 4. Block diagram of efficiency optimization controller.

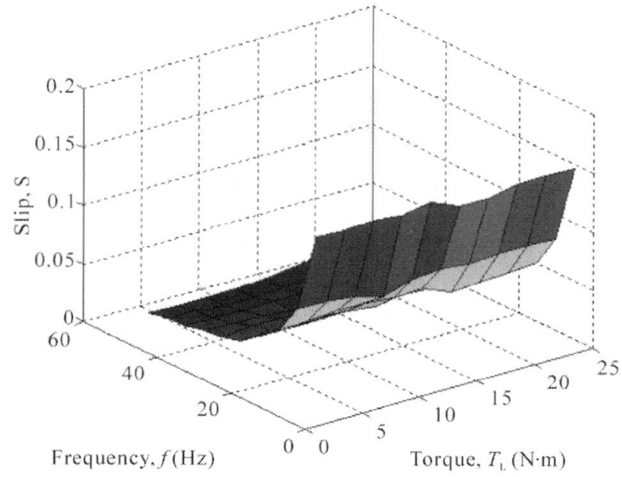

Figure 5. The relationship between slip, torque and frequency for the optimized system.

In this case, the error in slip will be gained and integrated to get the value of compensation voltage that should be added to the voltage in the optimized system. The value of compensation voltage V_{com} can be calculated by the following equation:

$$V_{com} = K_i \int \left(s_{ref} - s \right) \mathrm{d}t \tag{5}$$

where K_i = integrating constant.

Equation (5) can be realized by adding an integrator as a feedback loop into the optimized drive system.

The MATLAB Simulink model of optimization controller with slip compensation is shown in Figure 6.

ANALYSIS OF OPTIMIZED DRIVE SYSTEM

For quantitative analysis, the efficiency of the optimized system with slip compensation was compared with that in the original system with V/f = constant controller under the same operating conditions by using the simulation results of the MATLAB Simulink model of the drive system, shown in Figure 7.

Figures 8-10 are examples of the system behavior under different operating conditions. Figure 8 shows that the efficiency optimization controller increases the efficiency of the drive system, and makes it approximately constant for all the range of load. Slip compensation reduces the efficiency of about 4%.

Figure 9 shows the relationship between efficiency and frequency at load torque = 10 N.m, form which, it is clear that optimization has not significant effect on the drive system performance.

Figure 10 shows the slip response at load torque = 15 N.m and frequency = 5 Hz, from which it is clear that optimization and slip compensation have significant effect on slip.

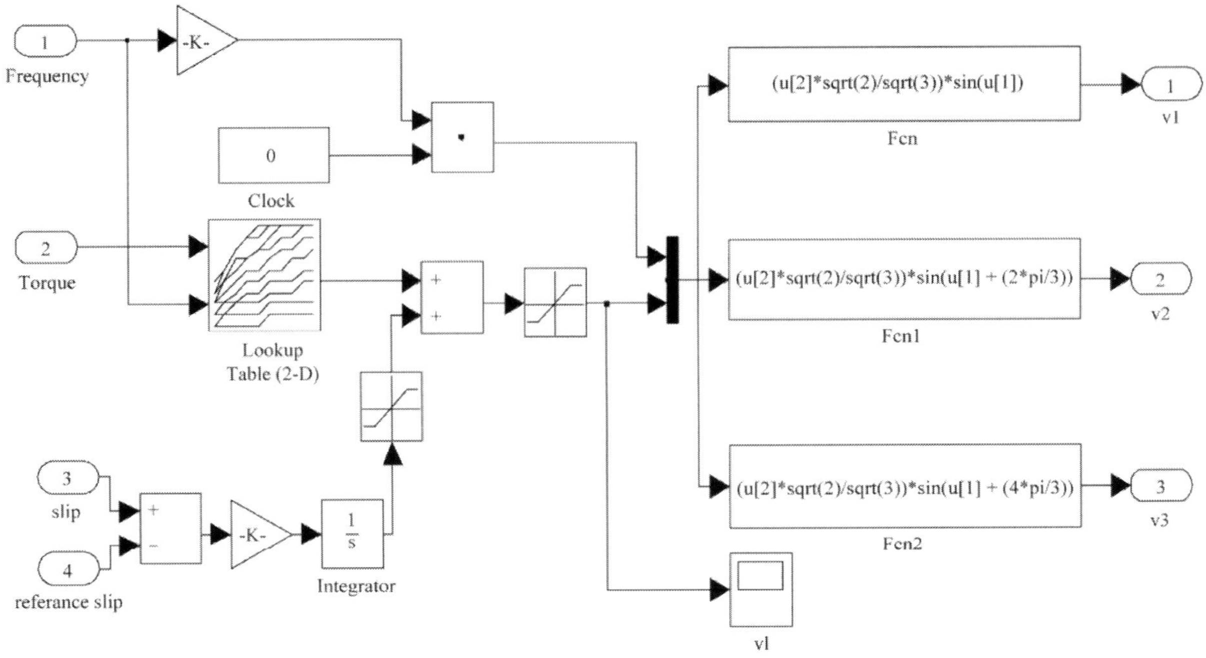

Figure 6. The MATLAB Simulink model of optimization controller with compensation.

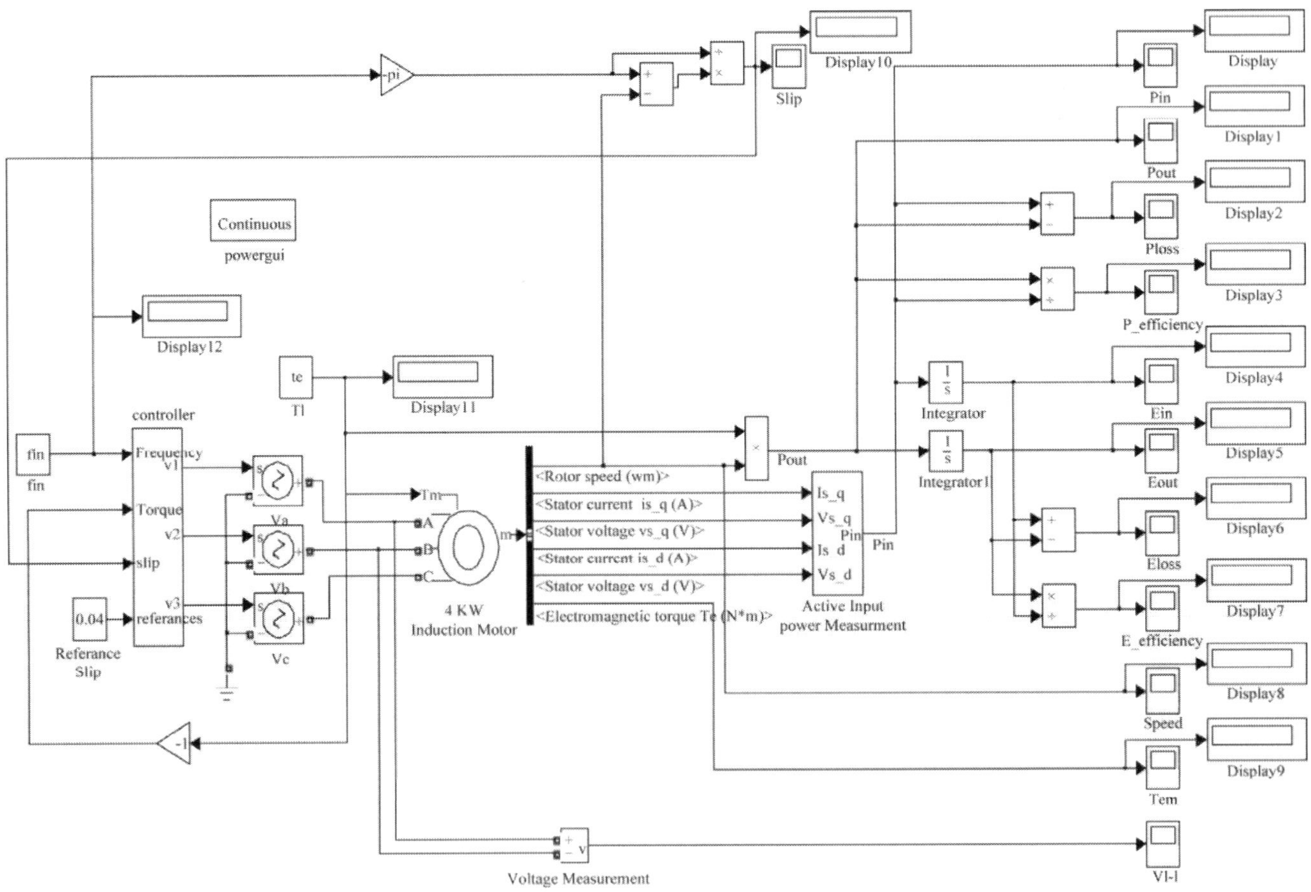

Figure 7. MATLAB Simulink model of optimized drive system with slip compensation.

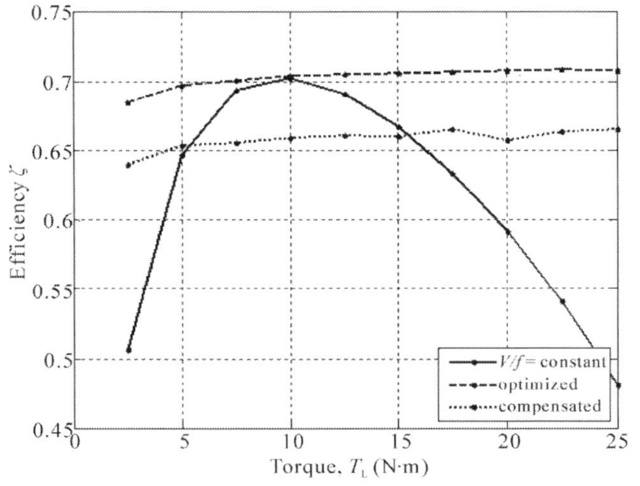

Figure 8. The relationship between efficiency and torque at frequency = 10 Hz.

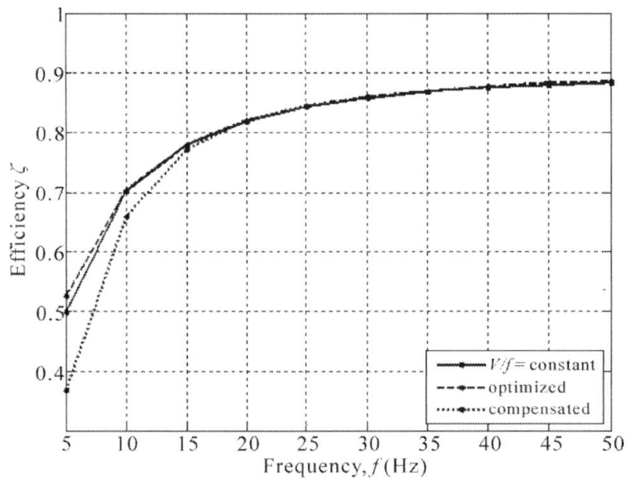

Figure 9. The relationship between efficiency and frequency at torque = 10 N·m.

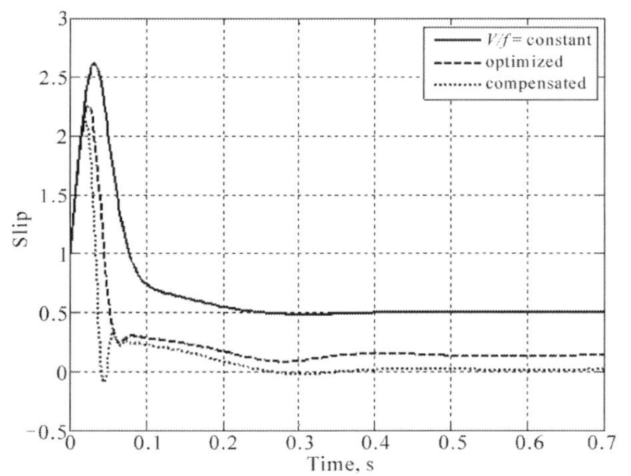

Figure 10. The slip response at torque = 15 N·m and frequency = 5 Hz.

CONCLUSIONS

In this paper, an efficiency optimization controller, based on search method has been developed. The proposed controller manipulates the value of stator voltage that maximizes the efficiency at any given operating point. To reduce the slip to a certain value, a slip compensator has been inserted into the drive system. The suggested technique can be used in variable frequency, variable load electrical drive systems. Based on simulation analysis, it was noticed that the compensated optimized system gives significant results at low frequencies and light loads.

REFERENCES

1. W. Leonhard, "Controlled AC Drives, a Successful Transfer from Ideas to Industrial Practice," *Control Engineering Practice*, Vol. 4, No. 7, 1996, pp. 897-908. doi:10.1016/0967-0661(96)00087-1
2. G. C. D. Sousa, B. K. Bose and J. G. Cleland, "Fuzzy Logic Based on-Line Efficiency Optimization Control of an Indirect Vector-Controlled Induction Motor Drive," *IEEE Transactions on Industrial Electronics*, Vol. 42, No. 2, 1995, pp. 192-198. doi:10.1109/41.370386
3. H. R. Andersen, C. B. Rasmussen, E. Ritchie and J. K. Pedersen, "Efficiency Comparison of Electrical Motors for Energy Optimized Variable Speed Low Power and Low Cost Household Equipment," *6th European Conference on Power Electronics and Applications*, Seville, 1995, pp. 3381-3386.
4. F. C. Lin and S. M. Yang, "On-Line Tuning of an Efficiency-Optimized Vector Controlled Induction Motor Drive," *Tamkang Journal of Science and Engineering*, Vol. 6, No. 2, 2003, pp. 103-110.
5. H. Sarhan and R. Issa, "Modeling, Simulation and Test of Inverter-Induction Motor Drive System with Improved Performance," *Journal of Engineering Sciences, Assiuit University*, Vol. 33, No. 5, 2005, pp. 1873-1890.
6. F. Abrahamsen, J. K. Pedersen and F. Blaabjerg, "State-of-the-Art of Optimal Efficiency Control of Low Cost Induction Motors Drives," *The Power Electronics and Motion Control Council*'96, Budapest, 2-4 Septemper 1996, pp. 163-170.
7. I. Kioskeridis and N. Margaris, "Loss Minimization in Induction Motor Adjustable-Speed Drives," *IEEE Trans-actions on Industrial Electronics*, Vol. 43, No. 1, 1996, pp. 226-231. doi:10.1109/41.481429
8. I. Kioskeridis and N. Margaris, "Loss Minimization in Scalar-Controlled Induction Motor Drives with Search Controllers," *IEEE Transactions on Power Electronics*, Vol. 11, No. 2, 1996, pp. 213-220. doi:10.1109/63.486168
9. A. K. Sharma, R. A. Gupta and Laxmi Srivastava. "Implementation of Neural Network in Energy Saving of Induction Motor Drives with Indirect Vector Control," *Journal of Theoritical and Applied Information Technology*, Vol. 5, 2008, pp. 774-779.
10. F. Abrahamsen *et al.*, "Efficiency-Optimized Control of Medium-Size Induction Motor Drives," *IEEE Transactions on Industry Applications*, Vol. 37, No. 6, 2001, pp. 1761-1767. doi:10.1109/28.968189
11. S. M. Yang and F. C. Lin, "Loss-Minimization Control of Vector-Controlled Induction Motor Drives," *Journal of the Chinese Institute of Engineers*, Vol. 26, No. 1, 2003, pp. 37-45. doi:10.1080/02533839.2003.9670752

12. A. A. Ansari and D. M. Deshpande, "Induction Motor Efficiency Optimization Using Fuzzy Logic," *International Journal of Advanced Engineering and Applications*, Vol. 1, 2010, pp. 177-180.

13. O. I. Okogo, "MATLAB Simulation of Induction Machine with Saturable Leakage and Magnetizing Inductances," *The Pacific Journal of Science and Technology*, Vol. 5, No. 1, 2003, pp. 5-15.

CHAPTER 11 A DIGITAL PHASE LOCKED LOOP SPEED CONTROL OF THREE PHASE INDUCTION MOTOR DRIVE: PERFORMANCES ANALYSIS

Ben Hamed Mouna and Sbita Lassaâd

Keywords: Digital Phase Locked Loop (DPLL), Induction Motor, Scalar Strategy, Speed Drives and Load Appliance

INTRODUCTION

IMs are used in many industry sectors as the leading element to convert electrical energy into mechanical one. The main challenge is based on robustness and low cost. The performance of the motor speed control required in the industrial drives depends on the application specifications. In some applications, an open loop speed variation of the drive motor is enough. In others, feedback control is required for better speed regulation performances. The performances of the speed regulation depend on the performances of the speed controller. In light of available literature, important aspects of the motor drive system, it is considered worthwhile to review it critically in order to find the gaps existing knowledge on the IM drive systems. Researchers have used various types of closed loop controllers for the IM rotor speed. PI controllers are widely and still used in the outer speed loop [1,2]. They have a simple structure and can offer a satisfactory performance for a wide range of operation. Due to the continuous variation in the plant parameters and the non linearities conditions, PI controllers may become unable to provide the required control performance [3,4]. These inherent disadvantages of the PI controllers have encouraged the replacement of this controller with adaptive [5,6] and robust [7-9] control laws. Adaptive control laws impose a very computation burden while H∞ robust control requires the knowledge of the disturbances boundaries [7]. Sliding mode control is also used in the speed loop [10]. Chattering in the steady state is the main drawback of this strategy [11,12]. Recently, speed controllers based on artificial intelligence techniques such as fuzzy logic and neural networks based controllers have been proposed [13,14]. Since these approaches do not rely on any detailed mathematical plant, the algorithm would remain robust despite of parameter deviations and noise measurement [15,16]. However, the computation expenses and the requirement of expert knowledge for the system setup have seriously restricted their applications in practice. In [17], an internal model controller (IMC) is developed for a scalar controlled IM feedback. This IMC controller has the advantage of robustness, ease of design and good responses. In some applications, a precise speed is required. Excellent speed regulation can be achieved with a DPLL. In [18], Moore found that this technique has significant ability to obtain precise speed regulation. When the feedback signal of the IM speed is synchronized with a reference signal, perfect signal speed regulation can be realized about 0.02% to 0.1% of the steady state accuracy which is difficult to obtain by conventional closed loop system. In the re-search work carried out in [19], the speed control of a three phase squirrel cage IM fed by employing triacs devices in the lines is described. The speed closed loop regulation of the system is then investigated by employing DPLL scheme. In [20], the study of the speed control of the IM drive using DPLL is discussed. The IM is fed with thyristors power inverter. In this, the DPLL is established using digital and analog circuits. In [21], the authors have developed an inverter fed IM drive using the PLL speed controller. While more literature has been discussing

Citation: B. Mouna and S. Lassaâd. "A Digital Phase Locked Loop Speed Control of Three Phase Induction Motor Drive: Performances Analysis," Energy and Power Engineering, Vol. 3 No. 1, 2011, pp. 61-68. DOI:10.4236/epe.2011.31009.

the design of such control loop, very few articles show the experimental results obtained at a specific target speed under a particular load appliance and none of these articles studies the PLL performance under a wide range of speeds and loads. Therefore, the aim of this paper is to highlight the high performance of the DPLL in experimentation for a wide range of speeds and shaft load appliance.

BASIC OF PLL IM

PLL is a feedback loop where a voltage controlled oscillator can be automatically synchronize to a periodic input signal [22-24]. The basic PLL has three components connected in a feedback loop [25-27] as shown in the block diagram of Figure 1: a voltage controlled oscillator (VCO), a phase detector (PD) and a low pass filter (LPF).

The VCO is an oscillator whose frequency f_{osc} is proportional to input voltage. The voltage at the input of the VCO determines the frequency of the periodic signal at its output. The VCO output and a periodic input signal (the reference) are inputs to the phase detector. When the loop is locked on the input signal, the frequency of the VCO output is exactly equal to that of a reference. It is also said that the PLL is in the locked condition. PD produces a signal proportional to the phase difference between the reference signal and the VCO output signal. The output of the PD is altered through a low pass filter. The loop is closed by connecting the filter output to the input of the VCO. The output voltage of the filter is used to control the VCO frequency. A PLL basic property is that it attempts to maintain the frequency lock between the VCO output and reference signal even if the frequency of the input signal varies in time [26]. Suppose that the PLL is in the locked condition, and that the frequency of the incoming signal increases slightly. The phase difference between the VCO signal and the in-coming signal will begin to increase in time. As a result, the filter output voltage increases, and the VCO output frequency increases until it matches the reference frequency, thus keeping the PLL in the locked condition. The range of frequencies from minimal to maximal values where the locked PLL remains in the locked condition is called the lock range of the PLL [25-26]. If the PLL is initially locked, and input signal frequency be-comes smaller than f_{min} or if input signal frequency exceeds f_{max}, the PLL fails to keep the VCO frequency equal to input signal one and the PLL becomes unlocked.

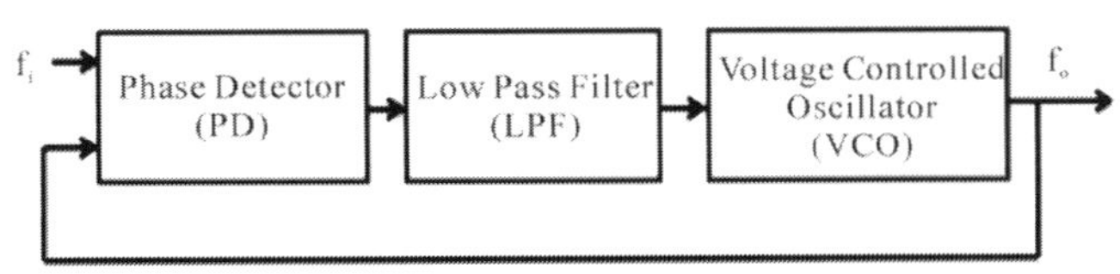

Figure 1. Bloc diagram of the PLL.

When the PLL is unlocked, the VCO oscillates at the frequency f_o called the center frequency, or the VCO free-running frequency. The lock can be established again if the incoming signal frequency gets close enough to f_o. The range of frequencies $f_i = f_o - f_c$ to $f_i = f_o + f_c$ such that the initially unlocked PLL becomes locked is called the capture range of the PLL [24]. The capture range 2fc depends on the characteristic of the loop filter [25]. For the used filter, an approximate implicit expression for the capture range can be founded as:

$$fc \cong \frac{VDD}{2} \frac{K_0}{\sqrt{1+\left(\frac{f_c}{f_p}\right)^2}} \quad (1)$$

If the capture range is much larger than the cut off frequency of the filter, $f_c f_p^{-1} > 1$, the expression for the capture range is simplified as:

$$2fc \cong \sqrt{2V_{DD} K_0 f_p} \quad (2)$$

K_0 is the VCO gain or the frequency sensitivity. Thanks to the rapid development of technology, the PLL is implemented using the integrated circuit systems. The most used integrated circuit is the well-known 4046 chip [23].

SYSTEM CONFIGURATION OF THE DRIVE SYSTEM AND EXPERIMENTAL RESULTS

The overall system block diagram of 4046 PLL IM speed control is given in Figure 2. This system is roughly classified into two parts: the control and the power parts. The control part is consisting mainly of 4046 PLL and a scalar control unit. A simple positive supply voltage is needed for the chip. The positive supply voltage VDD is connected to pin 16 and the ground is connected to pin 8. The reference square wave signal goes to the input of an internal amplifier at the pin 14 of the chip. The reference signal must be a square wave with a 50% duty cycle. Therefore, to obtain this target signal, a direct voltage is fed to pin 9 (the input of the 4046 VCO) to give at its output (pin 4), a square wave signal representing the target rotor speed. The VCO requires one external capacitor C_1 and one or two external resistors (R_1 or R_1 and R_2). Resistor R_1 and capacitor C_1 determine the frequency range of the VCO and resistor R_2 enables the VCO to have a frequency offset if required.

The output motor speed is sensed by a tacho generator. The motor speed in square waveform is needed. A second 4046 is added and the motor speed in square waveform is received at its VCO output. The obtained signal is feedback at pin 3 and compared to the target frequency received at pin 4 by the phase detector (PD) [27]. The used PD in this application is the comparator I of 4046. It is simply XOR logic gate with logic down output when the two inputs are either high or both low and the logic up otherwise. The PD output is proportional to the phase error between two inputs [25-27]. The obtained PD output is then filtered by an external low pass filter [27]. The obtained filtered voltage is proportional to the synchronous (stator supply) frequency which feeds the scalar control unit. The investigated DPLL IM speed regulation is built and tested using a 1kw — 2.65 A squirrel cage IM. Figure 3 shows the experimental set up drive system of the hardware configuration. The IM is powered by an insulated gate bipolar transistor (IGBT) voltage source inverter (VSI). It utilizes a thyristor rectifier as a dc bus voltage link. The pulse width modulation (PWM) signals to control the power modules are generated by the scalar control card equipped with a digital processor. A tacho generator delivering 1v for 1000 rpm is mounted on the rotor shaft to measure the rotor speed.

The DPLL is built around the 4046 integrated circuit system. A DC generator (1 kw - 6.2 A) is coupled with the IM which feeds the resistor bank used as a load. Investigated system is tested for various ranges of target rotor speeds under various range of loads. For each test, the rotor speeds, waveforms of real and target speeds and the output of the LPF as control signal are together illustrated. Obtained experimental results are given in Figures 4 to 18. Excellent steady state frequency has been obtained especially at high and low speeds proving the high performances of the rotor speed regulation using the DPLL. At no load, excellent steady state speed regulation is obtained. At a time of load appliance, the square root of the stator phase current increases to compensate the effect of the load appliance. Thus, the square root of stator phase voltage decreases due the voltage drop caused by the stator resistance. The fact that decreases the rotor speed. As, the PLL is in the locked condition, and that the frequency of the incoming signal decreases slightly. The phase difference between the VCO signal and the incoming signal will begin to increase in time. As a result, the filter output voltage increases, and the VCO output frequency increases until it matches the reference frequency, thus keeping the PLL in the locked condition. As, it is illustrated by the obtained experimental results under load appliance. At a time of load elimination, the square root of the stator phase current decreases. Thus, the square root of stator phase voltage increases. The fact that increases the rotor speed. As, the PLL is in the locked condition, and that the frequency of the incoming signal increases slightly. The phase difference between the VCO signal and the incoming signal will begin to decrease in time. As a result, the filter output voltage decreases, and the VCO output frequency decreases until it matches the reference frequency, thus keeping the PLL in the locked condition. As, it is illustrated by the obtained experimental results under load elimination. The digital scope screen images are illustrating the frequencies and speeds signals which are calibrated as follow: 1v for 1000rpm.

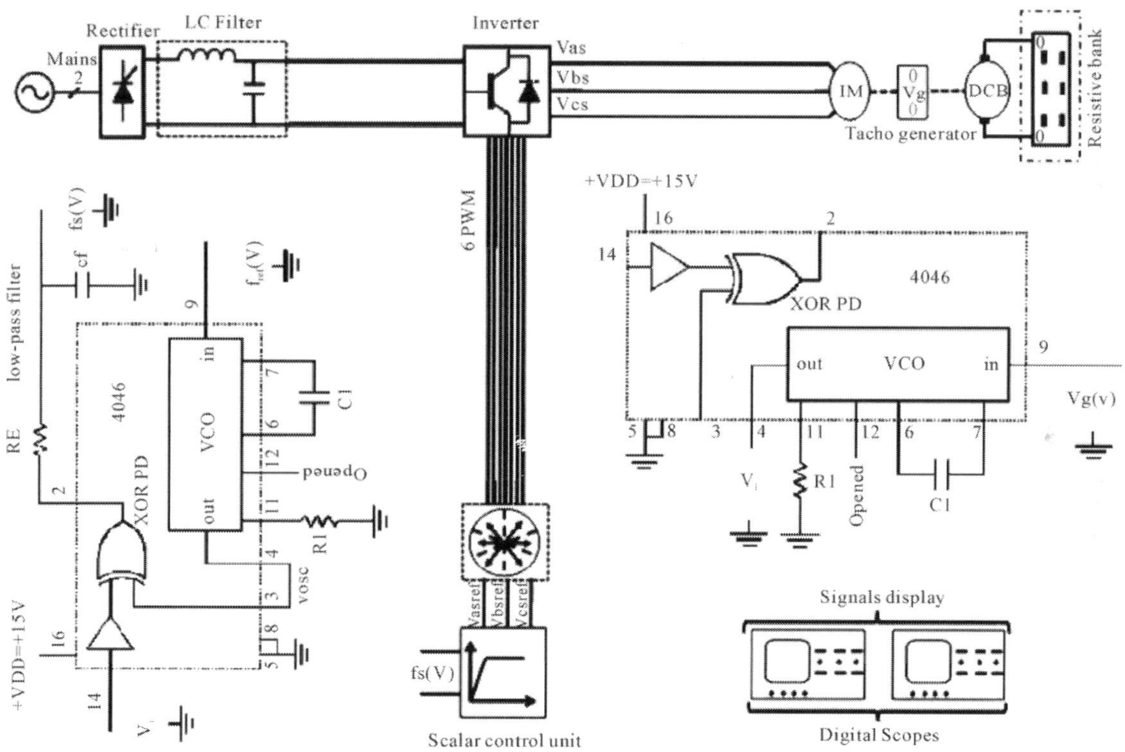

Figure 2. Bloc diagram of the DPLL IM speed control.

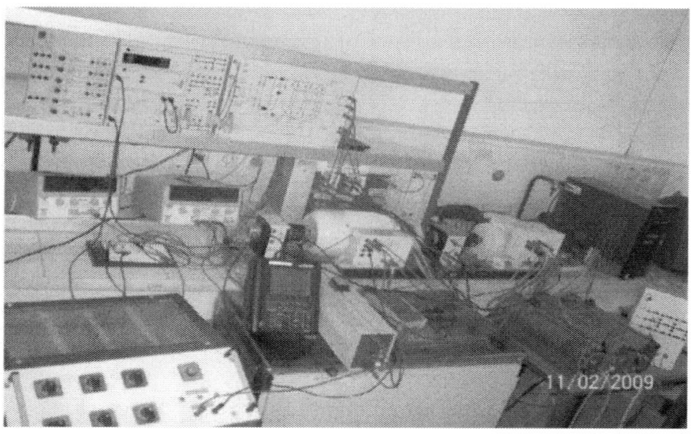

Figure 3. A photo of the experimental set up.

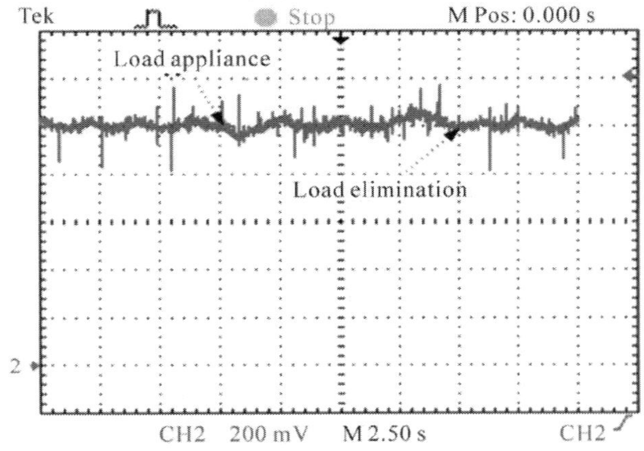

Figure 4. IM speed for a 1000rpm target speed under load appliance of 35% of its rated value.

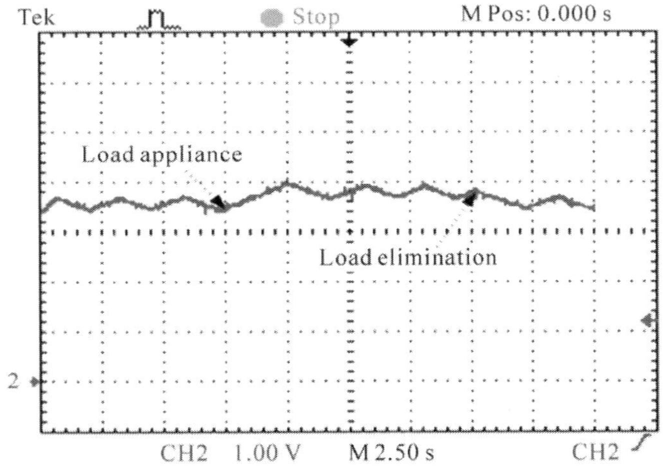

Figure 5. Control law for a 1000rpm target speed under load appliance of 35% of its rated value.

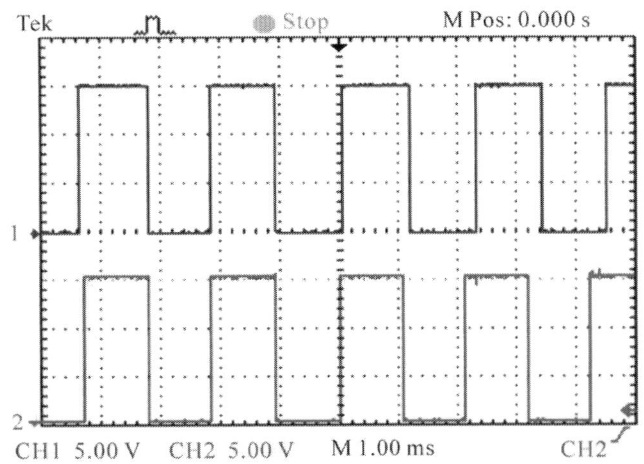

Figure 6. IM speed for a 1000rpm target speed under load appliance of 35% of its rated value.

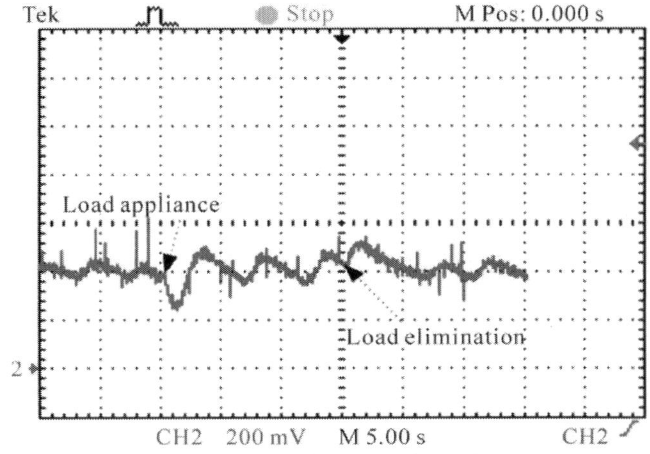

Figure 7. IM speed for a 400rpm target speed under load appliance of 40% of its rated value.

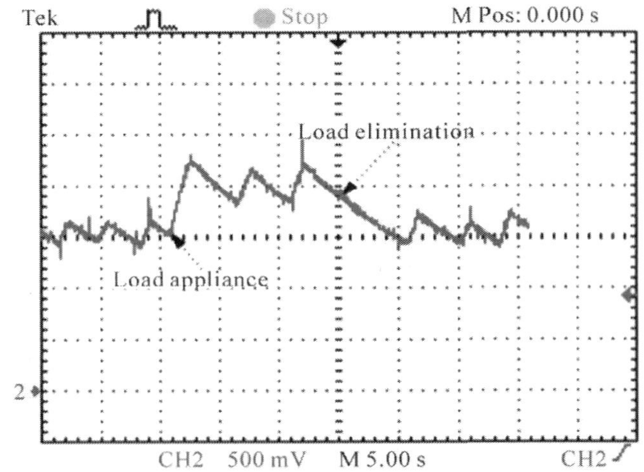

Figure 8. Real and target square waveforms for a 1000rpm target speed under load appliance of 40% of its rated value.

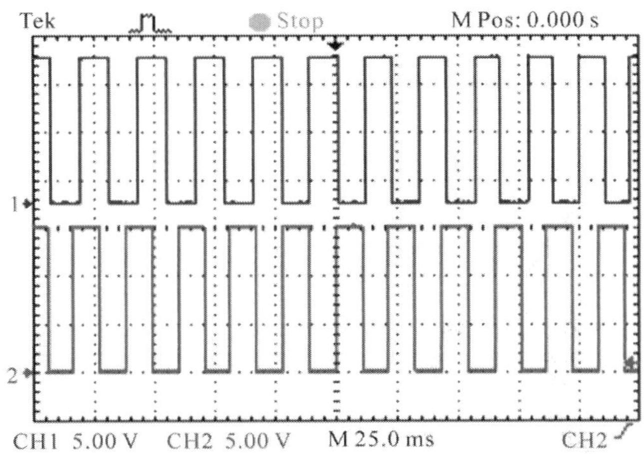

Figure 9. Real and target square waveforms for a 400rpm target speed under load appliance of 40% of its rated value.

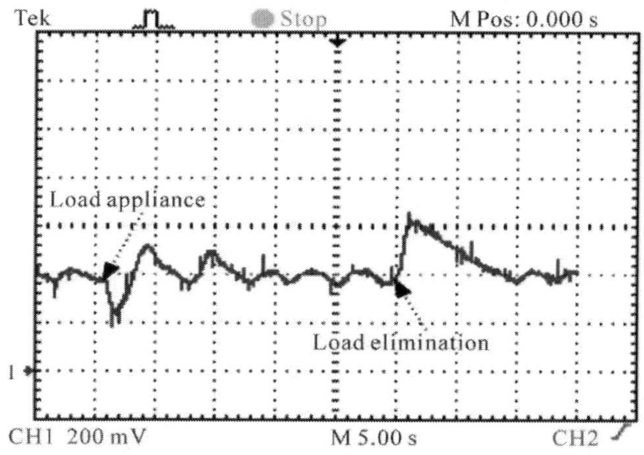

Figure 10. IM speed for a 400rpm target speed under load appliance of 50% of its rated value.

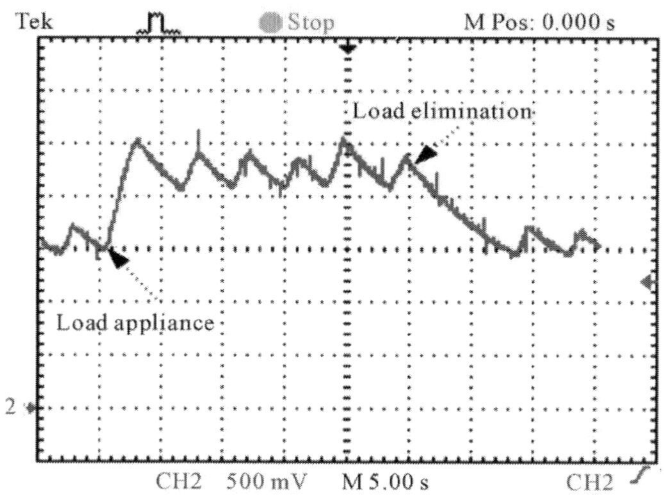

Figure 11. Control law for a 400rpm target speed under load appliance of 50% of its rated value.

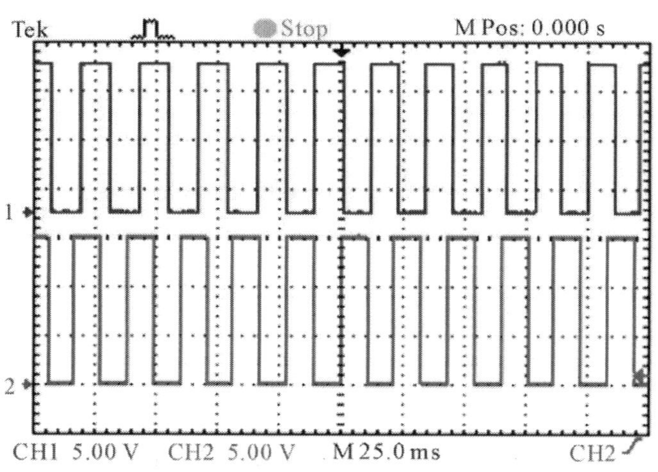

Figure 12. Real and target square waveforms for a 400rpm target speed under load appliance of 50% of its rated value.

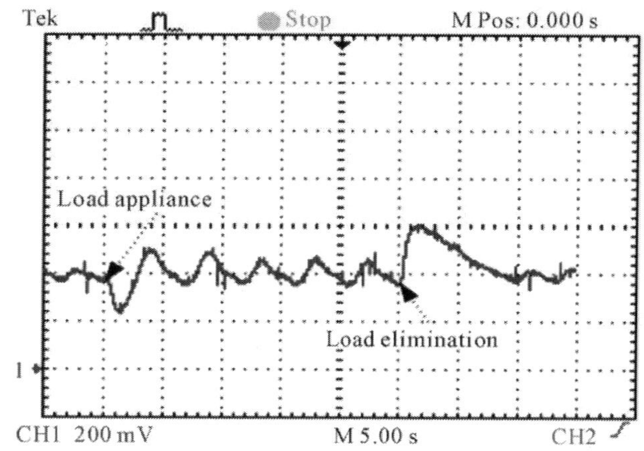

Figure 13. IM speed for a 400rpm target speed under load appliance of 60% of its rated value.

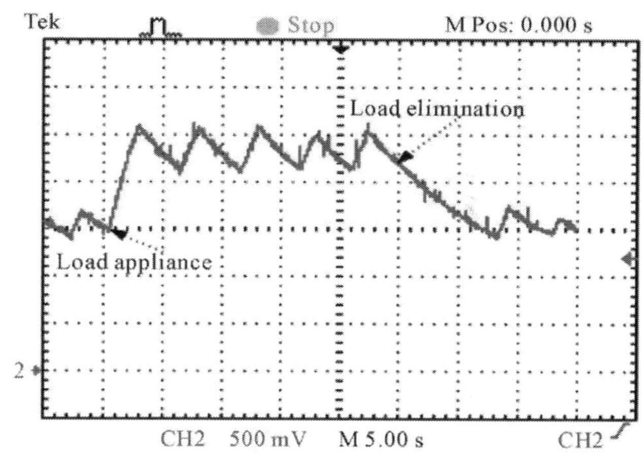

Figure 14. Control law for a 400rpm target speed under load appliance of 60% of its rated value.

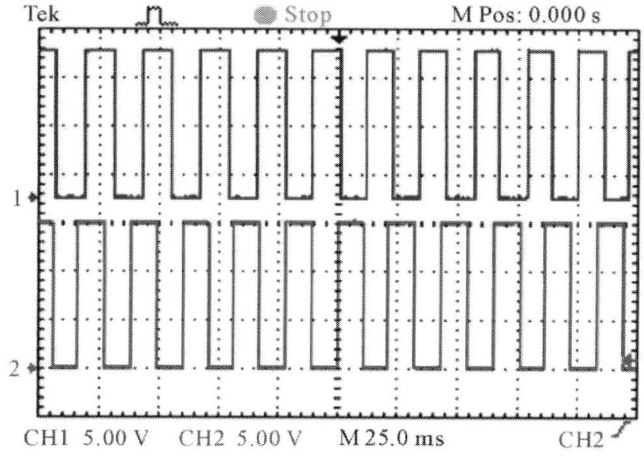

Figure 15. Real and square waveforms for a 400rpm target speed under load appliance of 60% of its rated value.

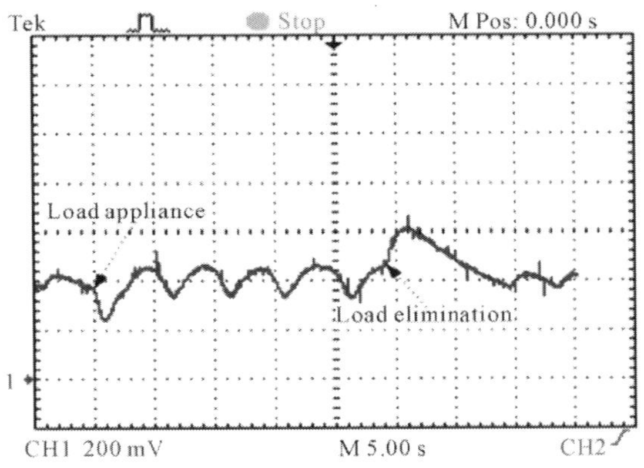

Figure 16. IM speed for a 400rpm target frequency under load appliance of 65% of its rated value.

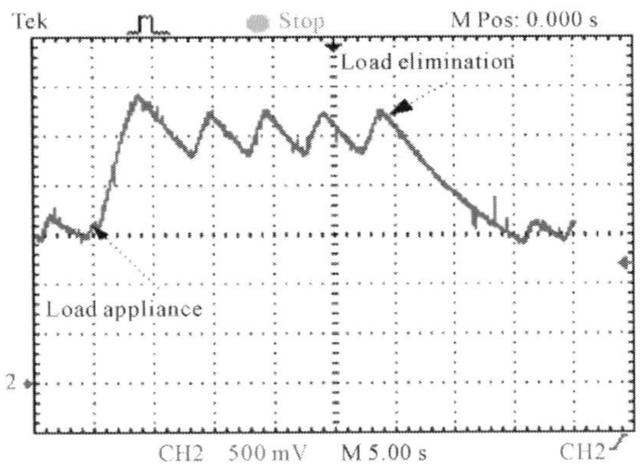

Figure 17. Control law for a 400rpm target speed under load appliance of 65% of its rated value.

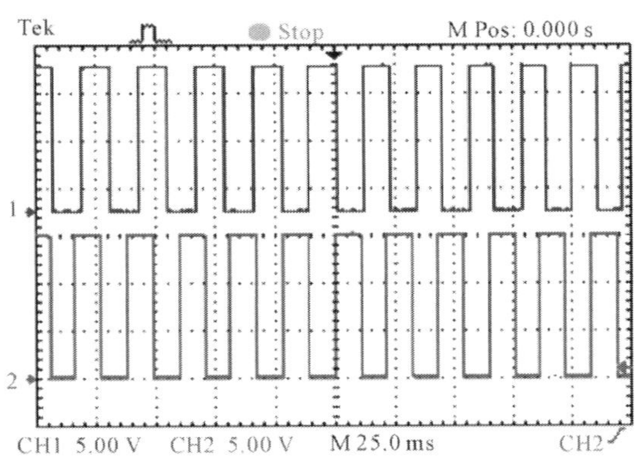

Figure 18. Real and target square waveforms for a 400rpm target speed under load appliance of 65% of its rated value.

CONCLUSION

In this paper, an experimental set up DPLL IM speed controller is designed and achieved. The overall system with three phase 1kw squirrel cage IM has been designed and tested in the laboratory over the entire range of speeds and loads. Obtained experimental results show that a precise speed regulation is achieved at steady state. Compared with the other control laws, the present scheme is in somewhat simpler, cheaper and provides speed control with high precision. All high performances scalar controlled IM drives require accurate rotational speed information for feedback control. This information is mainly provided by tackogenerator. The use of this sensor implies more electronics and high cost. To over-come these problems, a speed observer can be included in the control loop. Thanks to high performance of the PLL technique, it can be used to reconstruct the IM speed. The PLL sensorless novel scalar strategy IM speed controller will be the subject for a follow up further studies.

REFERENCES

1. S. Cong and Y. Liang, "PID-Like Neural Network Nonlinear Adaptive Control for Uncertain Multivariable Motion Control Systems," *IEEE Transactions on Industrial Electronics*, Vol. 56, No. 10, October 2009, pp. 3872-3879. doi:10.1109/TIE.2009.2018433
2. A. Rubai, M. J. Castro-Stitiriche and A. R. Ofli, "DSP Based Laboratory Implementation of Hybrid Fuzzy PID Controller Using Genetic Optimization for High Performance Motor Drives," *IEEE Transactions on Industry Applications*, Vol. 44, No. 6, November-December 2008, pp. 1977-1986. doi:10.1109/TIA.2008.2006347
3. M. Missiala, B. Vafakhah, J. Salmon and A. M. Knight, "Fuzzy Self-Tuning Speed Control of an Indirect Field-Oriented Control Induction Motor Drive," *IEEE Transactions on Industry Applications*, Vol. 44, No. 6, November-December 2008, pp. 1732-1740. doi:10.1109/TIA.2008.2006342
4. H. P. Huang, J. L. Yong and T. H. Chang, "Development and Fuzzy Control of a Pipe Inspection Robot," *IEEE Transactions on Industrial Electronics*, Vol. 57, No. 3, March 2010, pp. 1088-1095. doi:10.1109/TIE.2009.2031671
5. T. Orlowska-Kowalska, M. Dybkousi and K. Szabat, "Adaptive Sliding Mode Neuro-Fuzzy Control of Two Mass Induction Motor Drive without Mechanical Sensors," *IEEE Transactions on Industrial Electronics*, Vol. 57, 2009, pp. 553-564. doi:10.1109/TIE.2009.2036023
6. K. Szabat, T. Orlowska-Kowalska and M. Dybkousi, "Adaptive Sliding Mode Neuro-Fuzzy Control of Two Mass Induction Motor Drive without Mechanical Sensors," *IEEE Transactions on Industrial Electronics*, Vol. 56, 2009, pp. 4038-4042. doi:10.1109/TIE.2009.2022514
7. S. Katayama, K. Yubai and J. Hirai, "Iterative design of the reduced order weight and controller for the H_∞ loop shaping method under open loop magnitude constraints for SISO systems," *IEEE Transactions on Industry Electronics*, Vol. 56, 2009, pp. 3854-3863. doi:10.1109/TIE.2009.2017099
8. K. K. Hoong, C. Y. Chung and S. Lan, "Model Based H_∞ Control of Unified Power Quality Conditioner," *IEEE Transactions on Industrial Electronics*, Vol. 56, 2009, pp. 2493-2504. doi:10.1109/TIE.2009.2020705
9. I. Yalsh and U. Shaked, "Neuro adaptive H_∞ Estimation and Its Application to Improved Tracking in GPS Re-ceivers," *IEEE Transactions on Industrial Electronics*, Vol. 56, 2009, pp. 642-647. doi:10.1109/TIE.2008.2008343
10. B. Betron, T. Ahmed-Ali and M. Benbouzid, "High Order Sliding Mode Control of Variable Speed Wind Turbines," *IEEE Transactions on Industrial Electronics*,

Vol. 56, No. 9, September 2009, pp. 3314-3321. doi:10.1109/TIE.2008. 2006949

11. Y. Xinghuo and O. Kaynak, "Sliding Mode Control with Soft Computing: A Survey," *IEEE Transactions on Industrial Electronics*, Vol. 56, No. 9, September2009, pp. 3275-3285. doi:10.1109/TIE.2009.2027531

12. U. V. Utkin, "Sliding Mode in Control and Optimization," Springer, Berlin, 1992.

13. M. B. Hamed, L. Sbita and W. Abboud, "Neural Networks for Controlled Speed Sensorless Direct Field Oriented Induction Motor Drives," *International Journal of Electrical Engineering*, Vol. 8, No. 11, 2008, pp. 84-96.

14. L. Sbita and M. B. Hamed, "Fuzzy Controller and ANN Speed Estimation for Induction Motor Drives," *Proceeding of the* 4*th IEEE International Conference*, 2007.

15. X.-D. Sun, K.-H. Kho, B.-G. Yu and M. Matsui, "Fuzzy Logic Based V/f Control of an Induction Motor for a DC Grid Power Leveling System Using Flywheel Energy Storage Equipment," *IEEE Transactions on Industrial Electronics*, Vol. 56, 2009, pp. 3161-3168. doi:10.1109/TIE.2009.2021679

16. L. Y. Wang, T. Y. Chai and L. F. Zhai, "Neural Network Based Terminal Sliding Mode Control of Robotic Manipulators Including Actuator Dynamics," *IEEE Transactions on Industrial Electronics*, Vol. 56, No. 9, September 2009, pp. 3296-3304. doi:10.1109/TIE.2008.2011350

17. M. B. Hamed and L. Sbita, "Internal Model Controller of an ANN Speed Sensorless Controlled Induction Motor Drives," *Journal of Applied Sciences*, Vol. 7, No. 11, 2007, pp. 1456-1466. doi:10.3923/jas.2007.1456.1466

18. A. W. Moore, "Phase locked loops for motor speed control," *IEEE Spectrum*, 1973, pp. 61-67. doi:10.1109/MSP EC.1973.5216851

19. W. L. Kenly and B. K. Bose, "A Triac Speed Control of Three Phase Induction Motor with Phase Locked Loop Regulation, " *IEEE Transactions on Industrial Electronics*, Vol. IAI-12, 1976, pp. 492-498.

20. R. Moffat, P. Sen and R. Younker, "Digital Phase Locked Loop for Induction Motor Speed Control," *IEEE Trans-actions on Industry Applications*, Vol. IAI-15, No. 2, March1979, pp. 176-182. doi:10.1109/TIA.1979.4503634

21. P. C. Sen and M. Macdonald, "Stability Analysis of Induction Motor Drives Using Phase Locked Loop Control System," *IEEE Transactions on Industrial Electronics*, Vol. IECI-27, No. 3, August 1980, pp. 147-155. doi:10.1109/TIECI.1980.351668

22. N. M. H. Ismail and M. Othman, "Low Power Phase Locked Loop Frequency Synthesizer for 2.4 GHz Band Zigbee," *American Journal of Engineering and Applied Sciences*, Vol. 2, 2009, pp. 337-343. doi:10.3844/ajeassp.2009.337.343

23. G. C. Hseih and J. C. Hung, "Phase-Locked Loop Techniques: A Survey," *IEEE Transactions on Industrial Electronics*, Vol. 43, No. 6, December 1996, pp. 609-615. doi:10.1109/41.544547

24. A. Takano, "Quick-Response Torque-Controlled Induction Motor Drives Using Phase-Locked Loop Speed Con-trol with Disturbance Compensation," *IEEE Transactions on Industrial Electronics*, Vol. 43, No. 6, December 1996, pp. 640-646. doi:10.1109/41.544551

25. R. A. Al-Taha and S. R. Jassim, "Stability analysis of a linear discrete data induction motor drive using phase locked loop (PLL)," *Proceedings SSD Conference*, 2005.

26. T. Lagutere, "Conceptions et modélisations d'oscillateurs et de leurs boucles verrouillage de phase associes pour des applications de radiocommunications mobiles professionnelles," Ph.D. Dissertation, Poitier University, 2008.

27. H. Y. Aleksandar, "Test Generator for Examination of PLL and DDS Systems," *Electronics*, 2008, pp. 72-76.

CHAPTER 12

FUZZY-LOGIC BASED SPEED CONTROL OF INDUCTION MOTOR

Mohammad Abdul Mannan, Asif Islam, Mohammad Nasir Uddin, Mohammad Kamrul Hassan, Toshiaki Murata, and Junji Tamura

Keywords: Indirect-Field-Oriented Control, PI Controller, Fuzzy Logic Controller, Induction Motor, Speed Control

INTRODUCTION

Induction motors have been considered for many industrial applications since its discovery [1-3] because of its low maintenance, robustness, low cost, high efficiency, good self starting, simplicity of design, absence of the collector brooms system and small inertia. The most problematic things of induction motor are its complex, nonlinear, multivariable mathematical model, and it is not inherently capable of providing variable speed operation [4,5]. The problems of induction motor can be solved through the use of intelligent control and adjustable speed controllers, such as scalar and vector control drive [6,7].

The scalar control methods like the volt-hertz control performance is good only for steady state condition but it leads to poor dynamic performance. Since Blashke [6] and Hasse [8] have developed the new technique known as vector control or field oriented control (FOC), the use of the induction machine becomes more and more frequent because FOC provides better performance for steady state as well as transient conditions. This control strategy can provide the same performance as achieved from a separately excited DC machine, and is proven to be well adapted to all type of electrical drives associated with induction machines [9].

The FOC schemes are classified into two groups: the direct method of field orientation proposed by Blaschke [6] and the indirect method of field orientation proposed by Hasse [8]. The direct method requires flux acquisition, which is mostly obtained by computation techniques using machine terminal quantities, where as indirect method avoids the requirement of flux acquisition by using known motor parameters to compute the appropriate motor slip speed to obtain the desired flux position. The scheme of indirect is simpler to implement than the direct method of FOC hence, indirect field oriented control (IFOC) method has become more popular [1,10-13].

In order to achieve the same performance of DC machine from induction motor, the rotor flux and torque of are decoupled in terms of stator current components neglecting the core loss [6-16]. The core loss of induction motor is very common and may not be negligible to achieve precise industrial application [17]. The decoupling control of rotor flux and torque in terms of stator current components of induction motor drive taking core loss into account is complicated [18-21]. Therefore, an IFOC method of induction motor drives taking core loss into account in terms of magnetizing current components has been proposed with PI speed controller loop in [22]. The extension work [22] of literature [18] has been done to regulate speed and rotor flux based on PI controller. In [22], the PI controller gains were changed to obtain desired speed under the variation of load torque. One set of fixed PI controller gain is unable to track the desired speed. Moreover, the steady state error cannot be minimized to zero by the conventional PI controller [16,21]. An IFOC of induction motor drive to regulate speed based on fuzzy-logic has been proposed in [16] neglecting core loss. To overcome the previously discussed disadvantages of conventional PI controller and the complication effects of core loss of induction motor drive, it would be desirable to design a well controller. The fuzzylogic control (FLC) is seemed to be a suitable controller in terms of high dynamic response under the variation of load torque and parameters [16].

In this paper, a fuzzy-logic-based high-performance control of IM drive including core loss is proposed. The FLC is used to regulate rotor speed in terms of magnetizing q-axis current where the reference d-axis current is kept constant at

Citation: M. Mannan, A. Islam, M. Uddin, M. Hassan, T. Murata, and J. Tamura. "Fuzzy-Logic Based Speed Control of Induction Motor," Intelligent Control and Automation, Vol. 3 No. 3, 2012, pp. 229-235. DOI: 10.4236/ica.2012.33026.

rated value. The proposed controller can track properly the desired speed under the variation of load torque and parameters. The effectiveness of the proposed controller with space vector PWM inverter-fed IM drive is shown by computer simulation which done by Matlab/Simulink. The simulation is also done for PI speed controller of induction motor taking core loss into account. The simulation study shows the superiority of the proposed FLC over the conventional PI controller.

IFOC OF INDUCTION MOTOR DRIVE TAKING CORE LOSS INTO ACCOUNT

Figure 1 shows the equivalent circuit of an induction motor taking core loss into account in the d-q axes synchronously rotating reference frame. The voltage equations are given by:

$$\left.\begin{aligned} v_{sd} &= R_s i_{sd} + p\Phi_{sd} - \omega_e \Phi_{sq} \\ v_{sq} &= R_s i_{sq} + p\Phi_{sq} + \omega_e \Phi_{sd} \end{aligned}\right\} \qquad (1)$$

$$\left.\begin{aligned} 0 &= R_r i_{rd} + p\Phi_{rd} - \omega_s \Phi_{rq} \\ 0 &= R_r i_{rq} + p\Phi_{rq} + \omega_s \Phi_{rd} \end{aligned}\right\} \qquad (2)$$

$$\left.\begin{aligned} R_c i_{cd} &= p\Phi_{md} - \omega_e \Phi_{mq} \\ R_c i_{cq} &= p\Phi_{mq} + \omega_e \Phi_{md} \end{aligned}\right\} \qquad (3)$$

where p is differential operator; v_{sd} and v_{sq} are the stator dand q-axis input voltages; i_{sd} and i_{sq} are the stator dand q-axis currents; i_{rd} and i_{rq} are the rotor dand q-axis currents; i_{cd} and i_{cq} are the core loss dand q-axis currents; F_{sd} and F_{sq} are the stator dand q-axis fluxes; F_{rd} and F_{rq} are the rotor dand q-axis fluxes; F_{md} and F_{mq} are the magnetizing dand q-axis fluxes; R_s, R_r and R_c are stator, rotor and core loss resistances; w_e and w_s are primary and slip angular frequencies.

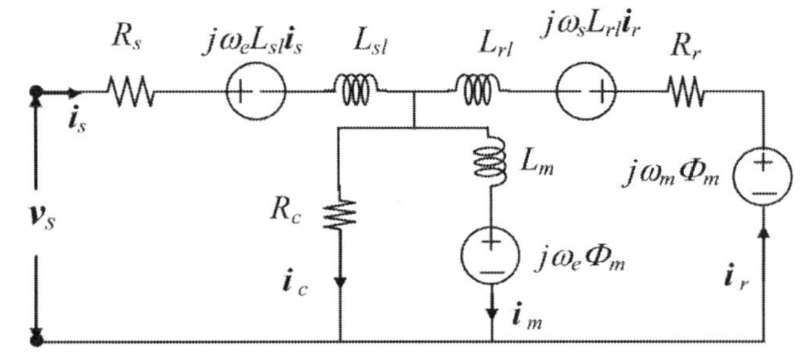

Figure 1. Equivalent circuit of IM taking core loss into account in a synchronously rotating reference frame.

The current equations are written as

$$\left.\begin{aligned} i_{sd} + i_{rd} &= i_{cd} + i_{md} \\ i_{sq} + i_{rq} &= i_{cq} + i_{mq} \end{aligned}\right\} \qquad (4)$$

where, i_{md} and i_{mq} are the magnetizing dand q-axis currents. The flux equations are given as

$$\Phi_{sd} = L_{sl} i_{sd} + \Phi_{md}; \quad \Phi_{sq} = L_{sl} i_{sq} + \Phi_{mq} \qquad (5)$$

$$\Phi_{rd} = L_{rl} i_{rd} + \Phi_{md}; \quad \Phi_{rq} = L_{rl} i_{rq} + \Phi_{mq} \qquad (6)$$

$$\Phi_{md} = L_m i_{md}; \quad \Phi_{mq} = L_m i_{mq} \qquad (7)$$

where, L_{sl} is stator leakage inductance; L_{rl} is rotor leakage inductance; L_m is magnetizing inductance; L_s is stator self-inductance; L_{rl} is rotor self-inductance.

The mechanical dynamic system and developed torque are given as

$$p\omega_m = -(D/J)\omega_m + (P_n/J)(T_e - T_L) \qquad (8)$$

$$T_e = P_n (L_m / L_{rl})(i_{mq}\Phi_{rd} - i_{md}\Phi_{rq}) \qquad (9)$$

where, w_m is rotor angular frequency; T_e and T_L are electromagnetic and load torques; P_n is number of pole pair; J is moment of inertia; and D is damping factor.

The main feature of IFOC method is given as follows:

$$\Phi_{rd} = \text{Constant}; \quad \Phi_{rq} = 0 \qquad (10)$$

According to the FOC constrains of Equation (10), the core loss current rotor flux, torque and slip can be obtained

$$i_{cd} = -L_m \omega_e i_{mq} / R_c; \quad i_{cq} = L_m \omega_e i_{md} / R_c \qquad (11)$$

$$\Phi_{rd} = L_m (i_{sd} - i_{cd}) \qquad (12)$$

$$T_e = (P_n L_m \Phi_{rd} / L_r)(i_{sq} - i_{cq}) \qquad (13)$$

$$\omega_s = (R_r L_m / L_r \Phi_{rd})(i_{sq} - i_{cq}) \qquad (14)$$

Figure 2 shows the steady state calculation of stator current components for different torque at rated speed with and without consideration of core loss.

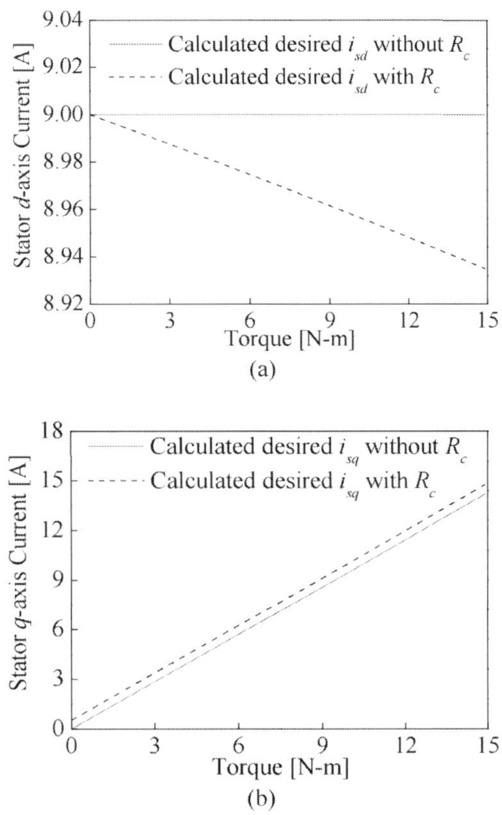

Figure 2. Steady state stator current for different torque at rated speed. (a) Stator d-axis current; (b) Stator q-axis current.

This calculation has done by using the parameters which are given in Table 1. The calculation of stator d-axis current to control flux and stator q-axis flux to control torque are not same of practical one. The decoupling control of IM in terms stator currents will not be met as shown in Equations (11)-(13) and Figure 2, because in practical IM core loss currents i_{cd} and i_{cq} are not zero. The rotor flux, torque and slip in terms of magnetizing current are [18].

$$\Phi_{rd} = L_m i_{md} \quad (15)$$

$$T_e = P_n \left(L_m / L_{rl} \right) i_{mq} \Phi_{rd} \quad (16)$$

$$\omega_s = \left(R_r L_m / L_{rl} \right) \left(i_{mq} / \Phi_{rd} \right) \quad (17)$$

From Equations (15)-(17), it is seen that the decoupling control of torque and rotor flux of IM taking core loss into account can be designed easily in terms of magnetizing current components instead of stator current components. PI based speed controller is proposed in [18] to obtain i_{mq} for regulating rotor speed where the reference rotor d-axis flux is kept constant at rated value.

DESIGN OF A PI CONTROLLER

The use of PI control has a long history in control engineering and is acceptable for many real industrial applications. Hence, PI control is still the commonly used control technique in the industrial applications despite the development of the advanced control techniques. The main reasons for this popularity are that PI controllers are often effective and are easy to maintain by plant personnel. The structure of PI controller is relatively simple which can be easily understood and implemented in practice.

Table 1. Ratings and parameters of induction motor.

3 hp, 1430 rpm, 14.96 N-m, 4 pole, 50 Hz
$R_s = 0.55\ \Omega$, $R_r = 0.75\ \Omega$, $R_c = 320\ \Omega$, $L_{sl} = 5$ mH, $L_{rl} = 5$ mH, $L_m = 63$ mH, $J = 0.0179$ Kg·m^2, $D = 0.001$ N-m-s

Due to these merits of PI controller, it has applied to control speed, torque and current of IM control in Industrial applications [2,3].

In this paper, discrete-time PI controller is designed to achieve the desired speed. The discrete-time PI controller can be defined by the following expression to calculate the first difference of desired q-axis magnetizing current components.

$$\Delta i^*_{mq}(k) = K_{I\omega} e_\omega (k) - K_{P\omega} \Delta \omega_m (k) \quad (18)$$

where,

$$e_\omega (k) = \omega^*_m (k) - \omega_m (k),$$

$$\Delta \omega_m (k) = \omega_m (k) - \omega_m (k-1),$$

$$\Delta i^*_{mq}(k) = i^*_{mq}(k) - i^*_{mq}(k-1)$$

where, k is sampling instant, K_{Pw} and K_{Iw} indicate the proportional and integral constant respectively for speed controller loop. Superscript * indicates the desired or reference value.

But the PI controller, where the controller gains are chosen by trial and error method, is not adequate to follow the reference speed without steady state error. Also, the PI controller is not robust under the variation of load torque and parameters of IM drive. The disadvantages of PI controller are clarified in the simulation results section. Therefore, we proposed speed controller of IM drive based on FLC to eliminate the steady state error and increase the robustness.

DESIGN OF A FUZZY LOGIC CONTROLLER

The FLC is proposed to find out the desired magnetizing q-axis current according to reference speed under the variation of load torque and parameters. The magnetizing d-axis current is kept constant which is found out from the constant rated rotor d-axis flux.

It is seen from Equation (14) that the desired magnetizing q-axis current is a function of error of speed and change of speed error. To adopt Δi^*_{mq}, the design of FLC is explained in the following.

Fuzzification

To design the proposed FLC, the first assumption is that the difference of the actual speed from the desired or reference speed, $e_w(k)$ and the difference of speed error, $De_w(k)$ are the only available controller input linguistic variables. The first difference of magnetizing q-axis current, Δi^*_{mq}, is considered as the output linguistic variable. For convenience, the inputs and output of the FLC are scaled with different coefficients K_w, K_e and K_i respectively. The scaling factors K_w, K_e and K_i are chosen for error of speed, change of error of speed and first difference of magnetizing q-axis current respectively as shown in Figure 3. These scaling factors can be constants or variables and play an important role for FLC design in order to achieve a good response in both transient and steady states. In this work, these scaling factors are considered as constant and are selected by trial and error. The triangular membership functions with overlap used for the input and output fuzzy sets are shown in Figure 4 in which the linguistic variables are represented by NB (Negative Big), NM (Negative Medium), NS (Negative Small), Z (Zero), PS (Positive Small), PM (Positive Medium) and PB (Positive Big). The grade of input membership functions can be obtained as follows.

$$\mu(x) = \left[w - 2|x - m| \right] / w \quad (19)$$

where m(x) is the value of grade of membership, w is the width and m is the coordinate of the point at which the grade of membership is 1, x is the value of the input variable.

Rule Base

The fuzzy mapping of the input variables to the output is represented by fuzzy IF-THEN rules of the following form:

IF $\langle e^n_\omega$ is N$\rangle$ and $\langle \Delta e^n_\omega$ is N$\rangle$ THEN $\langle \Delta i^n_{mq}$ is S$\rangle$

IF $\langle e^n_\omega$ is P$\rangle$ and $\langle \Delta e^n_\omega$ is Z$\rangle$ THEN $\langle \Delta i^n_{mq}$ is B$\rangle$

The entire rule base is given in Table 2. There are total 49 rules to achieve desired speed trajectory.

Inference and Defuzzification

From the rule base in Table 2, the inference engine provides fuzzy value of Δi^{n}_{mq}, and then crisp numerical value of Δi^{*}_{mq} is obtained by using defuzzification procedure. The most popular method of inference and defuzzification is Mamdani's max-min (or sum-product) composition with center of gravity method [23]. In this work, Mamdani type fuzzy inference [23] is used. The center of gravity method [23] is used for defuzzification to obtain Δi^{n}_{mq}. The normalized output function is given as

$$\Delta i^{*}_{mq}(n) = \sum_{i=1}^{N} \mu_i C_i \Big/ \sum_{i=1}^{N} \mu_i \qquad (20)$$

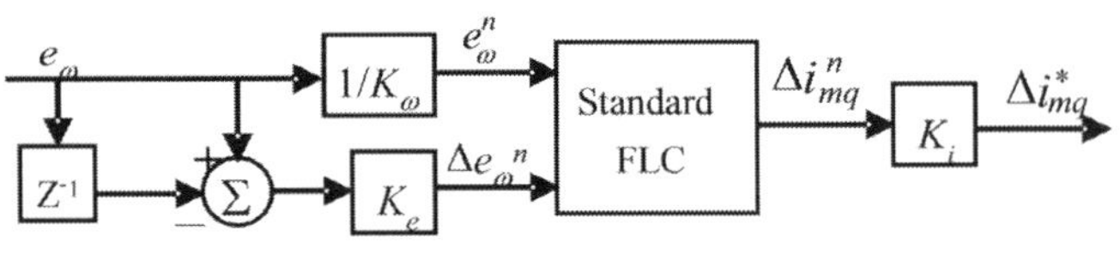

Figure 3. Scheme structure of the proposed FLC.

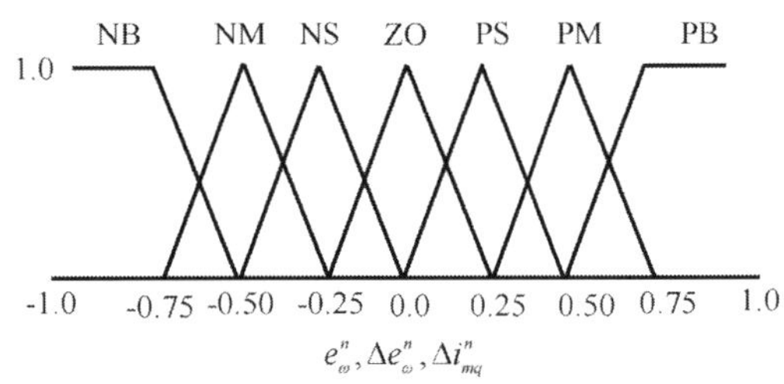

Figure 4. Fuzzy sets and their corresponding memberships functions.

Table 2. Fuzzy rules.

Δi^n_{mq}		Δe^n_ω						
		NB	NM	NS	ZO	PS	PM	PB
	NB	NB	NB	NM	NM	NS	NS	ZO
	NM	NB	NM	NM	NS	NS	ZO	PS
	NS	NM	NM	NS	NS	ZO	PS	PS
e^n_ω	ZO	NM	NS	NS	ZO	PS	PS	PM
	PS	NS	NS	ZO	PS	PS	PM	PM
	PM	NS	ZO	PS	PS	PM	PM	PB
	PB	ZO	PS	PS	PM	PM	PB	PB

where N is total number rules, m_i is the membership grade for ith rule and C_i is the coordinate corresponding to the maximum value of the respective consequent membership function [C_i Î {0.0, 0.9}]. After finding out Δi^{n}_{mq}, the actual desired first difference magnetizing current, Δi^{*}_{mq}, can be found out by product of scaling factor as shown in Figure 3.

SIMULATION RESULTS

In order to verify the performance of proposed fuzzylogic-based speed control of IM drive, computer simulations were performed using Matlab/Simulink. The ratings and parameters of IM drive used are given in Table 1. The sampling period was selected as T_s = 100 msec. In this work, the scaling factors for speed error and change of speed error were chosen as K_w = rated speed in rad/sec, K_e = 3 and K_i = rated value of i_{mq}. These scaling factors are selected by trial and error to achieve good response in both transient and steady states. We have conducted our simulation work based on Figure 5 where the IM is driven by space vector PWM inverter [22].

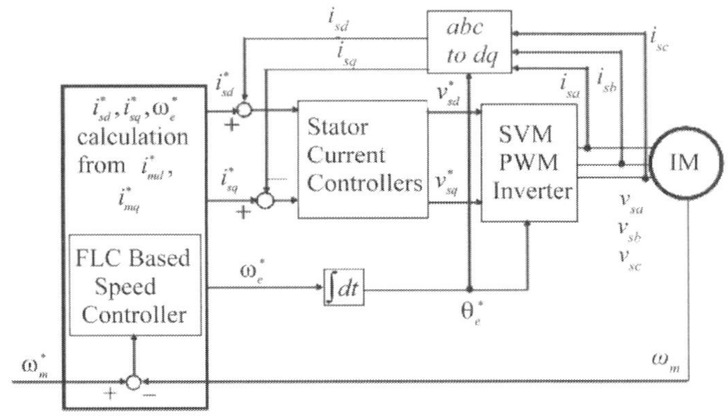

Figure 5. Proposed FLC-based IM drive.

In order to compare the performance of the proposed FLC with conventional PI controller, simulation tests were performed for both FLC and PI controllers. Figure 6 shows the simulation results based on PI controller for different values of PI controller gains. In these cases reference speed is changed from 0.0 rpm to rated 1000 rpm at 0.0 sec and load torque is changed from 6 to 12 N-m at 1.0 sec.

In Case I of Figure 6, the actual speed follow the reference speed without overshoot but the actual speed cannot follow the reference speed due to the change of load torque. In Case II of Figure 6, where the integral constant is increased, the steady state error is occurred between the actual speed and reference speed. In Case III of Figure 6, the actual speed follows the reference speed without steady state error, but overshoot is occurred due to the change of reference speed. It is clear from Figure 6 that the conventional PI controller is not suitable to achieve the desired speed of IM drive without overshoot or steady state error and not robust under the variation of load torque and parameters.

Figure 7 shows the transient responses for indirectfield-oriented control of IM drive using the proposed fuzzy-logic controller with consideration of core loss and without consideration of core loss.

In this simulation work, the step changed from 500 to 1000 rpm at 1.0 sec, load torque changed from 6 to 12 N-m at 2 sec, and again speed changed from 1000 to 500 rpm at 3 sec. It is evident from Figure 6 that the actual speed using proposed FLC follows the reference speed without any overshoot and steady state error. The design controller is also robust under the variation of disturbance of load torque as shown in Figure 6. If the rotor speed response of Figure 7 is compared with that of Figure 6, it can be clearly understand that the proposed fuzzy-logic controller is better than the conventional PI controller.

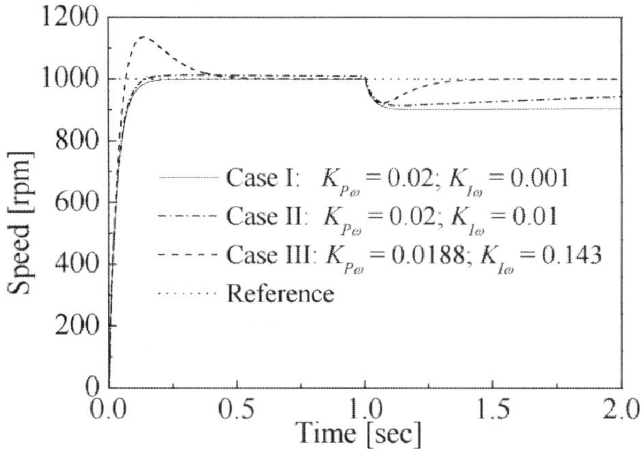

Figure 6. Transient responses for step change of reference speed and step change of load torque using PI controller with different sets of proportional and integral gains.

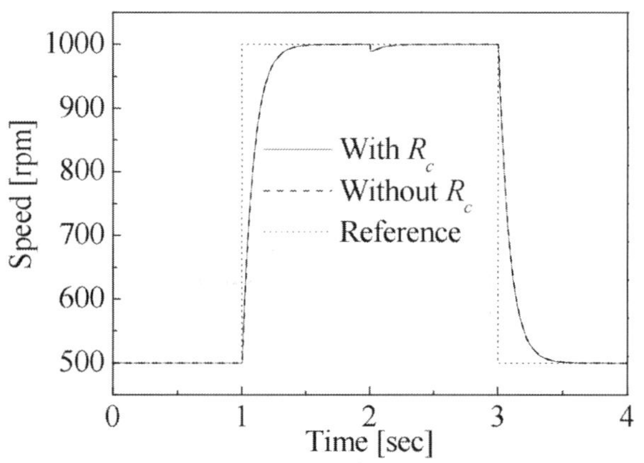

Figure 7. Transient responses of speed for step change of reference speed and step change of load torque using FLC.

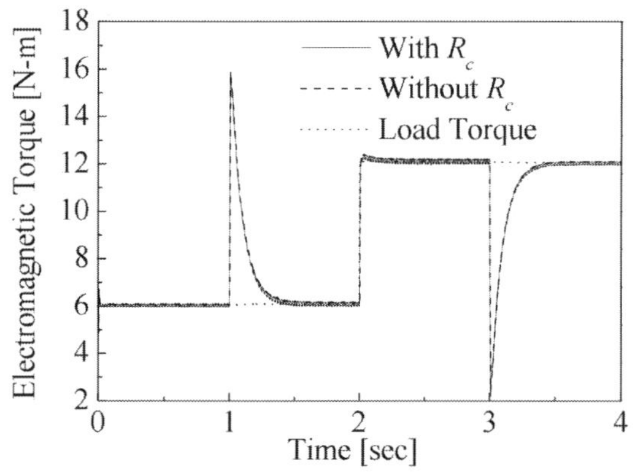

Figure 8. Transient response of torque using proposed fuzzy logic control.

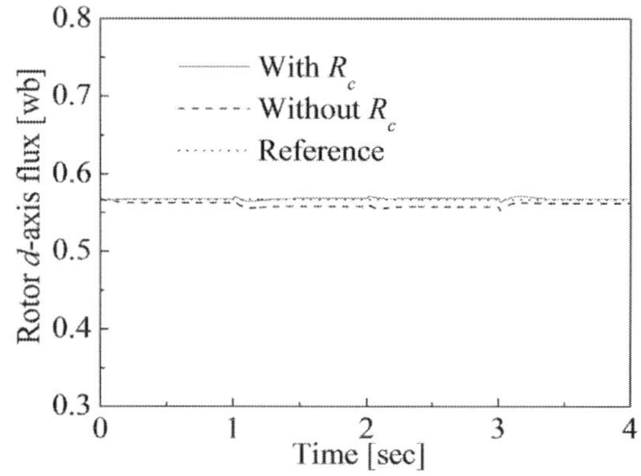

Figure 9. Transient response of d-axis rotor flux using proposed fuzzy logic control.

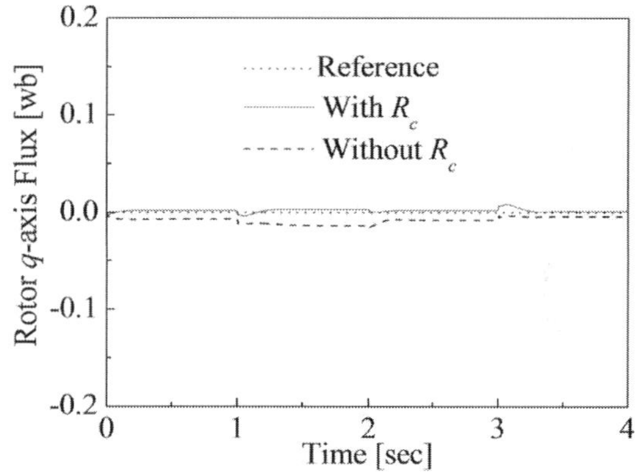

Figure 10. Transient response of q-axis rotor flux using proposed fuzzy logic control.

Figure 8 shows the transient response of torque. The variation of load torque can be compensated by changing the electromagnetic torque as shown in Figure 8. Figures 9 and 10 show the transient response of rotor flux due to the step changes of reference speed and load torque. It is notified from these figures that the rotor fluxes cannot be reached to its references values without consideration of core loss. If the field-oriented control is designed without taking core loss into account the constraints of field oriented control is not met. Thus the precise torque and flux control is possible by considering the core loss. The precise torque and flux control is possible by decoupling the magnetizing current components rather than the stator current components as shown in Figure 11. The transient response of stator current components is shown in Figure 12. The required stator current components is not reached to decoupling the torque and flux control which is cleared due to the deviations of required current and actual current. It is clear from Figures 10 and 11 the decoupling control of torque and rotor flux can be achieved in terms of d-axis and q-axis magnetizing current, respectively.

In order to verify the robustness of proposed fuzzy logic controller, the simulation work has been done by varying the change of mutual inductance and rotor resistance as shown in Figure 13. From Figures 7 and 13, it is cleared that the proposed fuzzy logic controller is robust under variations of disturbance and parameters.

In order to implement our proposed control system, at first we have to measured the value of core loss resistance. How the core loss resistance can be measured and included into the field-oriented control of IM has been discussed in detail in [17-21]. The proposed fuzzy logic controller can be implemented in practical applications based on DSP. The explanation of experiment to implement of FLC based on DSP has been discussed in [16,24].

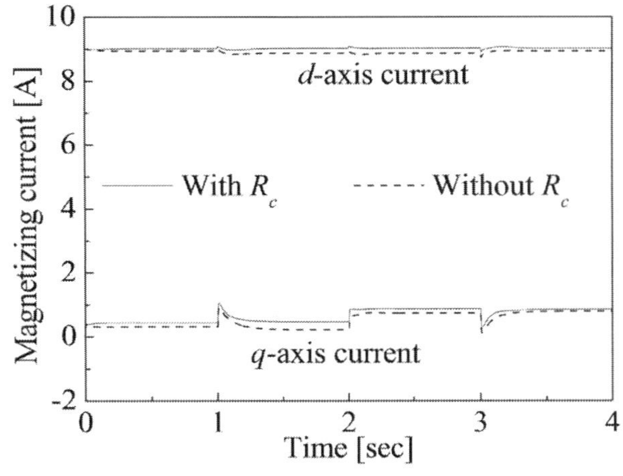

Figure 11. Transient response of magnetizing current using proposed fuzzy logic control.

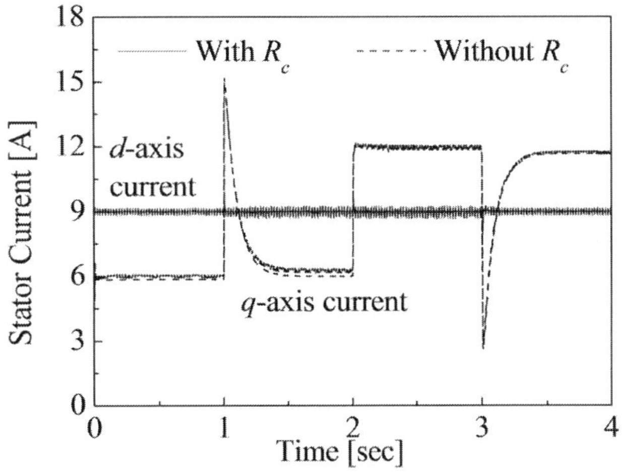

Figure 12. Transient response of magnetizing current using proposed fuzzy logic control.

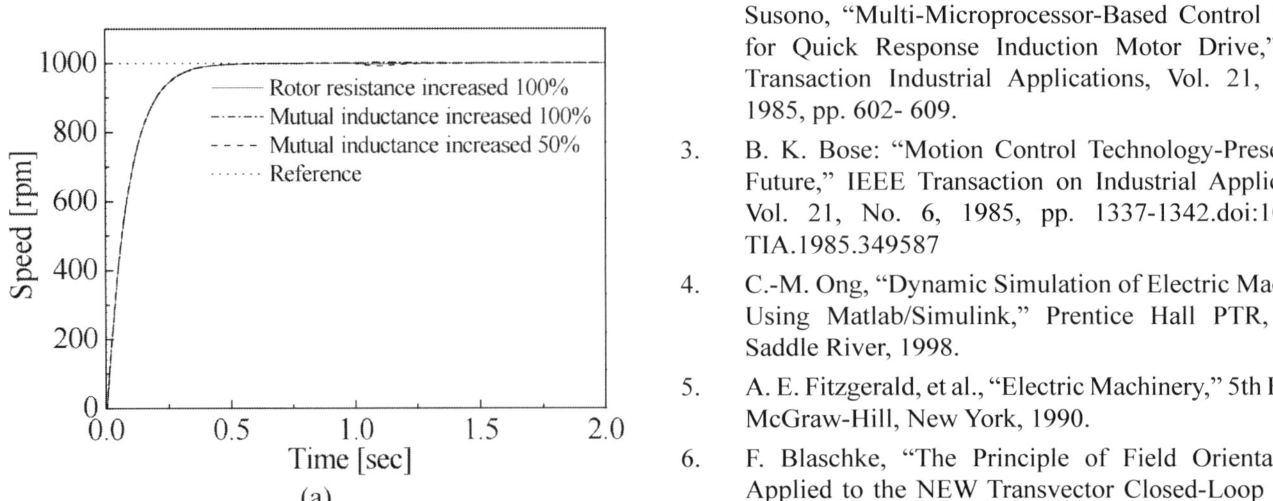

Figure 13. Transient response of speed with variations of parameters using proposed fuzzy logic control. (a) Response of speed; (b) Zoomed version of Figure (a).

CONCLUSION

The field oriented control of induction motor with core loss is no longer in terms of stator current components. So, the field oriented control of induction motor is designed in terms of magnetizing current. In order to improve the performance of an induction motor speed controller with consideration of core loss, a fuzzy-logic controller for indirect field oriented induction motor drive taking core loss into account is proposed. Due to the consideration of core loss, the FLC is designed to regulate the speed in term of magnetizing q-axis current component. In order to verify the effectiveness of the proposed FLC, we have conducted simulations using the equivalent circuit of IM in synchronously rotating reference frame. The simulation results based on conventional PI controller and the proposed FLC are demonstrated to compare both of those controllers. At last it can be concluded that the performances of the proposed FLC in both transient and steady states are better than those of conventional PI controller.

REFERENCES

1. P. Vas, "Electrical Machines and Drives: A Space-Vector Theory Approach," Clarendon Press, New York, 1992.
2. F. Harashima, S. Kondo, K. Ohnishi, M. Kajita and M. Susono, "Multi-Microprocessor-Based Control System for Quick Response Induction Motor Drive," IEEE Transaction Industrial Applications, Vol. 21, No. 3, 1985, pp. 602- 609.
3. B. K. Bose: "Motion Control Technology-Present and Future," IEEE Transaction on Industrial Applications, Vol. 21, No. 6, 1985, pp. 1337-1342.doi:10.1109/TIA.1985.349587
4. C.-M. Ong, "Dynamic Simulation of Electric Machinery Using Matlab/Simulink," Prentice Hall PTR, Upper Saddle River, 1998.
5. A. E. Fitzgerald, et al., "Electric Machinery," 5th Edition, McGraw-Hill, New York, 1990.
6. F. Blaschke, "The Principle of Field Orientation as Applied to the NEW Transvector Closed-Loop System for Rotating-Field Machines," Siemens Review, Vol. 34, No. 3, 1972, pp. 217-220. [Paper reference:2]
7. Y. Y. Tzou, "Fuzzy Tuning Current-Vector Control of a

Three-Phase PWM Inverter for High-Performance AC Drives," IEEE Transaction on Industrial Electronics, Vol. 45, No. 5, 1998, pp. 782-791. doi:10.1109/41.720335

8. K. Hasse, "Zum Dynamischen Verhalten der Asynchronmachine bei Betriek Mit Variabler Standerfrequenz und Standerspannung," ETZ-A, Vol. 89, 1968, p. 77. [Paper reference:2]

9. R. D. Lorenz and D. B. Lawson, "A Simplified Approach to Continuous On-Line Tuning of Field-Oriented Induction Machine Drives," IEEE Transaction on Industry Application, Vol. 26, No. 3, 1990. [Paper reference:1]

10. B. K. Bose, "Power Electronics and Variable Frequency Drives," IEEE Press Standard Publishers, 1997.

11. I. Boldea and S. A. Nasar, "Vector Control of AC Drives," CRC Press, London, 1992.

12. W. Leonhard, "Control of Electric Drives," Springer Verlag, Berlin, 1985.

13. R. Krishnan, "Electric Motors Drives Modeling Analysis and Control," Publication Prentice Hall of India, New Delhi, 2002.

14. A. Mechernene, M. Zerikat and M. Hachblef, "Fuzzy Speed Regulation for Induction Motor Associated with Field-Oriented Control," IJ-STA, Vol. 2, No. 2, 2008, pp. 804-817.

15. R. Marino, S. Peresada and P. Valigi, "Adaptive InputOutput Linearizing Control of Induction Motors," IEEE Transaction on Automation, Vol. 38, No. 2, 1993, pp. 208- 211.doi:10.1109/9.250510

16. M. N. Uddin, T. S. Radwan and M. A. Rahman, "Performances of Fuzzy-Logic-Based Indirect Vector Control for Induction Motor Drive," IEEE Transaction on Industrial Applications, Vol. 38, No. 5, 2002, pp. 1219-1225. [Paper reference:2]

17. I. Boldea and S. A. Nasar, "Unified Treatment of Core Losses and Saturation in the Orthogonal-Axis Model of Electrical Machines," Proceedings of Institute of Electrical Engineers, Vol. 134-B, No. 6, 1987, pp. 355-363. [Paper reference:1]

18. E. Levi, "Impact of Iron Loss on Behavior of Vector Controlled Induction Machines," IEEE Transaction on Industrial Application, Vol. 31, No. 6, 1995, pp. 1287-1296.doi:10.1109/28.475699 [Paper reference:3]

19. E. Levi, A. Boglietti and M. Lazzari, "Comparative Study of Detuning Effects in Indirect Vector Controlled Induction Motor Drives Due to Iron Losses," Proceedings of International Conference on Power Electronics and Drive Systems, Liverpool, 21-24 February 1995, pp. 633-638.

20. E. Levi, "Rotor Flux Oriented Control of Induction Machines Considering the Core Loss," Electric Machine and Power Systems, Vol. 24, 1996, pp. 37-50. doi:10.1080/07313569608955658

21. E. Levi, "Iron Loss in Rotor-Flux-Oriented Induction Machines: Identification, Assessment of Detuning, and Compensation," IEEE Transaction on Power Electronics, Vol. 11, No. 5, 1996, pp. 698-709. doi:10.1109/63.535402

22. M. A. Mannan, T. Murata, J. Tamura and T. Tsuchiya, "Indirect Field Oriented Control for High Performance Induction Motor Drives Using Space Vector Modulation with Consideration of Core Loss," Proceedings of PESC'03, Acapulco, 15-19 June 2003, pp. 1449-1454. [Paper reference:4]

23. D. Driankov, H. Hellendoorn and M. Reinfrank, "An Introduction to Fuzzy Control," Springer-Verlag, Berlin, 1993. [Paper reference:3]

24. S. Tunyasrirut, T. Suksri and S. Srilad, "Fuzzy Logic Control for a Speed Control of Induction Motor Using Space Vector Pulse Width Modulation," World Academy of Science, Engineering and Technology, No. 25, Topics 14, 2007, pp. 71-77.

CHAPTER 13 ETHERNET CONTROL AC MOTOR VIA PLC USING LABVIEW

Nader N. Barsoum and Pin Rui Chin

Keywords: LabVIEW, PLC, Ethernet, Induction Motor, OPC Servers, Inverter

INTRODUCTION

In the past, engineers had been designing the engineering systems that require a lot of hardwares. It is merely impossible to design distance control of the system as more hardwares and wiring were needed. In addition, if engineers wish to improve the design, all the unrelevant hardwares need to be scrap away, which is not sustain-able. With the rise of the technology, programmable logic controller (PLC) have eased the engineering design and lessen materials required, it is because the entire design is implemented in software programming paradigm. PLC had been commonly used in the industry, including con-trolling induction motor inverter fed variable drive sys-tem. Design distance control machinery is now possible, even by using Ethernet as the communication device between the computer and the PLC [1]. Apart of design the program structure by its own proprietary software, the convenience part of PLC is the accessibility and controllability by other softwares. Note that such softwares must have driver utility of the particular PLC. Therefore engineers can use LabVIEW [2], which has various types of industrial applications which are in virtual instrument (VI) instead of the real and heavy instrument, to control the PLC.

PROJECT DESIGN CONFIGURATION & PREP-ARATION

NI OPC Server

NI OPC Server has OMRON FINS Ethernet driver that allow the communication between OMRON CJ1M-CPU11-ETN21 PLC with LabVIEW. OMRON originally supplies their customers with FINS gateway, inter-facing software that communicates with the PLC and its proprietary software, OMRON CX-Programmer over the Ethernet network [3].

With the OMRON FINS Ethernet driver in NI OPC, users can setup the server by just a few simple setups and create variable tags that can be linked directly to the PLC's registers. These tags are named as OPC tags. The NI OPC Servers also have NI OPC Quick Client that enable users to monitor the status of the PLC in real-time. [3].

As long as the OPC tags had been created, the communication between the LabVIEW and PLC had been simplified as the driver can automatically apply the relevant FINS commands provided the tags are correctly configured [3]. Meanwhile in LabVIEW, the program can be design by using Shared Variables which is link to the OPC tags.

LabVIEW

LabVIEW is the acronym for Laboratory Virtual Instrumentation Engineering Workbench and is a graphical development environment for generating flexible and scalable design, control, and test applications rapidly at minimal cost. With LabVIEW, engineers and scientists are able to interface with real-world signals, analyse data for meaningful information, and share results through intuitive displays, reports, and the Web. Regardless of programming experience, LabVIEW makes development fast and easy for all users [3].

The programming style used in LabVIEW is G programming, which is abbreviation for graphical programming. It is also known as dataflow programming as it is depending on the structure of the graphical block diagram to execute the user-designed program. Com-pared to text-based programming, LabVIEW is user friendly as the user can design the program by simply arrange and wiring the relevant icons together [3]. LabVIEW programs are named as virtual instruments, or VI, because their appearance and operation

Citation: N. Barsoum and P. Chin. "Ethernet Control AC Motor via PLC Using LabVIEW," Intelligent Control and Automation, Vol. 2 No. 4, 2011, pp. 330-339. DOI: 10.4236/ica.2011.24038.

mimic the physical instruments, such as oscilloscope and multi-meters [4]. Similar to other conventional programming, LabVIEW has standard features such as looping structures, data structures, event-handling, object-oriented programming. LabVIEW also has an extensive library of math functions similar to MATLAB libraries and also formula nodes that allow text-based programming for certain sections of the code that require complex logical structures. Besides that, LabVIEW also has networking library functions that can easily allow users to reference [3].

Compared to other softwares like Microsoft Visual Basic, LabVIEW is a better option as it comes together with a library of functions included Shared Variables Project Library, which is bound to the OPC tags, that allow server and client communication by connecting relevant icons with the Shared Variables. If Microsoft Visual Basic (VB) is used, the OMRON FINS Ethernet driver must be developed using the MS Comm function and this would require more time to develop the code [3].

LabVIEW has front-end interface applications that al-low user to design and then use for controlling systems. In general LabVIEW has three main elements: the front panel, the block diagram and the connector panel. The front panel allows the user to build the controls and indicators. The controls are including knobs, push buttons, dials, and other input mechanism. Indicators are graphs, LEDs, and other output displays. Meanwhile, the block diagram let user to add code using VIs and structures to control the front panel objects. The connector panel al-lows user to represent a single VI as a sub VI icon that can be called in another VI. The elements are illustrated in Figure 1.

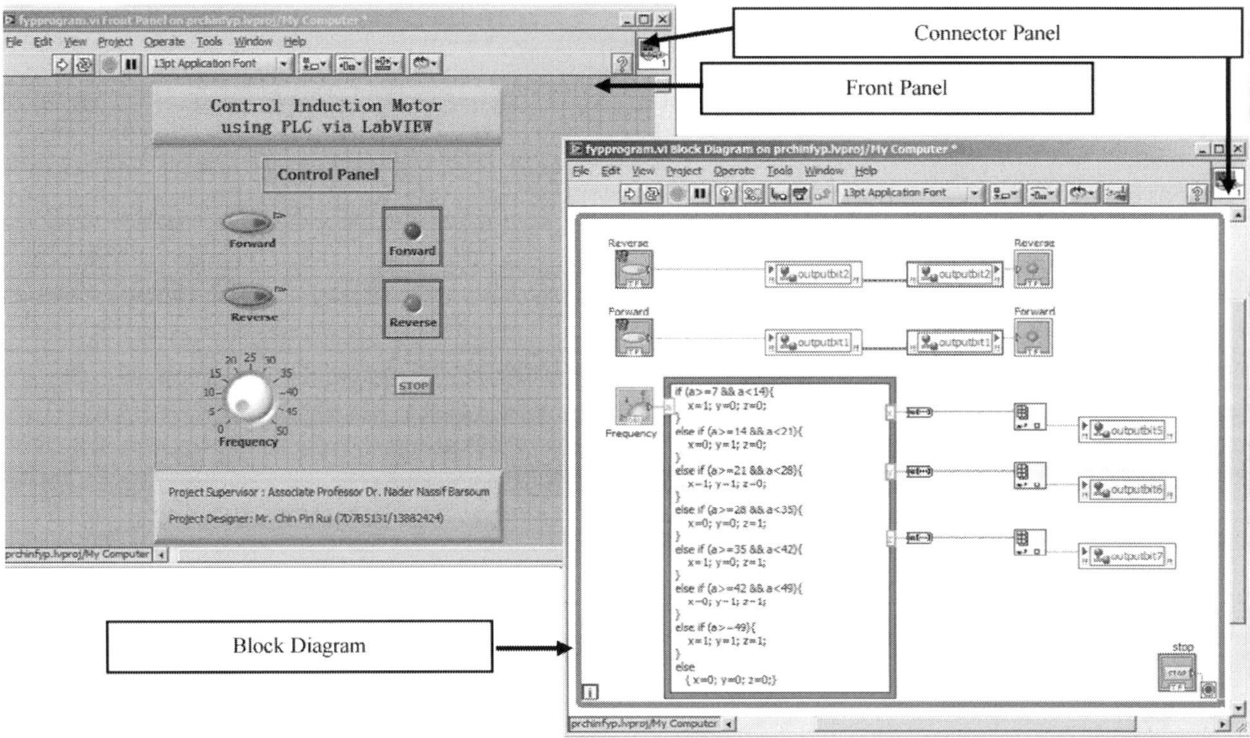

Figure 1. Three main elements of LabVIEW software.

Shared variable is a library function variable that allows sharing of data between applications or different data sources across a network. There are many existing data sharing method in LabVIEW, such as UDP/TCP, LabVIEW queues, and Real-Time FIFO. Compared to Datasocket Communication, using shared variable is user friendly as the configuration can be done by simple setup in the library instead of writing URL and perform more wiring in Datasocket Communication [3,5].

Programmable Logic Controller

The PLC used for the implementation is OMRON CJ series. There are 4 units used in this PLC, namely Power Supply Unit (CJ1W-PA202), CPU Unit with Ethernet function (CJ1M-CPU11- ETN21), Basic Input Unit (CJ1W-ID211), and Basic Output Unit (CJ1W-OC211).

There is an End Cover for the PLC unit. All these units can be connected by assembled them together and lock the sliders by moving them towards the back of the units. The End Cover must be connected on the far right hand side of the PLC. Otherwise fatal error will occur [6].

Figure 2 demonstrates the arrangement of the PLC units.

In order to let the PLC to operate through the Ethernet, the PLC must be given an IP address with a destination node number (DA1) that is not a duplicate of DA1s of other IP addresses in the network. The destination node number is also known as the last octet of an IP address.

The PLC node number is configured by turning the dials on the CJ1W-ETN21 module as illustrated in Figure 3 the PLC must also be given a unit number in the network besides a destination node address (DNA) [3].

Node No $\times$ 16^1 is the most significant bit (MSB) of the node address whereas $\times 16^0$ is the least significant bit (LSB) of the node address. IP addresses are in decimal form therefore any reference made to the DA1 of the PLC Ethernet module in software configurations must be in decimals. To convert any hexadecimal number to decimal, the following example can be applied [3]:

MSB LSB

2 E

$$14 \times 16^{0}$$
$$+\ 2 \times 16^{1}$$
$$46$$

The PLC is configured with the IP address 10.1.136.46, for the implementation described in this paper. The router is for personal home network. Therefore the IP addresses associated with the devices in the network is classified under Class C.

Variable Frequency Drive

OMRON SYSDRIVE 3G3MV-A2007 inverter is a variable frequency drive can be used to alter the frequency of the electrical power supplied to the motor so in order to change the motor speed [6].

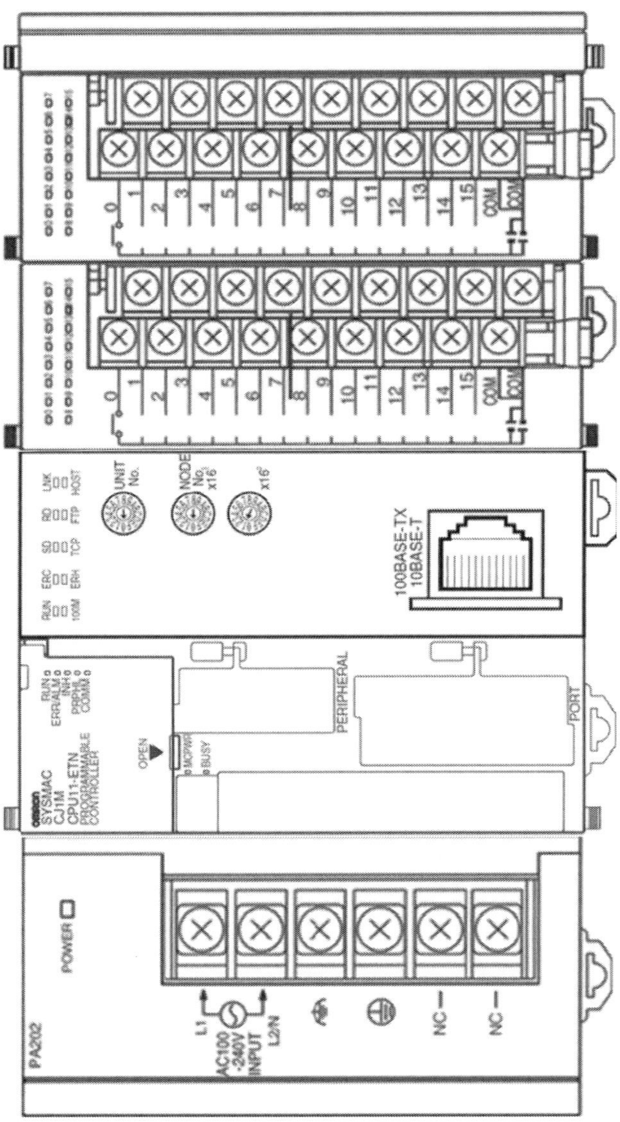

Figure 2. Connection for PLC units in this project [7,8].

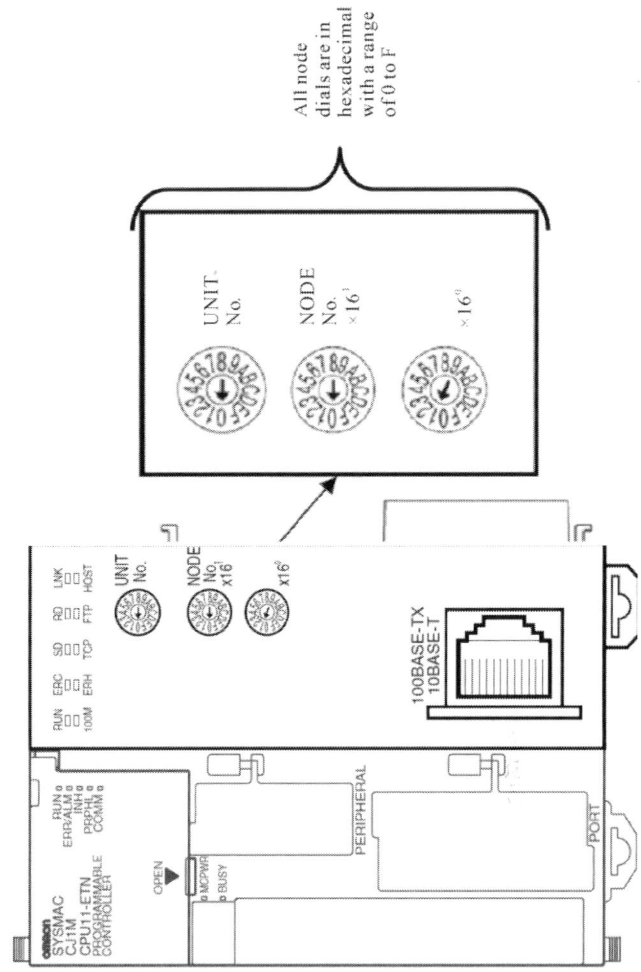

Figure 3. CJ1W-CPU11-ETN21 hardware configuration.

The SYSDRIVE 3G3MV-A2007 inverter is suitable for a variety of applications as it incorporates many convenient controls and I/O functions that are easy-to-use as well as open loop vector control function. The advantages of vector control function are that it ensures a torque output that is 150% of the rated motor torque at an output frequency of 1 Hz, allows powerful revolution at low frequencies and restrains the revolution fluctuation caused by the load [6].

The list of function parameters of the 3G3MV inverter need to be configured in order to control the 3 phase squirrel cage induction motor effectively.

IMPLEMENTATION

Implementation Process

The Ethernet control systems presented in this paper is to control squirrel cage three phase induction motor. In order to achieve the objectives, the establishment of the communication between PLC and LabVIEW is crucial as LabVIEW is 3rd party software instead of using the software implemented in the PLC itself. Thus, the implementation used LabVIEW to perform the start and stop operation of the motor, either in forward or reverse direction, and varying the speed by changing the frequency of the motor. However, this system is not a supervisory control and data acquisition (SCADA) since there is no practical data measurement acquire from the actual output of the motor.

The system has three-layer network architecture illustrated in Figure 4.

As illustrated in Figure 5, user gets the authority to control through the host computer, which is a laptop. Then the input data by the user will convert into Boolean data and send to CJ1M-CPU11-ETN21 Programmable Logic Controller through Ethernet cable and router. Once the Boolean data have been processed by the PLC, the relevant address in the basic output of the PLC will be turned on. This process allow the 3G3MV Inverter/ Variable Frequency Drive to operate the Three Phase Squirrel Cage Induction Motor according to the input data given by the user. In addition, 3G3MV Inverter/ Variable Frequency Drive also serve as inverter between the power supply unit and the motor as the input power is single phase power, while the Squirrel Cage Induction Motor is operate in three phase power.

Implementation of VI Design

The objective of the VI program in this paper is to allow user to make the decision of the start and stop operation of the motor, either in forward or reverse direction, and varying the speed by changing the frequency of the motor, by perform two simple step. Firstly select the turning direction by press the push button in the VI Front Panel, which is either forward or reverse. Secondly, vary the speed by turning the Frequency knob to the value of the frequency that the user desire. Figure 6 illustrate the VI Front Panel.

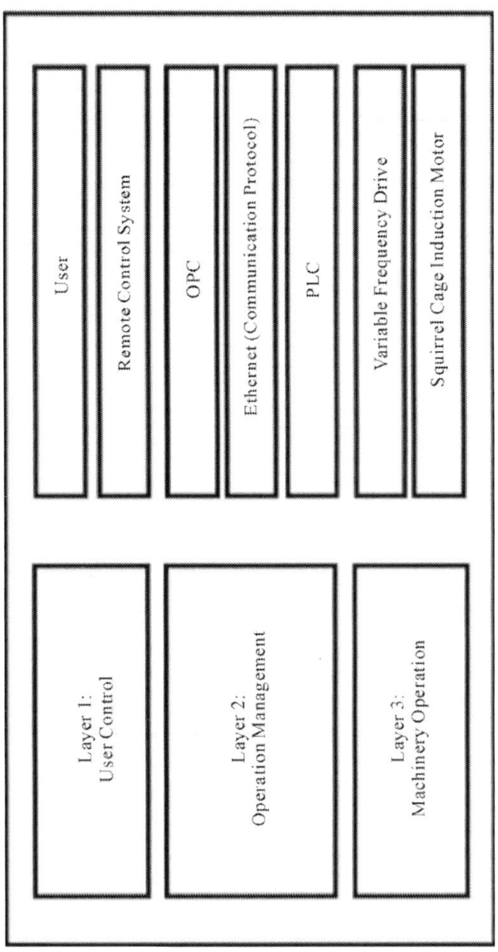

Figure 4. Three layer network architecture.

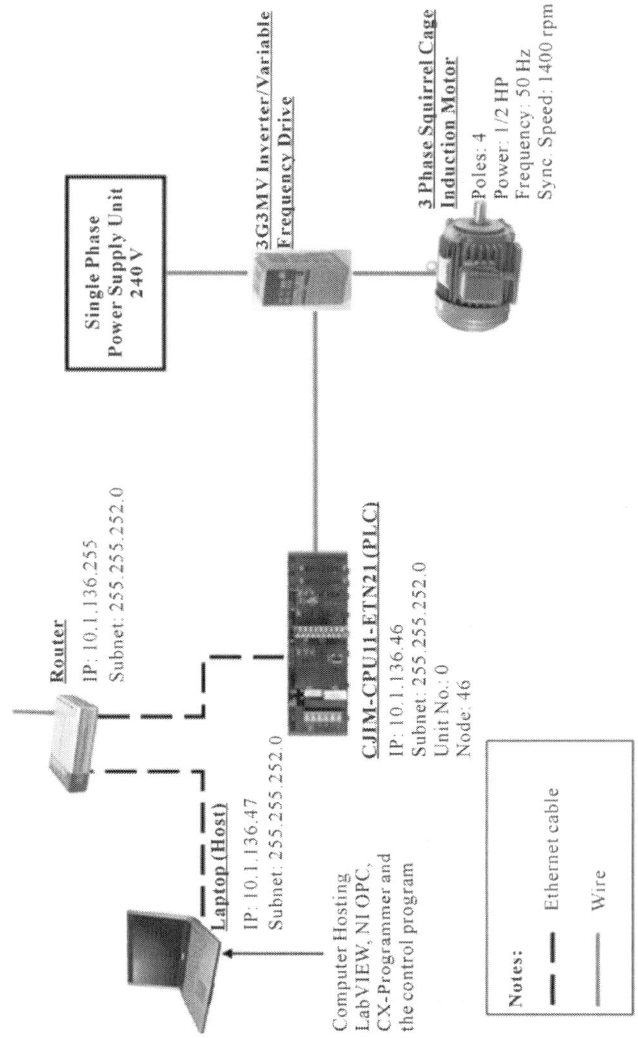

Figure 5. Actual network configuration.

Before running the VI, make sure all the hardwares have been switched on and configured correctly, and launch the NI OPC Quick Client so that the OPC tags can be browsed by the shared variables in this VI.

There are 5 OPC tags was created and used in the implementation, and the details of the tags have been tabulated in Table 1.

By referring to Figure 6, the Green push button is the switch to determine the motor is turning in forward direction. Meanwhile, the Orange push button is the switch to determine the motor is turning in reverse direction. Both Green and Orange light indicators at the right hand side show whether the push buttons have been switched ON. The knob labelled as "Frequency" is the key pro-gram to control the frequency as well as the speed of the motor. The push button labelled as "Stop" function to stop the program execution.

Figure 7 illustrate the VI Block Diagram, which the programming part of the VI. By referring to the Figure 7, the Red colour square boxes with dotted lines are the VI components that are visible in both Front Panel and Block Diagram. For Example the push buttons, knob, light indicator. Meanwhile, the Red colour square boxes without dotted lines are the VI component that is visible in Block Diagram but they are not visible in Front Panel, which are essential VI to structure the program. For example, shared variables, formula node, number to Boolean converter (consist of number to Boolean array and index array) and the While loop. The While loop, which is similar concept with the While loop in C Programming, is used in this VI to ensure the program execute continuously until the stop button is triggered.

Table 1. The details of OPC tags and its connection to inverter.

OPC Tag name	PLC Address	3G3MV Inverter	
		Connected Terminal	Terminal Name
Outputbit 1	CIO0001.01	S1	Multi-function input 1 (Forward/Stop)
Outputbit 2	CIO0001.02	S2	Multi-function input 2 (Reverse/Stop)
Outputbit 5	CIO0001.05	S5	Multi-function input 5 (Multi-step speed reference 1)
Outputbit 6	CIO0001.06	S6	Multi-function input 6 (Multi-step speed reference 2)
Outputbit 7	CIO0001.07	S7	Multi-function input 7 (Multi-step speed reference 3)

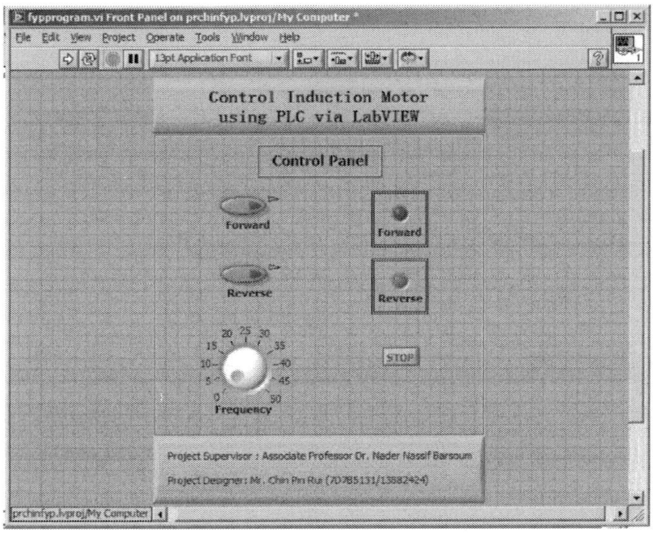

Figure 6. VI front panel of the project.

The VI block diagram consists of two parts of the program. First part of the program is to allow user to switch on either Forward or Reverse direction of the squirrel cage induction motor. The second part of the program is to vary the frequency of the motor by changing the knob value.

By referring to Figure 8, if the user has switched on the Forward button, the signal will be transmitted to the "outputbit1" shared variable, which is in write mode, *i.e.* the signal will be written into the PLC. Then once the signal has been successfully written to the PLC, the light indicator of the output address 1.01 of the PLC, and the "Forward" direction indicator in the LabVIEW's Front Panel will light. This process identical to the reverse direction as well.

By referring to the Figure 9, the Red colour square boxes indicate the VI named formula node, where the language used in the formula node is C. After the user set the value of the frequency by turning the "Frequency Variable" knob, the value will be sent to the formula node, where it is denoted as "*a*".

The output of the formula node has been divided into x, y and z, in which the number is either 0 or 1. In order to be readable by the shared variables, it is necessary to convert the number into Boolean format. When the signal is in Boolean format and sent to the shared variable, the information will be written into the PLC, and the light indicator of the output address 1.05, 1.06, 1.07 of the PLC will light up according to the desired output.

Table 2 below describes the relationship between the multi-step speed references 1 through 3 and frequency references 1 through 8. The concept of frequency varying in Table 2 is crucial as the PLC communicates with the variable frequency drive in Boolean format. The value of each frequency reference can be set in the function parameters of the variable frequency drive. In this paper, the value of each frequency reference has been tabulated in Table 3.

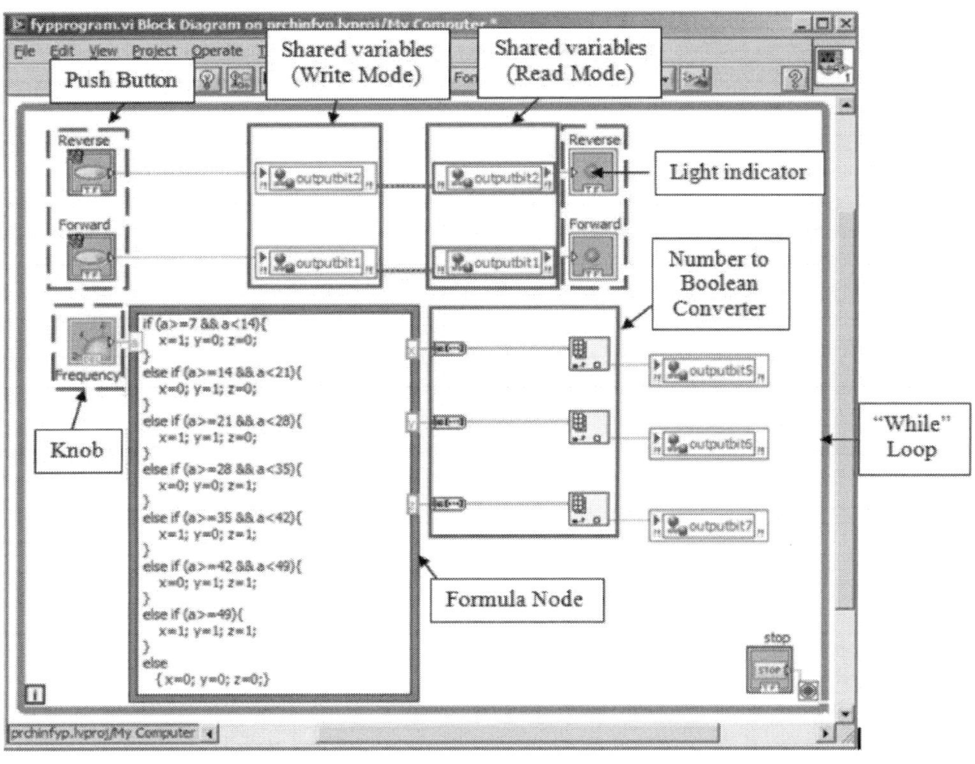

Figure 7. VI Block Diagram of the project.

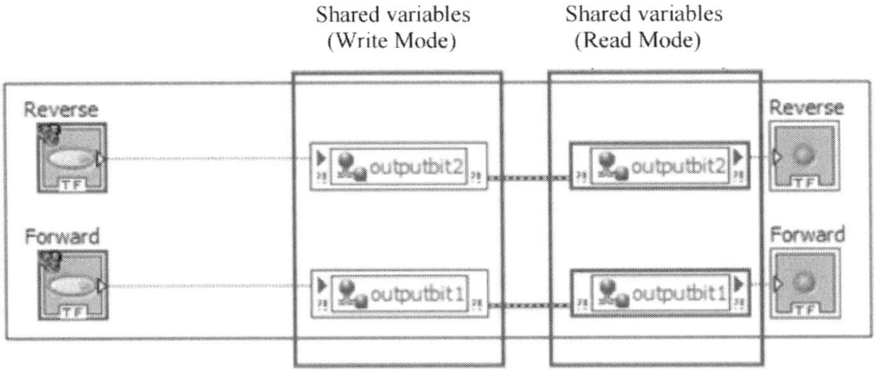

Figure 8. VI block diagram program (motor orientation switching part).

By referring to Table 3, as long as the user has set the input frequency according to any of the condition, the number will be converting into Boolean in which it will arrange according to the relevant Multi-step speed reference. Then this Boolean number will trigger the variable frequency drive and determine the frequency and as well as the speed of the squirrel cage induction motor according to the output condition listed in Table 3. For example, if the user set the input as 8 Hz, in which the input range of this value is within 7 *x* 14. Therefore, the Boo-lean number of Multi-step speed reference 1 is 1, and the rest are 0. Then, this signal triggers the variable frequency drive to deliver the frequency of 7 Hz.

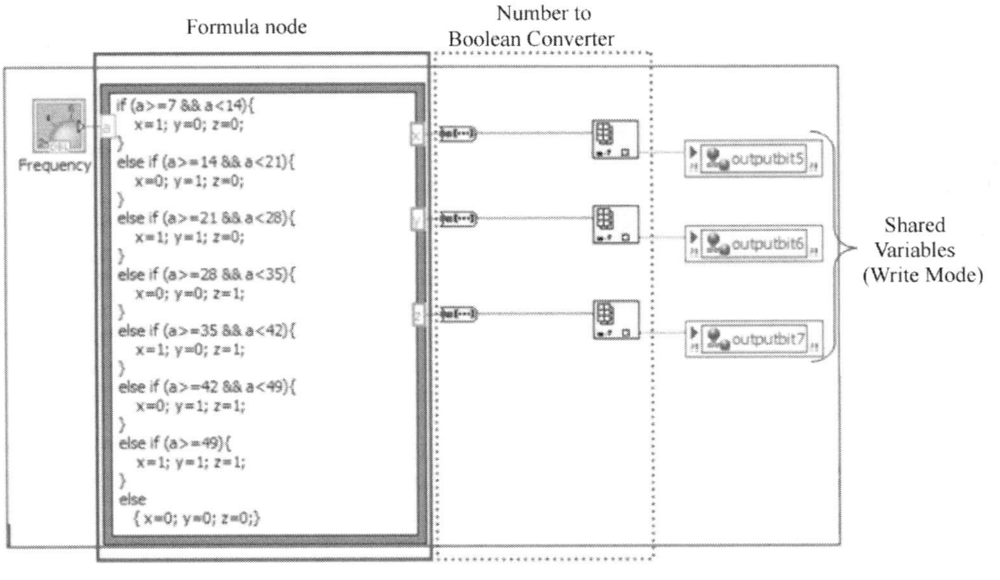

Figure 9. VI block diagram program (motor frequency varying part).

Table 2. Relationship between multi-step speed references and frequency reference [6].

Frequency reference	Multi-step speed reference 1 (Set value: 6)	Multi-step speed reference 2 (Set value: 7)	Multi-step speed reference 3 (Set value: 8)
Frequency ref. 1	OFF	OFF	OFF
Frequency ref. 2	**ON**	OFF	OFF
Frequency ref. 3	OFF	**ON**	OFF
Frequency ref. 4	**ON**	**ON**	OFF
Frequency ref. 5	OFF	OFF	**ON**
Frequency ref. 6	**ON**	OFF	**ON**
Frequency ref. 7	OFF	**ON**	**ON**
Frequency ref. 8	**ON**	**ON**	**ON**

Table 3. Conditions of input frequency in LabVIEW for varying frequency reference.

Frequency input range in LabVIEW (Hz)	Multi-step speed reference 1 (Outputbit 5)	Multi-step speed reference 2 (Outputbit 6)	Multi-step speed reference 3 (Outputbit 7)	Output frequency of squirrel cage induction motor (Hz)
x < 7	0	0	0	0
7 ≤ x < 14	1	0	0	7
14 ≤ x < 21	0	1	0	14
21 ≤ x < 28	1	1	0	21
28 ≤ x < 35	0	0	1	28
35 ≤ x < 42	1	0	1	35
42 ≤ x < 48	0	1	1	42
x > 48	1	1	1	49

TESTING AND VERIFICATION

After the implementation, test is performed in order to verify that the system is working smoothly. Figure 10 shows the front view of the system.

Figure 11 shows the PLC Ethernet Module light indicators that light up when the project VI is executing. And Figure 12 shows the snapshot of the Squirrel Cage Three Phase Induction Motor when it is running.

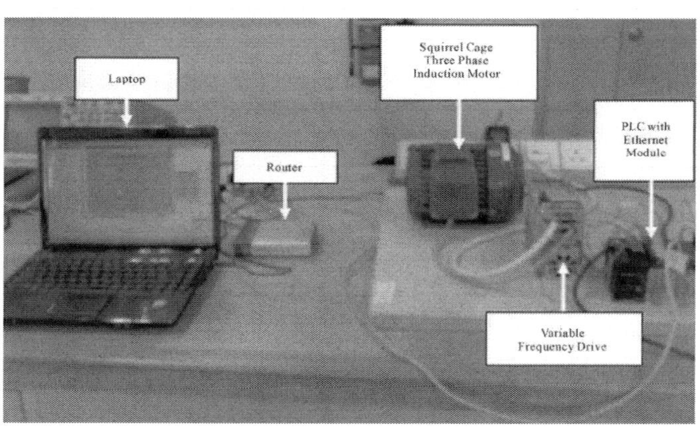

Figure 10. Front view of the project.

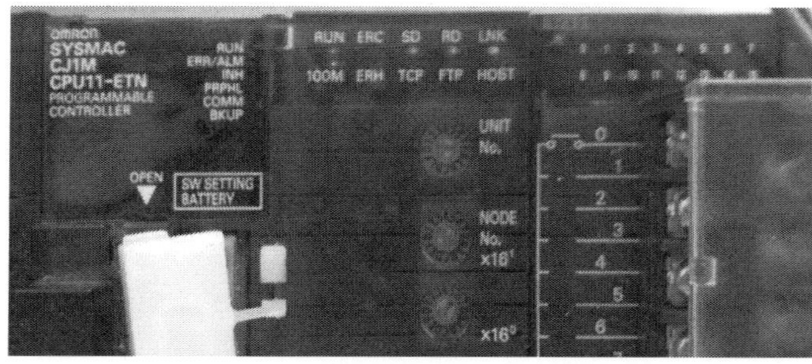

Figure 11. PLC ethernet module light indicator status.

Figure 12. Induction motor running in forward direction.

CONCLUSIONS

In conclusion, the objective, scope and fundamental requirements of the project had been achieved. But there are some defects in this project. Firstly, there is a delay of time during the execution, which every decision made by the user through LabVIEW will take about 30 seconds in average to send the signal to the PLC. The delay mainly caused by the poor specification of the Laptop used for implementation that leads the LabVIEW to take long time to execute the program and generate the Boo-lean data to send to the PLC.

Secondly, the Ethernet connectivity between the Lap-top and the PLC is easily disconnected. This problem arises also due to the poor specification of the Laptop used. This problem happen when the Laptop had used up too much RAM, causing the laptop lag and unable to use temporary. Therefore user cannot make too many actions and changes at the same time as it will crash the program. Thirdly, there is limitation in the speed control as there are only 3 multi-step references in the VFD. The frequency changes in this project cannot perform in smaller steps unless there are additional multi-step references provided.

In overall, choosing LabVIEW as the human machine interface for the implementation is a proper decision as it has various types of applications and functions that are easy to understand and use. Additionally, this approach is more economical as the objectives of the system implementation have been achieved with only basic functionality of the LabVIEW toolkits used, which are shared variables and NI OPC Servers.

OMRON CJ1 series PLC is easy to install and setup. Both hardware and software configuration can be easily done. It can carry out additional functions by simply add more units with various functions, like the Ethernet unit.

The 3G3MV inverter is a user friendly VFD that al-lows the user to configure the function parameter easily and the circuit wiring can be completed easily.

ACKNOWLEDGEMENTS

Special thanks to Daniel Wong and Jacintha Roland for their remarkable advice throughout the design and im-plementation of the system.

REFERENCES

1. W.-F. Chang, Y.-C. Wu and C.-W. Chiu, "Design and Implementation of a Web-Based Distance PLC Labora-tory," *Proceedings of the 35th Southeastern Symposium on System Theory*, Morgantown, 16-18 March 2003, pp. 326-329. doi:10.1109/SSST.2003.1194584
2. M. Rodrigues, J. Mendes and J. Fonseca, "Application of a Web-Based Monitoring and Control System in Plastic Rotational Moulding Machine," *IEEE International Con-ference on Industrial Technology*, Hammamet, 8-10 De-cember 2004, pp. 819-823
3. N. N. Barsoum and J. A. Roland, "Ethernet LabVIEW Control," Undergraduate Thesis, Curtin University Sara-wak Campus, Sarawak,, 2010.
4. National Instruments, "Getting Started with LabVIEW," Na-tional Instruments, Technique Report 373427F-01, 2009.
5. National Instruments, "Using the LabVIEW Shared Vari-able," 2010. http://zone.ni.com/devzone/cda/tut/p/id/4679
6. N. N. Barsoum and Y. Y. Ng, "PLC Humidity Control In-verter Fed Induction Motor," Undergraduate Thesis, Cur-tin University Sarawak Campus, Sarawak, 2009.
7. OMRON Industrial Automation, "SYSMAC CJ Series Programmable Controller," OMRON, Technique Report W393-E1-14, 2009.
8. OMRON Industrial Automation, "CJ1M CPU Units with Ethernet Functions," OMRON, Technique Report, 2005.

CHAPTER 14 INDUCTION GENERATOR WITHOUT AND WITH A VOLTAGE REGULATOR

M. L. Elhafyani, S. Zouggar, M. Benkaddour, and A. Aziz

Keywords: Wind Power, Self-Excited Induction Generator, Converters (Buck-Boost), Regulation

INTRODUCTION

The exploitation of the renewable energy sources raises a growing interest, not only in the Third World countries, but also in the industrialized countries, particularly for the developing countries and for any isolated site generally. The use of the mechanical converters (synchronous machine, asynchronous machine), is a subject of studies and research in the range of products of the electric power [1] -[12]. As far as the production of low power electrical energy (<50 kw) is concerned, the three-phase self-excited induction generator (SEIG) with cage or wound-rotor, competes as a generator with the synchronous machine. This is due primarily to its sturdiness, robustness, its simplicity of excitation (simple connection of condensers at the stator boundaries) and its relatively modest cost. During the isolated use of the asynchronous machine [1] [10] -[12], a variation of the capacity, load or diver speed involves an imbalance of the stator output voltage. This problem can be solved by operating, the condensers, the load or the drive speed by the intermediary of a regulator. In this paper, we present the capacity variation effect, speed of drive and load as function of the wind system without and with analogical regulator.

BEHAVIOR OF THE MACHINE WITHOUT THE VOLTAGE REGULATOR

Modelling of the Wind System

Equivalent Circuit of the Wind System

The Figure 1 represents the synoptic diagram of an asynchronous machine (1.5 KW) related to on the energy converter (AC-DC) and feeds a resistive load (R_S):

The Induction Asynchronous Generator does not excite without external power given to its stator, the rotor windings are shorted. Rotor winding will have current only when there is connection of voltage from Grid/battery operated AC source.

The initiation of the generator is possible only from a speed-dependent threshold value of the capacitances and the magnetic state of the machine (linear portion of the magnetization curve) (Figure 2).

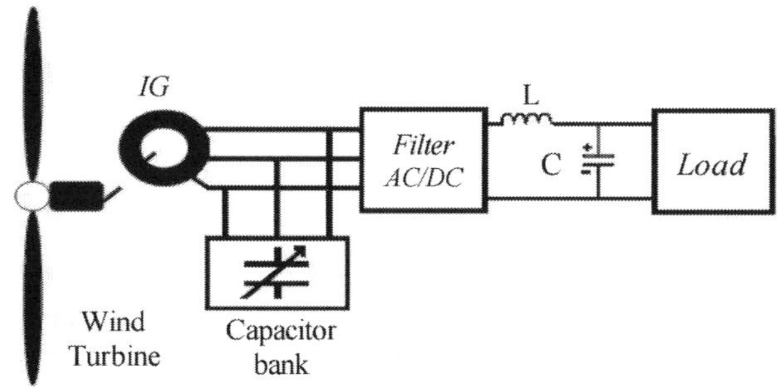

Figure 1. Synoptic diagram of a SEIG with a rectifier dependent has a resistive load.

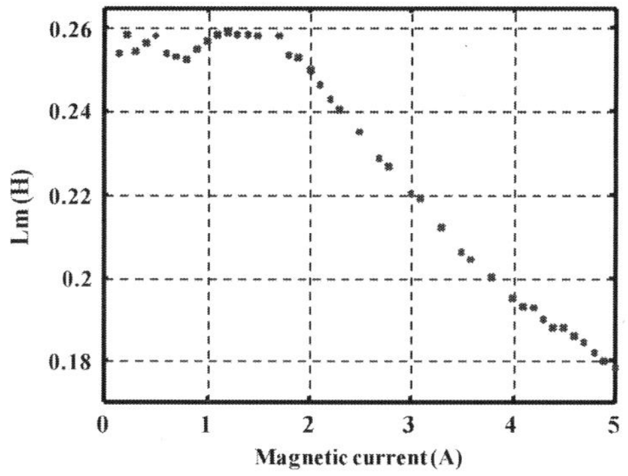

Figure 2. Experimental curve of the magnetizing inductance [4].

Citation: Elhafyani, M., Zouggar, S., Benkaddour, M., and Aziz, A. (2014). Induction Generator without and with a Voltage Regulator. Smart Grid and Renewable Energy, 5, 207-219. DOI: 10.4236/sgre.2014.59020.

Modelling of the Wind-Turbine

By applying the theory of momentum and Bernoulli's theorem, we can determine the incident power (theoretical power) due to wind:

$$P_{inci} = \frac{1}{2} \cdot \rho \cdot A \cdot V_w^3 \quad (1)$$

The wind turbine acts like a primary strength to train the induction generator [7]. The power produced by the wind turbine part is given by the following formula

$$P_{wt} = \frac{1}{2} \cdot C_P(\lambda) \cdot \rho \cdot A \cdot V_w^3 \quad (2)$$

where:

ρ: the air density (ρ = 1.225 kg/m³ at atmospheric pressure);

A: the swept area (m^2);

C_p: the coefficient of wind turbine;

V_w: the wind velocity (m/sec);

$C_p(\lambda)$ is called the power coefficient, which expresses the aerodynamic efficiency of the turbine. It depends on the ratio λ, which represents the ratio between the speed at the end of the blades and the wind speed. The ratio λ can be expressed by the following relation:

$$\lambda = \frac{\Omega \cdot R}{V_w} \quad (3)$$

Symbolization of the Mechanical Equation (See Figure 3)

The implementation and the simulation of the self-excited asynchronous machine were achieved by using the Orcad-Pspice tool whose libraries are rich in electrical and analogical components, and by elaborating symbolization procedures [13] -[15].

In electrical technology, there is no software for simulating mechanical parameters directly. We did this simulation on the Orcad/Pspice software-using the following analogies:

Mechanical	Electrical
Speed	Voltage
Torque	Current
Moment of inertia	Capacitance of a capacitor
Viscous friction	The inverse of resistivity

The mechanical equation of the induction machine is given by the following formula:

$$J\frac{d\omega_{mr}}{dt} = T_e - B_m \cdot \omega_{mr} - T_r \quad \text{or} \quad \omega_{mr} = \frac{2}{p}\omega_r \quad (4)$$

$$T_e = \left(\frac{2}{p}\right) J \frac{d\omega_r}{dt} + \left(\frac{2}{p}\right) B_m \cdot \omega_r + T_r \quad (5)$$

The electromagnetic torque T_e is given by:

$$T_e = \sqrt{3}\left(\frac{2}{p}\right) Lm\left(i_{as} \cdot i_{\gamma r} - i_{cs} \cdot i_{\alpha r}\right) \quad (6)$$

With:

J: Moment of inertia;

T_e: Couple électromagnétique en N×m;

T_r: Load torque (N·m);

B_m: The constant friction (kg·m²/s);

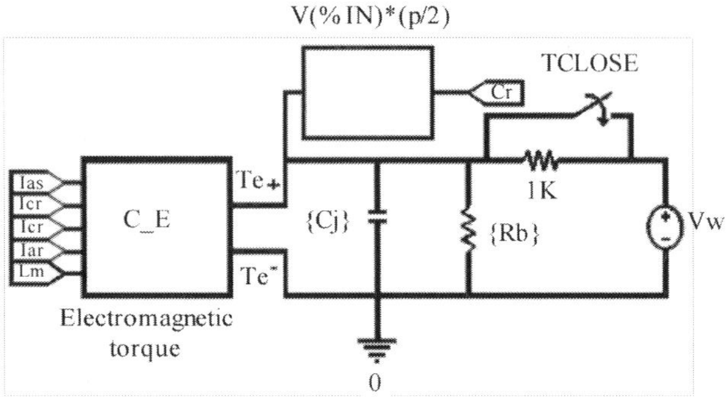

Figure 3. Electrical diagram of the mechanical equation.

ω_r et ω_{mr}: Electrical and mechanical speed (rad/s).

Applying the analogy between mechanical and electrical variables, we have:

$$J\frac{d\omega}{dt} = \sum_k T_k \Rightarrow C\frac{dV}{dt} = \sum_j i_j$$

$$\frac{1}{B_m} \Rightarrow R \quad (7)$$

Simulation Results without the Voltage Regulator

Speed Variations Effect

Figure 4 represents the stator voltage variation at the output of the rectifier according to the drive speed of the machine for a capacity of C = 80 μF and a load R_S = 70 Ω. We note that the voltage increases with the speed (the nominal voltage $220\sqrt{2}$ corresponds to a speed of the w_r = 298 rad/s).

Load Variations Effect

Figure 5 illustrates the evolution of the stator tension on the outlet side of the rectifier according to the load at different values of the drive speed, and at the excitation capacitor C = 80 mF. One fixes shows that a variation speed or load involves a variation of the tension (the nominal voltage $220\sqrt{2}$ V corresponds to a load R_S = 70 Ω, a C = 80 mF capacity and a speed w_r = 298 rad/s).

Capacity Variations Effect

Figure 6 represents the voltage waveform evolution of a stator phase (Figure 1) according it's of the self excitation capacitor C. At several values of the speed, one observes that the voltage waveform increases with the capacity value knowing well that the induction generator cannot start unless the excitation capacitor has reached a certain threshold.

The objective of these tests is to study the behavior of the asynchronous machine (SEIG) during variation of the speed, capacity and load, and to determine the exploitation limits the generator without resorting to any form of regulation.

MODELING AND SIMULATION THE WIND SYSTEM IN THE ORCAD-PSPICE ENVIRONMENT

Behaviour of the Machine with the Voltage Regulator

The synoptic diagram equivalent of the system (asynchronous machine, energy converters (AC-DC, DC-DC, regulator, and load) is given in Figure 7. The voltage regulator is intended to control the output voltage of the reversible inverter Buck-Boost in spite of the variations of speed, load and capacity.

This diagram is implemented and simulated in the Orcad-Pspice environment.

Generally the conception and the modeling of a wind system (self-excited induction machine, static converters (AC/DC and DC/DC)) require the use of performing simulators on electronics and electrical engineering to implement each block of the wind system and in particular the model of the self-excited induction generator in transient state. In our research framework, the Orcad/Pspice software can largely meet our need in electric simulations.

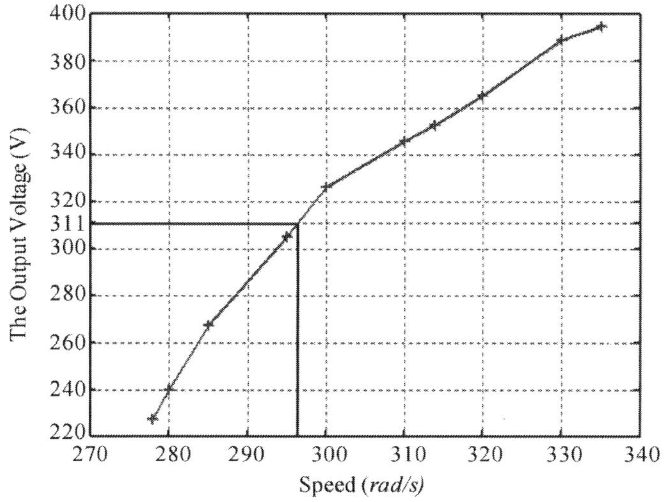

Figure 4. The output voltage evolution of the system (Figure 1) as a function of the drive speed at (C = 80 mF and R_S = 70 Ω).

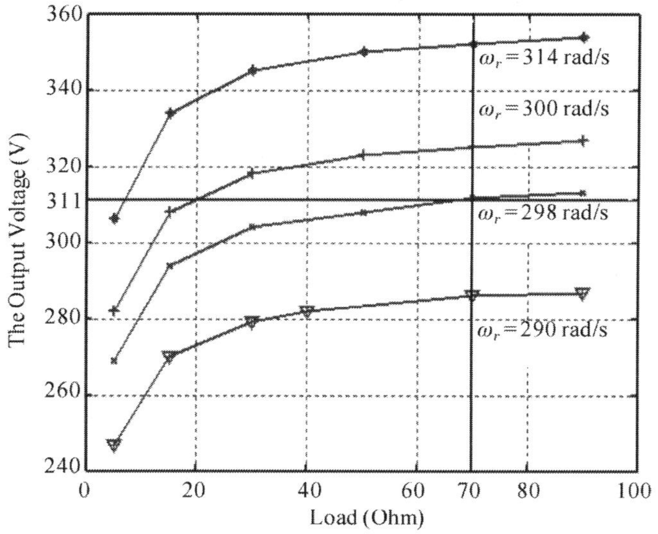

Figure 5. Variation of the output tension of the rectifier as a function of the load for different speeds (C = 80 mF).

The main advantages of this software are:

- Orcad/Pspice simulation software is the most widely used in the industry.
- It is based on the industry standard SPICE and therefore gives access to libraries of models developed by industrial manufacturers, and the availability of a large number of libraries on the market.
- It allows the combination of digital and analog components without any problems.
- It allows the ability to integrate Orcad electrical schematics in Matlab/Simulink tool especially with the version 16 of Orcad/Pspice.
- It is a very complete software because it allows to simulate all aspects of systems that we find in electrical engineering: power electronics, analog and digital electronic control, slaving, and mechanics.

Modelling of the Voltage Regulator

The model of the voltage regulator is given in Figure 8.

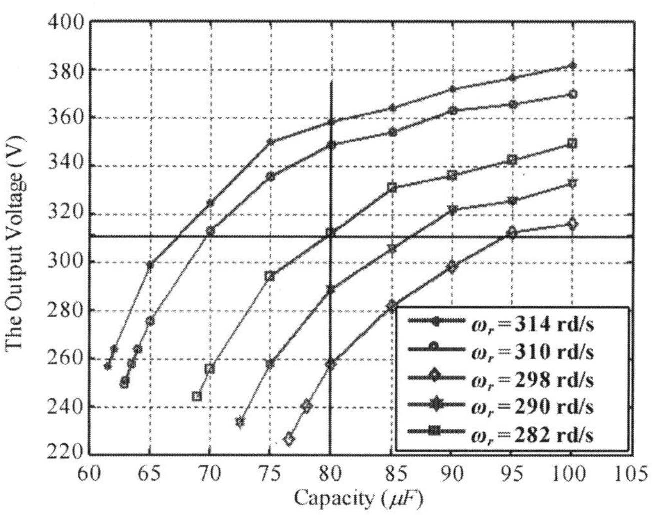

Figure 6. Evolution of the output tension of the system (Figure 1) according to the capacity for different speeds at (R_S = 70 Ω).

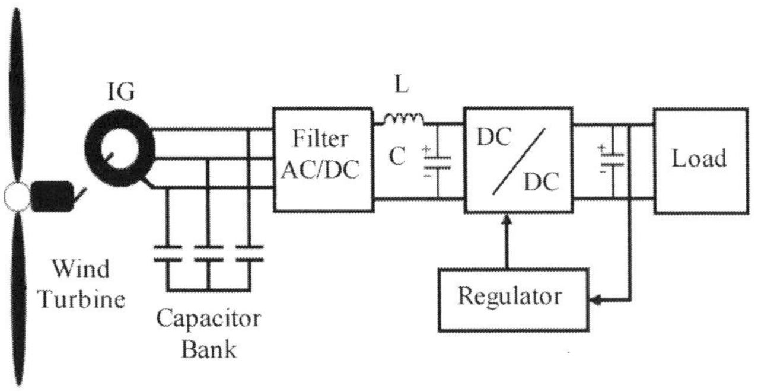

Figure 7. Descriptive diagram of the system.

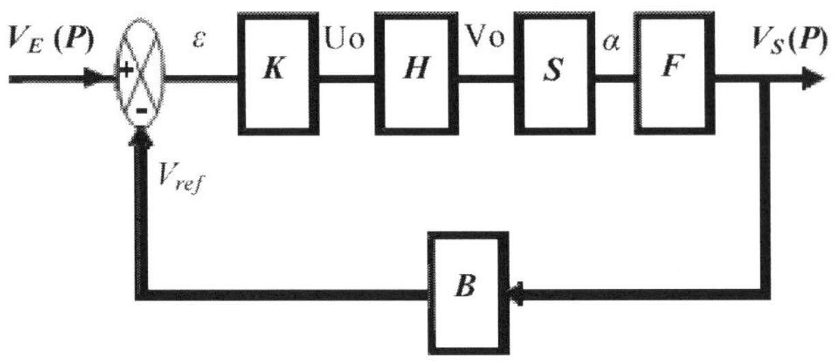

Figure 8. Equivalent circuit of the voltage regulator.

The different transfer functions of the various elements of Figure 8 are as follows:

B: divider of return:

$$B = \frac{V_{ref}}{V_s} \quad (8)$$

K: Corrector Proportional-Integrator PI

$$K(P) = \frac{U_0(P)}{\varepsilon(P)} = \frac{A_0}{1+\tau p} \quad (9)$$

H: Corrector Integrator

$$H(P) = \frac{V_0(P)}{U_0(P)} = \frac{1}{\tau_1 p} \quad (10)$$

S: Comparator circuit Puls Wave Modulation (PWM):

$$S(P) = \frac{\alpha(P)}{V_0(P)} \quad (11)$$

The transfer functions of the inverter:

$$F(P) = \frac{V_s(P)}{\alpha(P)} = \frac{V_E}{k \cdot p(1+\tau_2 p)} \quad (12)$$

with: α = Cyclic ratio P: Laplacian operator The total function of transfer of the system is as follows:

$$Ht(p) = \frac{Vs}{Ve}$$

$$Ht(p) = K(p) \cdot S(p) \cdot F(p) \cdot H(p) \cdot B \quad (13)$$

with:

$A_o = 2.2 \times 10^{-3}, \tau = 12.69, \tau_1 = 0.0423\ et\ S = 1.3 \times 10^{-3}$.

The Analog Voltage Control Algorithm

When you submit your final version, after your paper has been accepted, prepare it in two-column format, including figures and table We have implemented in the environment Orcad/Pspice the wind system of Figure 7 without the Analog Voltage Control (AVC) with taking into account the parameters of the converter given in the previous paragraph. Then we have analyzed the adaptation of the generator SEIG (LEROY SOMER: LA3 loop asynchronous machine 1.5 KW) and studied the effect of the variation of the resistive load (R_S), the excitation capacity, the drive speed and the duty cycle to the output system.

The IRF450 switch of the converter is controlled by a pulse generator outputting a square wave of high frequency 100 kHz and a variable duty ratio.

Influence of the Load and Duty Cycle of the Voltage

In Figure 9, we have shown the output voltage of the converter according to the R_S load and the duty cycle a for an excitation capacity C = 70 μF and a driving speed w_r = 298 rad/s. From these curves of the figure below we deduced that the output voltage of the converter increases with increasing the R_S load and the duty cycle.

Influence of the Capacity and the Duty Cycle on the Output Voltage

Figure 10 illustrates the voltage at the output of the Buck/Boost converter as a function of the capacity and the duty ratio a for a resistive load $R_S = 70\,\Omega$ and a driving speed w_r around 314 rad/s. From these curves we have observed that the output voltage of the converter depends on the self-excitation capacity and the duty cycle.

The voltage required by our specifications $(V_S = 311\,V)$ is given by a capacitor $C = 85\,\mu F$ and a ratio a of 0.25.

Influence of the Speed and the Duty Cycle on the Output Voltage

Figure 11 shows the output voltage of the converter as a function of the load and the duty cycle a for an excitation capacity of 75 μF and a resistive load of 70 W.

From these curves we found that the output voltage of the converter also varies with the duty cycle when the speed varies.

As to the Buck/boost converter, the variations in the output voltage according to the duty ratio, obtained in the simulation Orcad/Pspice, are consistent with the relation (14) and the proportionality of the duty ratio with the voltage at the variation in the mechanical and electrical quantities of the machine.

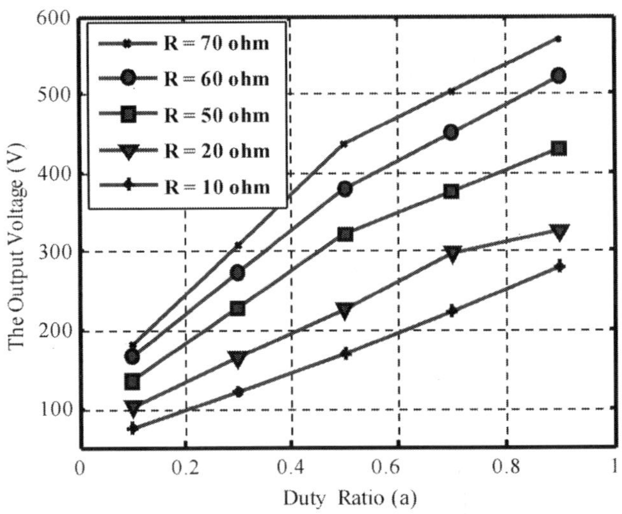

Figure 9. Typical simulation of the voltage V_S at the output of the buck/boost converter in function of the load and the duty ratio: C = 70 μF and w_r = 298 rad/s.

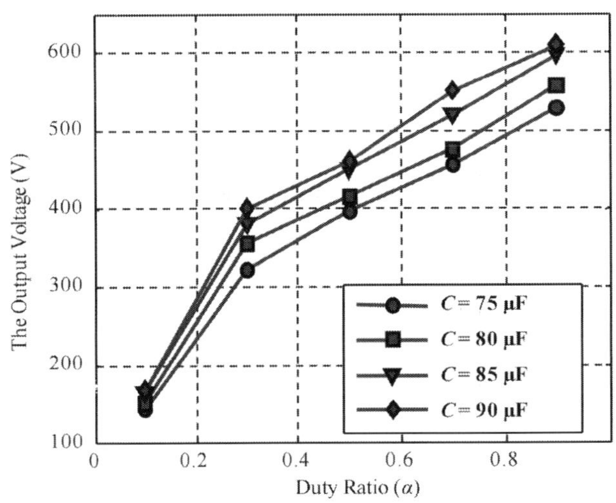

Figure 10. Typical simulation of the voltage V_S at the output of the buck/boost converter according to the capacity and the duty ratio for: $\omega_r = 314\ \text{rad/s}$ and $R_S = 70\ \Omega$.

$$|Vs| = \frac{\alpha}{1-\alpha} E \qquad (14)$$

Summary of Variations in the Parameters Load, Speed and Capacity

We saw in the previous section when the excitation capacity, the load and the driving speed varies, an adaptation of the system must be achieved by including between the generator (asynchronous machines SEIG + Rectifier) and the R_S load an adapter (Figure 1). This adapter must also include the DC-DC power converter, an AVC command to regulate the voltage in the operating conditions considered. The AVC command must meet our specifications so as to maintain a constant voltage on the terminals of the wind system studied.

So, for a given a wind system operating. The design of a command to adjust the duty cycle a, depending on weather changes (wind speed), changes of the capacity and of the load is needed in order to operate the wind system in optimal conditions.

Figure 12 shows the search algorithm of the duty cycle as a function of the output voltage whatever the variation of the SEIG characteristic.

In the following paragraph, we study the design and the operation in the Orcad/Pspice environment, of the Analog Voltage Control (AVC) linked to a converter (buck/boost) operating at high frequency (100 KHz).

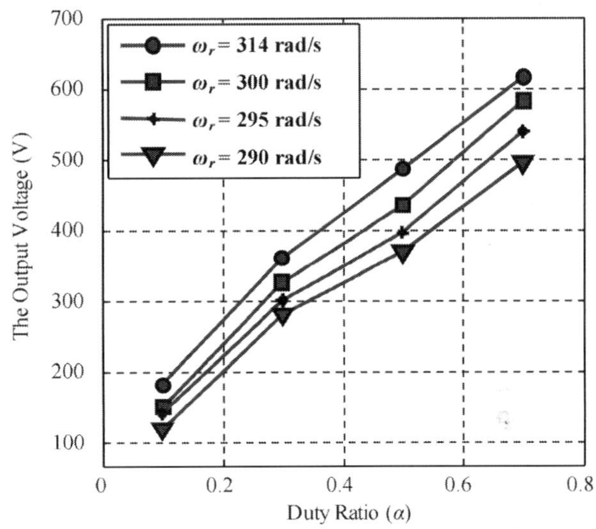

Figure 11. Typical simulation of the voltage V_S at the output of buck/boost converter according to the driving speed and the duty cycle for: C = 75 μF and R_S = 70 W.

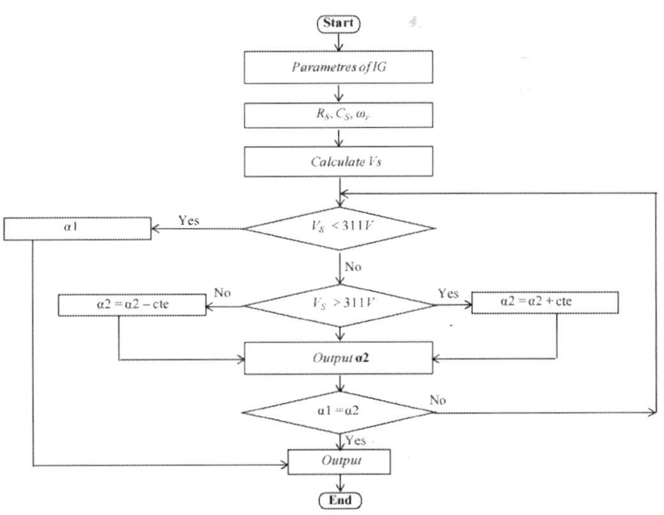

Figure 12. Algorithm for calculating the duty ratio.

RESULTS OF SIMULATION ANALOG VOLTAGE CONTROL ALGORITHM

Figure 13 represents the nominal output voltage of the inverter at fixed values of the capacity of self-excitation, load and wind speed.

Figure 14 illustrates the output voltage evolution of the inverter when the driver speed decreases by 8%. We note as light variation of the voltage at the time of the speed variation and then it is stabilized at its initial value (The wished value).

The voltage output variation of the inverter, when the self-excitation capacitor increases by 12%, is showed in Figure 15. We observe that the output voltage reaches the nominal value of stabilization after as light increase at the time of the capacity variation.

Figure 16 represents the output voltage evolution of the system (Figure 7) when the load increases by 22%. We note a slight variation of the voltage at the time of the load variation; then is stabilized at the nominal value fixed by our schedule of conditions.

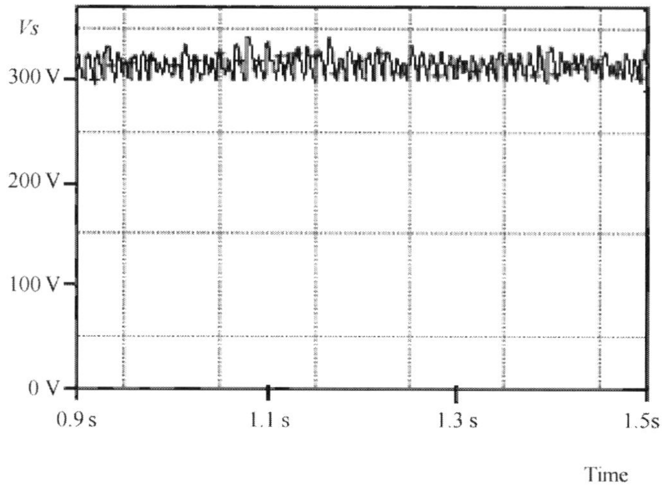

Figure 13. Variation of the output voltage of the chopper (Figure 5) at w_r= 298 rad/s, C = 80 mF and R_S = 70 W.

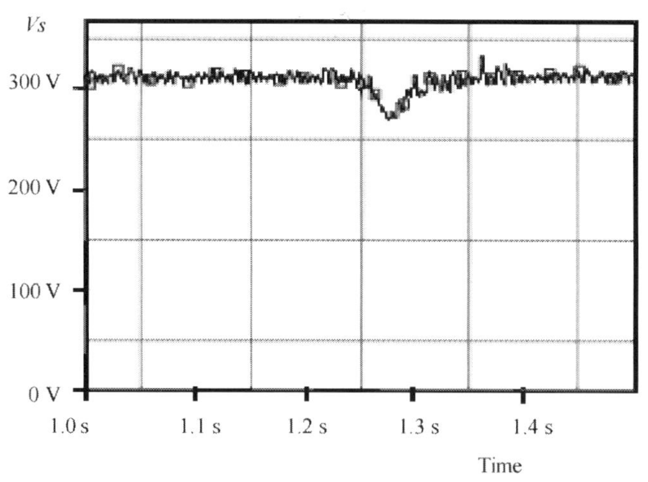

Figure 14. Variation of the output voltage of the chopper when the speed decreases by 8% at C = 80 mF and R_S = 70 W.

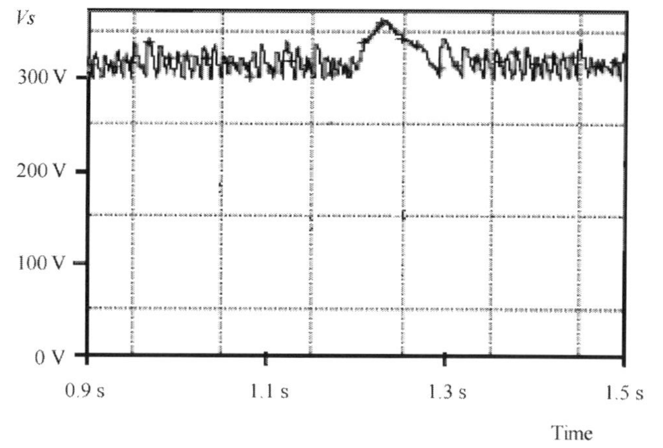

Figure 15. Variation of the output voltage of the chopper when the excitation capacitor increases by 12% at ω_r = 298 rad/s and R_S = 70 W.

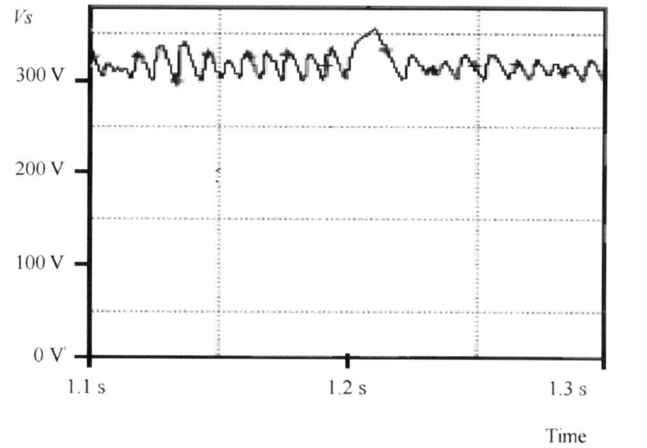

Figure 16. The output voltage evolution of the chopper when the load increases by 22% at ω_r = 298 rad/s and R_S = 70 W.

CONCLUSIONS

In this paper we proposed a robust law of the control circuit applied to a wind system with the aim of finding a constant value of tension $(220\sqrt{2}\ \mathrm{V})$ at the exit of the system provided that the drive speed does not exceed a maximum value w_r = 330 rad/s.

From the results of simulation of the studied system (asynchronous machine, converters (AC-DC and DC-DC) and load), it appears that:

- The voltage at the exit of the inverter is almost invariant for all the curves, even if one varies the characteristic parameters of the system (speed, capacity or load).
- The control designed makes the system oscillate around the desired tension.
- The disturbances which exist during the oscillations around the value $(220\sqrt{2}\ \mathrm{V})$ are due to the transient state of the asynchronous machine and the losses in the components of the system (Mosfet, diode and component of the regulator).

The SEIG has the following characteristics:

$P = 1.5$ kW, $f = 50$ Hz, $220/380$ V, $I_s = 4.4$ A, $p = 2$,
$r_s = 5.51\ \Omega$, $r_r = 2.24\ \Omega$, $Xr = 6.9\ \Omega$, $Xs = 6.9\ \Omega$,
$Xm = 38.81\ \Omega$.

REFERENCES

1. Elhafyani, M.L., Zouggar, S., Zidani, Y. and Benkaddour, M. (2006) Permanent and Dynamic Behaviours of Self-Excited Induction Generator in Balanced Mode. Journal of Physics: Condensed Matter, 7, 49-53. [Paper reference:2]
2. Premalatha, K. and Sudha, S. (2012) Self-Excitation and Voltage Control of an Induction Generator in an Independent Wind Energy Conversion System. International Journal of Modern Engineering Research (IJMER), 2, 454-461.
3. Marra, E.G. and Pomilio, J.A. (2000) Spice-Assisted Simulation of the αβγ Model of Cage-Rotor Induction Machines Including Saturation. IND2000_FDI05, 497-502.
4. Elhafyani, M.L., Zouggar, S. and Benkaddour, M. (2005) Comportement permanent et Transitoire d'une Génératrice asynchrone auto-excitée en régime équilibré. PREMME_05, Université My Ismail, FST-Errachidia, Maroc. [Paper reference:1]
5. Zidani, Y. and Naciri, M. (2001) A Numerical Analytical Approach for the optimal Capacitor used for the Self Excited Induction Generator. Proceedings of the 32nd IEEE PESC Power Electronics Specialists Conference, Vancouver, June 2001, 216-220.
6. Zidani, Y. and Naciri, M. (2001) A Steady State Analysis of the Self Excited Induction Generator Controlled by an Electronic Load Governor. Proceedings of the 4th IEEE PEDS Power Electronics and Drive Systems, Bali, October 2001, 903-907.
7. Mujadi, E., Sallan, J., Sanz, M. and Butterfield, P. (1999) Investigation of Self Excited Induction Generators for Wind Turbine Application. IEEE Industry Application Conference 34 IAS Annual Meeting, Arizona, October 1999, 3-7.
8. Seyoum, D., Grantham, C. and Rahman, F. (2001) Analysis of an Isolated Self-Excited Induction Generator Driven by a Variabe Speed Prime Mover. AUPEC 2001, Curtin University of Technology, Perth, September 2001, 23-26.
9. Mujadi, E., Gregory, B. and Bord, D. (1999) Self Excited Induction Generators for Variable-Speed Wind Turbine Generation. IEEE Industry Application Conference 34 IAS Annual Meeting, Arizona, October 1999, 343-352.
10. Barara, M., Akherraz, M., Abbou, A., Elhafyani, M.L. and Moutchou, M. (2013) Fuzzy Modeling Magnetizing Curve of a Self Excited Induction Generator and Effect of Mutual Inductance. 2013 International Renewable and Sustainable Energy Conference (IRSEC), Ouarzazate, 7-9 March 2013, 287-291.
11. Ouchbel, T., Zouggar, S., Elhafyani, M.L., Oukili, M., Sedik, M. and Rabhi, A. (2013) Regulation of the Excitation Reactive Power of the Asynchronous Wind Turbine at Variable Speed. Smart Grid and Renewable Energy, 4, 272-280.http://dx.doi.org/10.4236/sgre.2013.43033
12. Ouchbel, T., Zouggar, S., Elhafyani, M.L., Oukili, M., Sedik, M. and Aziz, A. (2014) Power Maximization of an Asynchronous Wind Turbine with a Variable Speed Feeding a Centrifugal Pump. Journal Energy Conversion and Management (EC & M), 78, 976-984. [Paper reference:2]
13. Elhafyani, M.L., Zouggar, S. and Benkaddour, M. (2006) Symbolization of the Asynchronous Machine in the Orcad/ Pspice environment. JER06, EST-Oujda, Morocco. [Paper reference:1]
14. Aziz, A., Kassmi, K., Olivié, F. and Sarrabayrouse, G. (2005) Symbolization of the Electric Diagram of the Marketing Solar Panels in the Orcad-Pspice Environment. Rapport LAAS (05270), Toulouse, 17.
15. Aziz, A., Kassmi, K., Maimouni, R., Olivié, F., Sarrabayrouse, G. and. Martinez, A. (2005) Integration of the New Components in the Library of the Orcad-Pspice Simulator. 2 Symbolization of the Electric Diagrams: Application to the Photovoltaic Systems. Rapport LAAS/CNRS (05271), Toulouse, 15.

CHAPTER 15

A NEW TYPE OF CAPACITIVE MACHINE

Arie Shenkman, Saad Tapuchi, and Dmitry Baimel

Keywords: Capacitive Machine, Electric Field, Torque, Synchronous Machine

INTRODUCTION

Today, most of the electrical machines operate on the principle of magnetic field effect at which the electric energy is converted into mechanical energy and vice versa. The magnetic field is produced by the current flowing in the stator and rotor coils.

On the other hand, it is also possible to construct the electrical machines that operate on the electric field principle, where the electrical energy is stored in an electric field [1]. It is known that the electric and the magnetic fields are always linked to each other and it is impossible to separate them. However, the electric field in the electromagnetic machines is much weaker than the magnetic field, and, practically, is negligible.

The electric machines operated on the electric field principle would have mechanical structure based on capacitors [2] - [4] while the mechanical structure of the electromagnetic machines is based on the inductors. While the development of the electromagnetic machines has reached the most advanced state, the electric field machines (capacitor machines) remained somehow ignored. The reason is low electric permittivity-ε of most dielectric materials. However, recently, new dielectric materials have been developed. These materials possess relatively high electric permittivity of up to ε_r=104. Therefore, the capacitor motors acquire an interest again.

The greatest difficulty in the development and production of capacitive machines is preventing friction between the stator and rotor plates which occurs when the dielectric materials are solid. A common method to prevent the friction is to ensure an air gap. However, in the capacitive machine this method is not suitable because even small air gap diminishes the advantages of using dielectric material with high dielectric permittivity ε.

Citation: Shenkman, A., Tapuchi, S. and Baimel, D. (2015). A New Type of Capacitive Machine. Energy and Power Engineering, 7, 31-40. DOI: 10.4236/epe.2015.72003.

One possible solution could be to use liquid electric materials with low density and high values of ε. Another possible solution might be changing the air gap to plasma, which can be accomplished by using high voltage and a narrow air gap.

The main factor determining the efficiency of electric power mechanisms for energy conversion is the density of the electric or magnetic energy stored in the electromagnetic or electric fields. The electromagnetic and electric field machines could be compared from the energy point of view [5] [6]. The energy density of the magnetic field in electromagnetic machines could be calculated by:

$$w_m = \frac{B^2}{2\mu_0\mu_r} \qquad (1)$$

where B is the magnetic flux density of the magnetic field, μ_0 is the magnetic permeability constant $(\mu_0 = 4\pi\times10^{-7}\ \text{H/m})$ and μ_r is the magnetic permeability of the air (in magnetic machines the process of energy conversion occurs within the air gap between the rotor and the stator so $\mu_r = 1$). By using the ferromagnetic materials (particularly electric steel), the magnetic field of the electromagnetic machines can have magnetic flux density of $B = 1-1.6\ \text{Wb/m}^2$. The density of the magnetic energy would be (for $B = 1.4\ \text{Wb/m}^2$):

$$w_m = \frac{1.4^2}{2\times4\pi\times10^7} = 800000\ \text{W/m}^3 \qquad (2)$$

The energy density of the electric field in the electric machines could be calculated by:

$$w_e = \frac{1}{2}\varepsilon_r\varepsilon_o E^2 \qquad (3)$$

where ε_0 is the electric permittivity constant $(\varepsilon_0 = 8.86\times10^{-12}\ \text{F/m})$ and E is electric flux density of the electrostatic field. Today's technology allows usage of very strong electrostatic fields, but because of low electric strength of most insulating materials, the electrostatic field is practically limited to 2 - 20 MV/m. Therefore, in the machine whose operation is based on the electric field principle and in which the process of energy conversion also takes place in the air $(\varepsilon_r = 1)$, the energy density of the electric field, for $E = 10^7$ V/m, could be determined by Equation (4):

$$w_e = \frac{8.86 \times 10^{-12} \times (10^7)}{2} = 450 \text{ W/m}^3 \quad (4)$$

Lately, new dielectric materials have been developed, such as ceramics based on a paratitanite bar, which has a relatively high electric permittivity of up to $\varepsilon_r = 10^4$ [F/m]. In electric machines, where the energy conversion process takes place in the new dielectric materials and not in the air, the density of the electric energy could be significantly increased.

For electric permittivity of $\varepsilon_r = 10^4$, the obtained density of the electric energy, for $E = 10^7$ V/m, would be:

$$w_e = \frac{10^4 \times 8.86 \times 10^{-12} \times (10^7)}{2} = 450 \times 10^4 \text{ W/m}^3 \quad (5)$$

Such a result is even greater than the result obtained with the electromagnetic machines, a fact that encourages development of new electric machines based on the electric field effect.

The electric field machines have several advantages upon the electromagnetic machines. The main advantage is that the electric field machines are much smaller in size and lighter in weight than the electromagnetic machines with the same output parameters. This could be explained by the fact that magnetic force lines of any electromagnetic machine are formed in a closed loop. The magnetic force lines are formed by the magnetic circuit, built from a very heavy material like electric steel. The magnetic circuit is used for the power transfer, while the energy conversion process takes place in the air-gap itself (between the stator and the rotor). The air-gap volume is relatively a small part of the volume of the whole machine.

In the electric field machines, the power lines of the electric field start and end in electric charges, i.e. the rotor and stator plates. The energy conversion process takes place in the gap between the stator and rotor, which takes up a sizable part of the volume of the whole machine. Assuming that the specific energy is equal for both electric field and electromagnetic machines, the machines based on the electric field effect will be significantly smaller. And since the dielectric materials (e.g. ceramics and plastics) are much lighter than the ferromagnetic materials, the electric machines based on the electric field effect will be also lighter. As a result of these two factors, the production of the machines based on the electric field effect will be cheaper.

Just like magnetic machines, the capacitive machines can be of two kinds, in dependence of the current/voltage used for their operation: direct current (DC) machines with commutator or alternative current (AC) machines: synchronous or asynchronous. Section II shows how the sectored capacitor could be used as a practical solution for the DC capacitive machines. Section III presents the proposed AC capacitive machine based on the sinusoidal capacitor. Section IV presents the simulation results of the proposed capacitive machine. Section V presents the conclusions of the paper.

THE SECTORED CAPACITOR AS A PRACTICAL SOLUTION FOR THE DC CAPACITIVE MACHINES

The simplest DC capacitive machine can be constructed from a round capacitor having two or more sectors, as shown in Figure 1. In this machine, two positively charged rotating sectors form the rotor and two negatively charged static sectors form the stator. In the position shown in Figure 1, the attractive force between the stator and rotor sectors establishes a moment, which rotates the rotor on a quarter of a circle until the sectors coincide. At the moment of a coincidence of the sectors, the applied voltage polarity is changed between the positively and negatively charged sectors. In other words, this machine should be constructed with commutate or (like regular DC machines). If the number of plates will be increased, the machine operation will be smoother. In order to ensure smooth operation, at least three couples of plates should be used.

The torques of the capacitive machine could be calculated by differentiation of the energy stored in the electric field of the capacitor [6]. In the case of angular movement, the differentiation is performed by an angle α:

$$M(\alpha) = \frac{d}{d\alpha}\left(\frac{C(\alpha)V^2}{2}\right) \quad (6)$$

where $C(\alpha)$ is varying capacitance of the capacitor plates as function of the rotor angle α and V is the voltage applied to the capacitor plates (this formula is adequate when the capacitor is connected to a power source). With the movement of the rotor, the capacitance between the rotor and the stator varies proportionally to the surface of coincidence between sectors. The surface of coincidence between the sectors is proportional to the rotor angle α and could be calculated by:

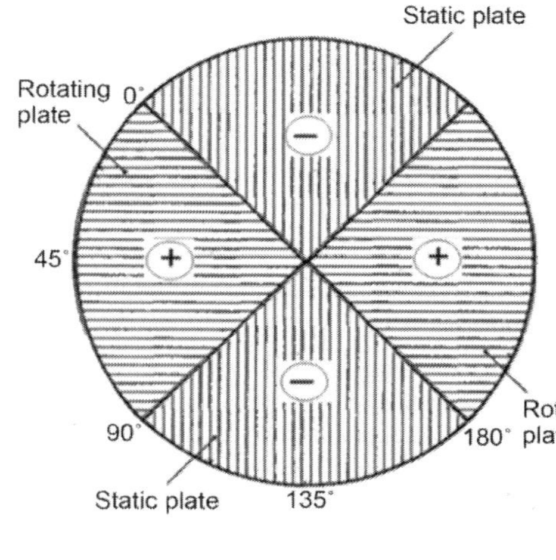

Figure 1. DC capacitive machine constructed from round capacitor with four sectors.

$$s = \rho^2 \frac{\alpha}{2} \quad (7)$$

where ρ is the radius of the plates. The capacitance of the round capacitor with four sectors (two couples of plates) could be calculated by:

$$C(\alpha) = 2\frac{\varepsilon\varepsilon_0\rho^2\alpha}{d} \tag{8}$$

where d is the distance between the plates.

The axial torque of the capacitive machine with three couples of plates could be calculated by substituting Equation (8) in Equation (6):

$$M(\alpha) = \frac{\varepsilon\varepsilon_0\rho^2V^2}{d} \tag{9}$$

Equation (9) shows that the moment of this capacitive machine is constant and independent of the rotor angle-α. The speed of capacitive motors is a function of the supplied power and the number of revolutions per minute could be calculated by $= 9.55 \times (P/M)$. For example, for the 130 W motor with insulating material which has dielectric permittivity of $\varepsilon = 10^4$, two couples of plates, the distance between the plates $d = 2$ mm, the radius of the plates $\rho = 10$ cm and applied voltage $V = 1$ kV, the moment will be $M = 0.44$ Nm. The motor speed will be $n = 955$ rpm.

THE PROPOSED AC CAPACITIVE MACHINE BASED ON A SINUSOIDAL CAPACITOR

This paper proposes novel synchronous machine operating on the principle of the electric field effect. The proposed machine is constructed from a variable capacitor having static and rotating plates.

The capacitance be haves as a sinusoidal function and depends on the plates movement angle. The proposed sinusoidal capacitor machine with rotating plates (rotor) above static plates (stator) is shown in Figure 2.

The plates are constructed in such a way that their radius-vector is dependent on the angle and their overlapping surface is a sinusoidal function of the angle. The stator plates are constructed as standard sectors. This form allows more efficient usage of the plate's surface and as a result, increases the maximum capacity of the capacitor. The variable surface of the plates is a function of the α angle and could be calculated by:

$$S(\alpha) = \frac{1}{2}\int_0^\alpha \rho^2(\alpha)\,d\alpha \tag{10}$$

On the other hand, the variable surface is sinusoidal and it could be also expressed as:

$$S(\alpha) = S_m\frac{1-\cos(2\alpha)}{2} \tag{11}$$

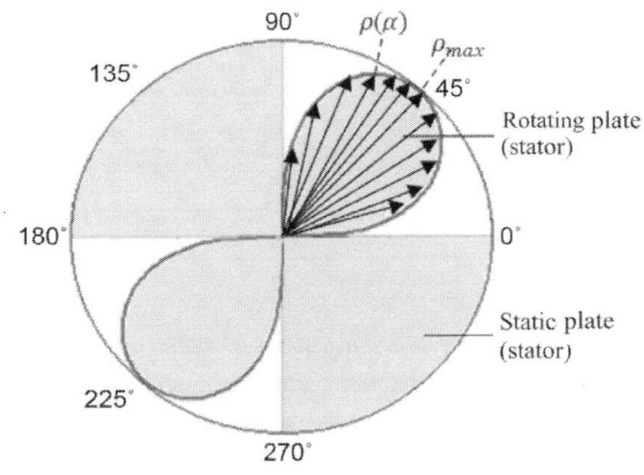

(a)

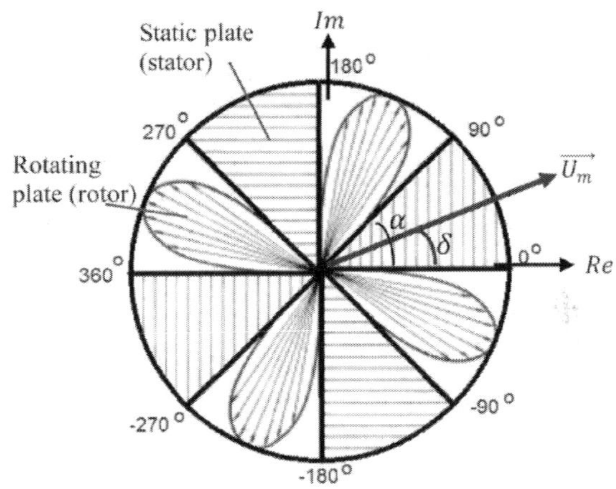

(b)

Figure 2. (a) The proposed capacitive motor with two rotating plates (rotor) and two static plates (stator); (b) The proposed capacitive motor with four rotating plates (rotor) and four static plates (stator).

where S_m is the maximum overlap surface of the capacitor (when the plates of the stator and the rotor are fully overlapped one above the other). The variable surface versus the rotating angle α is shown in Figure 3.

The radius of the capacitor plates (α) could be calculated from Equations (10) and (11):

$$\rho(\alpha) = \sqrt{2S_m\sin(\alpha)} \tag{12}$$

By using Equation (12), the diameter of the capacitor plates could be obtained:

$$D = 2\rho_m = 2\sqrt{2S_m} \tag{13}$$

where ρ_m is the maximal radius of the rotating plates.

When the rotor plates rotates above the stator plates, the capacity of the machine varies according to:

$$C(\alpha) = C_m\frac{1-\cos(2\alpha)}{2} = \varepsilon\varepsilon_0 k\frac{S_m}{d}\frac{1-\cos(2\alpha)}{2} \tag{14}$$

where d is the distance between the static and the rotating plates, ε is the relative electric permittivity of the insulating material between the plates, k is the number of pairs of the rotor/stator plates and C_m is the maximal capacitance of the machine with k couple of plates.

The torque of the capacitive machine could be calculated by the differentiation of the energy stored in the electric field of the capacitor. If the capacitor is supplied by the voltage-u, the axial torque of the capacitive machine could be determined by substituting Equation (14) into Equation (6):

$$M(\alpha)=\frac{C_m u^2}{2}\sin 2\alpha \tag{15}$$

As it seen from the Equation (15), if the proposed capacitive machine was supplied by a DC voltage source, the average torque would be zero. In order to obtain efficient average torque (different than zero), the applied voltage should be an alternating voltage $\left(u=U_m\sin(\omega t)\right)$ and the angular speed of the rotor should be equal to the angular frequency of the applied voltage.

The angular velocity could be defined as $\omega=\frac{d\alpha}{dt}$. Therefore, the rotor angle could be obtained by an integration of the angular velocity:

$$\alpha=\omega t+\frac{\delta}{2} \tag{16}$$

where δ is the angle between the starting line $\alpha=0°$ of the rotor plate and the phase angle of the alternating voltage (see Figure 3(b)). The meaning of the angle δ could be also explained by using a complex plane having "Real" and "Imaginary" axes. The alternating voltage applied to the sinusoidal capacitor could be represented by the rotating vector U_m, which rotates with the angular velocity ω on the complex plain. The capacitor rotating plates (rotor) are referred to the zero degree line. Therefore, the δ angle is the angle between voltage vector U_m and the rotor axis.

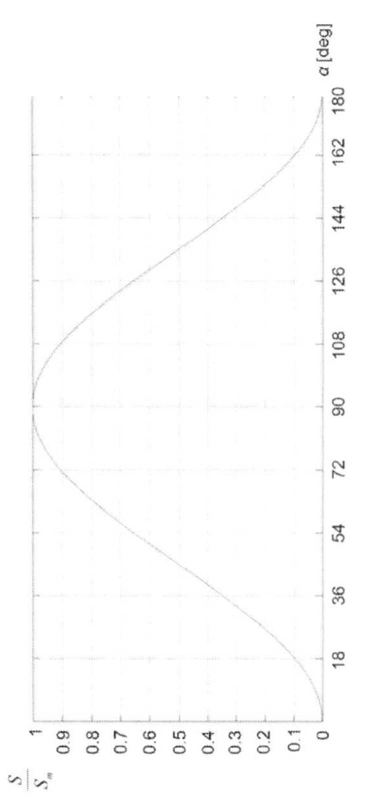

Figure 3. The variable surface of the capacitor versus the angle α.

The angular velocity of the rotor could be obtained by:

$$\omega_r=\omega/k \tag{17}$$

where k is the number of pairs of the rotor/stator plates. The speed of the machine in revolutions per minute (rpm) could be calculated according to the following formula:

$$n=\frac{60f}{k} \tag{18}$$

where f is the frequency of the applied AC voltage.

The torque of the proposed machine as a function of time could be calculated by substituting the Equation (16) into the Equation (15):

$$M(t)=C_m U_m^2 \sin(\omega t)^2 \sin\left(2(\omega t+\delta/2)\right) \tag{19}$$

$$M(t)=\frac{C_m U_m^2}{2}\left(1-\cos(2\omega t)\right)\sin(2\omega t+\delta) \tag{20}$$

$$M(t)=\frac{C_m U_m^2}{2}\left[\sin(2\omega t+\delta)-\frac{\sin\delta}{2}-\frac{\sin(4\omega t+\delta)}{2}\right] \tag{21}$$

The average torque, i.e., the effective torque, would be:

$$M_{avg}=\frac{1}{2\pi}\int_0^{2\pi} M(t)\,dt=-\frac{C_m U_m^2}{2}\sin\delta \tag{22}$$

The meaning of the minus sign in the formula of torque is that the torque resists motion, i.e. that machine operates in a generator mode and the conversion is from mechanical energy into electric energy. In the generator mode, the rotor precedes the alternating voltage (positive δ angle). In the motor mode, where the δ angle is negative, the rotor lags behind the alternating voltage. The generator and the motor operation modes are shown in the torque diagram, Figure 4.

The machine would develop maximal torque when the angle δ would be equal to $\pi/2$:

$$M_{max}=\frac{U^2 C_m}{2} \tag{23}$$

The obtained diagram is similar to the torque diagram of a synchronous machine. It can be seen from the presented analysis, that a rotating field is not needed for the capacitive machine operation as for the three-phase electromagnetic machines, and that a capacitor machine can also operate as a single-phase machine. There are no differences between the analysis of single-phase capacitor motor and the three-phase one.

SIMULATION OF THE PROPOSED AC CAPACITIVE MACHINE

The proposed AC capacitive machine was simulated by using the Matlab program. The parameters of the simulated machine are: 4 couples of plates ($k=4$), the maximal overlap surface of the rotor and stator plates is

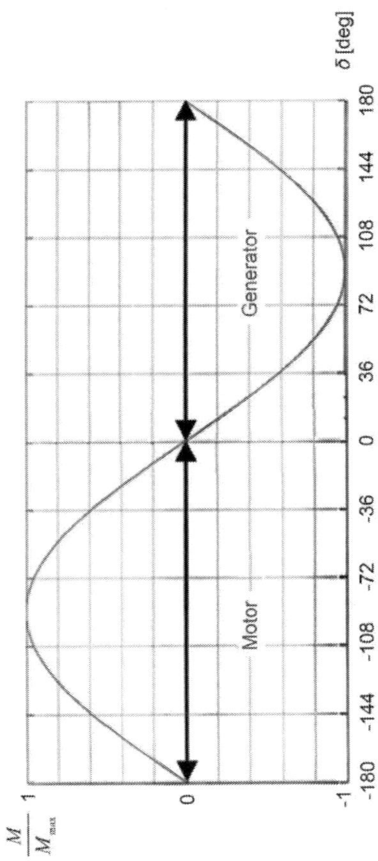

Figure 4. The torque diagram of the proposed capacitive machine.

$S_m = 50\ \text{cm}^2$, electric permittivity of the dielectric material is ε = 10^4, the distance between the plates is d = 3mm. The applied AC voltage is $U_m = 6\ \text{kV}$ and the frequency is f = 50 Hz $f = 50\ \text{Hz}$. The simulation results for δ = 30° (generator mode) are shown in Figure 5. It can be seen that the capacitance varies in a sinusoidal shape: $C(\alpha)$ = 0 for $\alpha = 0,\ \pi,\ 2\pi$; $C(\alpha) = 6\ \text{uF}$ for $\alpha = 0,\ 0.5\pi,\ 1.5\pi$. The obtained average torque is negative −54 Nm (see Figure 5(a)). Therefore, the machine operates in the generator mode. The spectrum of the instantaneous torque has three harmonics: DC, 100 Hz and 200 Hz. This result is logical because $M(t)$ consists of $-\frac{C_m U_m^2}{2}\frac{\sin\delta}{2}$ which is represented by DC harmonic, $\frac{C_m U_m^2}{2}\left[\sin(2\omega t + \delta)\right]$ which is represented by 100 Hz harmonic and $-\frac{C_m U_m^2}{2}\frac{\sin(4\omega t + \delta)}{2}$ which is represented by 200 Hz harmonic.

The simulation results for $\delta = 0°$ are shown in Figure 6. In this case, the machine operates on the boundary of generator and motor modes. The obtained average torque is zero. The spectrum of the instantaneous torque has

100 Hz and 200 Hz harmonics. The DC harmonic is zero because $\frac{C_m U_m^2}{2}\frac{\sin\delta}{2} = 0$.

The simulation results for δ = −30° are shown in Figure 7. In this case, the machine operates in the motor mode. The obtained average torque is 54 Nm.

The Matlab simulation results could be validated by numerical calculations. The maximal capacitance could be calculated by Equation (14):

$$C_m = \varepsilon\varepsilon_0 k\frac{S_m}{d} = 10^4 \times 8.86\times 10^{-12} \times 4\frac{50\times 10^{-4}}{3\times 10^{-4}} = 6\ \text{uF}$$

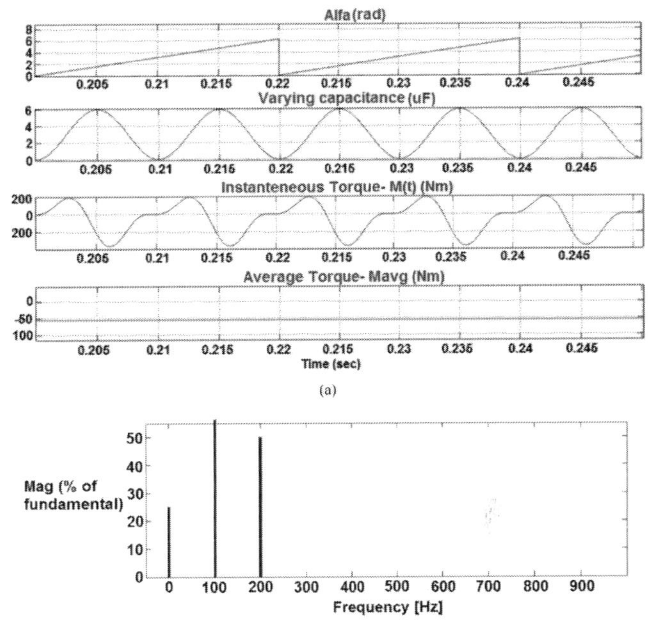

(a)

(b)

Figure 5. The angle δ = 30° (generator mode). (a) The simulated rotating rotor angle (varies from zero to 2π), varying capacitance $C(\alpha)$, instantaneous torque of the machine $M(t)$ and the average torque M_{avg}. (b) The spectrum of the instantaneous torque.

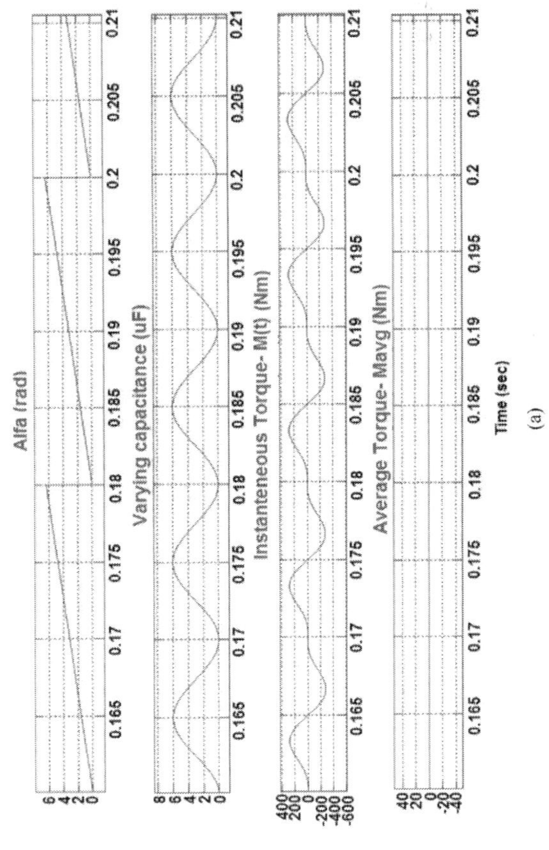

(a)

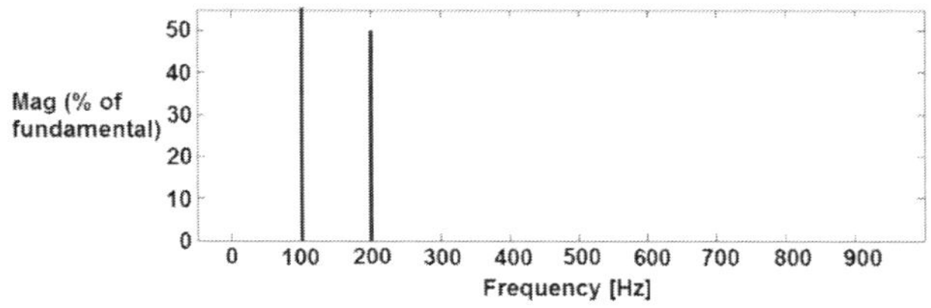

(b)

Figure 6. The angle $\delta = 0°$ (boundary between the generator and motor modes). (a) The simulated rotating rotor angle α (varies from zero to 2π), varying capacitance $C(\alpha)$, instantaneous torque of the machine $M(t)$ and the average torque M_{avg}; (b) The spectrum of the instantaneous torque.

For the case of $\delta = 30°$, the calculated average torque would be:

$$M_{avg} = -\frac{C_m U_m^2}{2}\sin\delta = -\frac{(6\times10^{-6})\times(6\times10^3)^2}{2}\times 0.5 = -54\ \text{Nm}$$

For the case of $\delta = -30°$, the calculated average torque would be:

$$M_{avg} = -\frac{C_m U_m^2}{2}\sin\delta = -\frac{(6\times10^{-6})\times(6\times10^3)^2(-0.5)}{2} = 54\ \text{Nm}$$

In this case, the motor output power would be:

$$P_{out} = M_{avg}\times\omega_r = M_{avg}\times\frac{\omega}{k} = \frac{54\times314}{4} = 4239\ \text{W}$$

The calculations show that the presented simulation results are correct.

CONCLUSIONS

The novel capacitive synchronous machine is proposed. The machine is operating on the principle of the electric field. By using new materials with high electric permittivity ε, the proposed machine is feasible from the theoretical and the practical points of view. The machine design is based on the sinusoidal capacitor whose capacitance varies according to a sinusoidal function. The proposed machine is actually electrical synchronous machine with characteristics similar to those of the magnetic synchronous machines.

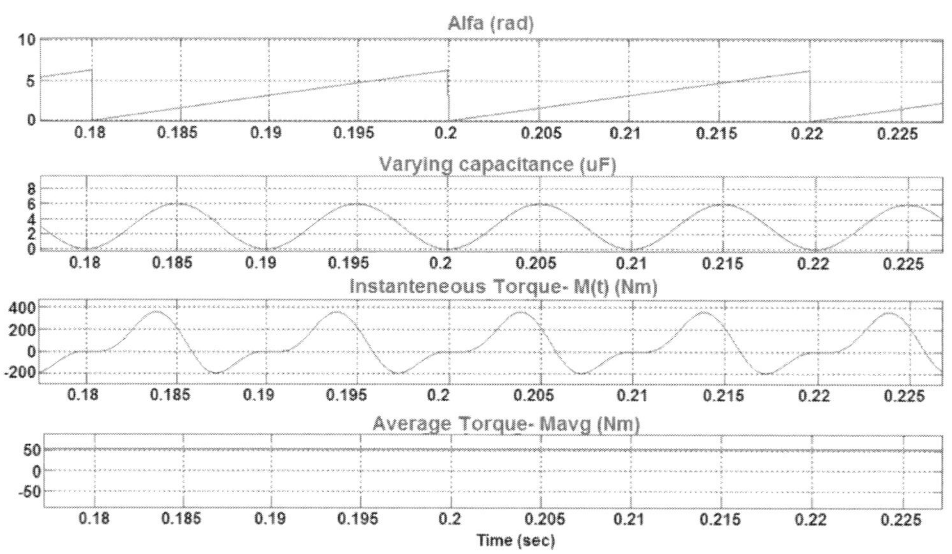

Figure 7. The angle $\delta = -30°$ (motor modes). The simulated rotating rotor angle α (varies from zero to 2π), varying capacitance $C(\alpha)$, instantaneous torque of the machine $M(t)$ and the average torque M_{avg}.

Extensive simulation results of the machine operated in the motor and the generator modes validate the theoretical analysis and show the practical feasibility of the proposed machine.

The proposed machine has a number of advantages upon the conventional electric machines based on the magnetic interaction. The main advantage is that the electric field machines are much smaller in size and lighter in weight than the electromagnetic machines with the same output parameters. Furthermore, the capacitor motor has very low inertial torque due to the low weight of the rotor. This fact allows rapid starting and stopping of the machine, as well as other variations during operation. This feature could be highly important in control systems. Another advantage is the simple structure of a single-phase capacitor motor.

Appendix: Example of Design of the Capacitive Motor

The proposed capacitive machine and the shape of the plates of both stator and rotor are shown inFigure 8. In order to increase the overall capacity of the motor, it is made of a large number of plates, and the insulation between the plates is of a material with a high ε value. This material takes the form of insulation washers. In order to ensure a good contact between the plates of the rotating rotor and the insulation washers, idling together with the stator plate, and in order to reduce friction, the lubricant should have also high ε value.

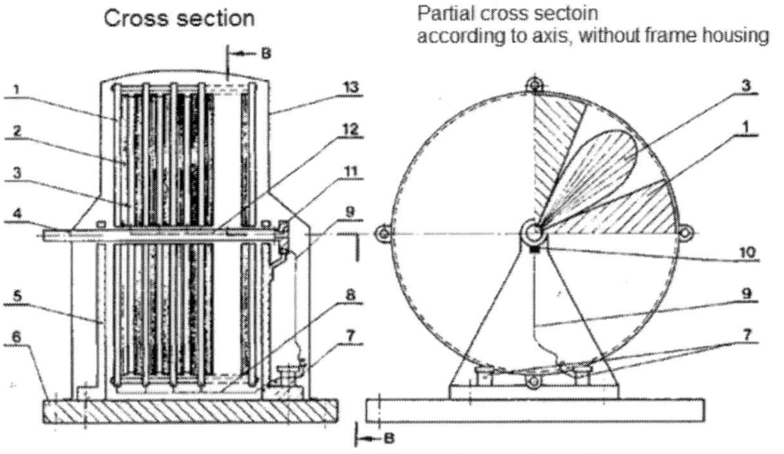

1—Stator plates; 2—Insulating washer; 3—Rotor plates; 4—Motor shaft; 5—Supports; 6—Bed plate; 7—Terminals; 8—Connecting lead of stator; 9, 12—Connecting lead of rotor; 10—Brush; 11—Slip rings; 12—Frame housing.

(a)

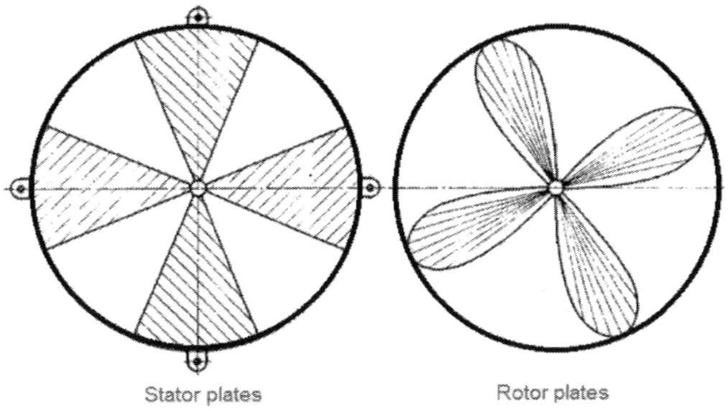

(b)

Figure 8. The proposed capacitive motor.

REFERENCES

1. Moore, A.D. (1973) Electrostatics and Its Applications. John Wiley & Sons, 133-145.
2. Hojjat, Y., Dadkhah, M., Modabberifar, M. and Ghanbari, M. (2007) Electrostatic Rotation of Plexiglas Disc Supported on Air Bearing. 14th International Conference on Mechatronics and Machine Vision in Practice, Xiamen, 4-6 December 2007, 130-135.
3. Santana-Martín, F.J., Montoya, S.G.-A., Monzón-Verona, J.M. and Montiel-Nelson, J.A. (2008) Analysis and Modeling of an Electrostatic Induction Micromotor. 18th International Conference on Electrical Machines, Vilamoura, 6-9 September 2008, 1-5.
4. Stranczl, M., Sarajlic, E., Fujita, H. and Yamahata, C. (2011) Modal Analysis and Modeling of Frictionless Electrostatic Rotary Stepper Micromotor. IEEE 24th International Conference on Micro Electro Mechanical Systems (MEMS), Cancun, 23-27 January 2011, 1257-1260.
5. Matsuzaki, K., Matsuo, T. and Mikuria, Y. (1994) Comparison of Electrostatic and Electromagnetic Motors Based on Fabrication and Performance Criteria. Proceedings of 5th International Symposium on Micro Machine and Human Science, Nagoya, 2-4 October 1994, 77-81.
6. Behjat, V. and Vahedi, A. (2004) Study Torque Ripple of the Side Drive Electrostatic Micromotor. 39th International Universities Power Engineering Conference, Bristol, 8 September 2004, 600-603.

CHAPTER 16

INDUCTANCE IN A SYNCHRONOUS LINEAR PERMANENT MAGNET GENERATOR

Boel Ekergård, Rafael Waters, and Mats Leijon

Keywords: Wave Power, Synchronous Linear Permanent Magnet Generator, Varying Inductance, ABC of Frame Reference

INTRODUCTION

Researchers all over the world are looking for ways to convert electricity from new renewable energy sources. Wave energy is at the present time a relatively unexploited energy source, but has a potential to make a significant contribution to electrical power generation [1]. Over the past 25 years a large variety of concepts for wave energy conversion have been proposed and a significant number of different technologies have been suggested [2].

The presented paper discusses a direct driven point absorber, with a linear permanent magnet generator placed on the ocean floor, see Figure 1. The main dimensions of the generator is presented in Table 1. The output characteristics are referring to a nominal load condition, when the translator is moving at a constant speed of 0.7 m/s. The three phases are Y-connected to a resistive load at 12 ohm. For more information and literature about the system, see [3].

To investigate the inductance terms of a synchronous machine, experimental tests, such as standstill frequency response or short circuit test, are usually performed [4]. These tests are, however, impossible to perform at the design stage, so to be able to estimate the unknown parameter the authors have chosen to work with both analytic calculations and a numeric two-dimension finite element model.

ANALYTICAL MODEL

If the generator is connected to a balanced three phase load, each phase flux linkage can be expressed as [5]:

Figure 1. Schematic figure of the linear generator implemented in the system.

Table 1. Main Dimensions of the Generator.

Output power	20 kW
Armature current	25 A
Phase to phase Voltage	450 V
Fundamental frequency	6 Hz
Stator length	2000 mm
Translator length	3000 mm
Stroke length	2000 mm

$$\lambda_a = L_{aa}\lambda_a + L_{ab}I_b + L_{ac}\lambda_c + \lambda_{PM,a} \quad (1)$$

$$\lambda_b = L_{bb}I_b + L_{ab}I_a + L_{bc}I_c + \lambda_{PM,b} \quad (2)$$

$$\lambda_c = L_{cc}I_c + L_{bc}I_b + L_{ac}I_a + \lambda_{PM,c} \quad (3)$$

L_{aa}, L_{bb} and L_{cc} represent the winding's self-inductance per phase, L_{ab}, L_{ac} and L_{bc} the winding's mutual inductance, I_a, I_b and I_c represent each phase current and λ_{PM} correspond to the flux linkage from the permanent magnet [5].

Citation: B. Ekergård, R. Waters, and M. Leijon. " Inductance in a Synchronous Linear Permanent Magnet Generator," Journal of Electromagnetic Analysis and Applications, Vol. 3 No. 5, 2011, pp. 155-159. DOI:10.4236/jemaa.2011.35025.

The winding's self-inductance, derived from Faraday's law of induction, is frequently expressed by the following expressions [5]:

$$L_{aa} = L_{bb} = L_{cc} = L_S + L_{leakage} \tag{4}$$

$$L_S = \frac{\mu_0 A N^2}{l_{air}} * N_{poles} \tag{5}$$

where A corresponds to the winding-turn area, N the number of turns, N_{poles} he number of poles in the generator, μ_0 the permeability of air and l_{air} represent the length of the magnetic flux in air, i.e., in the air gap.

To calculate the mutual inductance in a balanced three phase generator knowledge of the winding-design, i.e., how the different phase's linkage flux intersect, is required. In this case, the windings are separated by $\alpha = \pm \frac{2\pi}{3}$, and the mutual inductance can hence be calculated by [4]:

$$L_{ab} = L_{ac} = L_{bc} = L_S \cos\left(\pm \frac{2\pi}{3}\right) = -\frac{L_S}{2} \tag{6}$$

To identify the linkage flux from the permanent magnet, λ_{PM}, and thereby be able to calculate the mutual inductance between the stator and translator, L_{PM}, the value of the magnetic field density, B, in the air gap is utilized. The analytic expression for the term is:

$$B = \frac{B_r l_{PM}}{\mu_r \left(l_{air} + \left(\frac{l_{Fe} A_{air}}{\mu_{Fe} A_{Fe}} \right) + \left(\frac{l_{PM} A_{air}}{\mu_{r,p} A_{PM}} \right) \right)} \tag{7}$$

B_r and $\mu_{r,p}$ represent the remanence magnetization and the recoil permeability of the permanent magnet, l_{Fe}, l_{PM} and l_{air} represent the flux length in the highly magnetized part of the iron, in the permanent magnet and in the air gap and the terms A_{air} A_{Fe} and A_{PM} are the active cross sections areas [6].

The flux linkage and the mutual inductance between the stator and translator, L_{PM} can thereby be calculated according to

$$\lambda_{PM} = N = N \int B dA \tag{8}$$

$$L_{PM} = \frac{I_{PM}}{\lambda_{PM}} \tag{9}$$

The current term, I_{PM}, represent the permanent magnet's equal current density, given by B_r, l_{PM} and the permeability of air and the permanent magnet, μ_0 respectively μ_r, see Equation (10) [7].

$$I_{PM} = \frac{B_r l_{PM}}{\mu_0 \mu_r} \tag{10}$$

With the amplitude and known cycle of the term L_{PM} known, the induced voltage can be calculated by:

$$U = L_{PM} \frac{dI_{PM}}{dt} \tag{11}$$

Due to the relative motion between the stator and translator, the external surrounding for both the winding and the magnets changes continuously. To investigate how the different inductance terms are affected by the change of the external factors, for example the magnetic flux or the change of the reluctance in the magnetic circuit etc, simulations in a finite element simulation tool have been performed.

NUMERICAL MODEL

To be able to apply the finite element method, the area of investigation, in this case the two dimensional model of the generator, is divided into small regions, called elements. It is assumed that the unknown quantities, for example the vector potential A, in each element can be described by a simple function, i.e., a polynomial. The order of the polynomial determines the order of the element [8]. In the finite element simulation tool the authors have chosen to work with, each element is describe as a second order triangular, see Figure 2. Each element is thereby described by a second order polynomial, see Equation (12).

$$A = a + bx + cy + dxy + ex^2 + fy^2 \tag{12}$$

The xand yterm define the element's position in the two dimensional xy-plane, and the terms a to f are the unknown amplitudes of the investigated quantities.

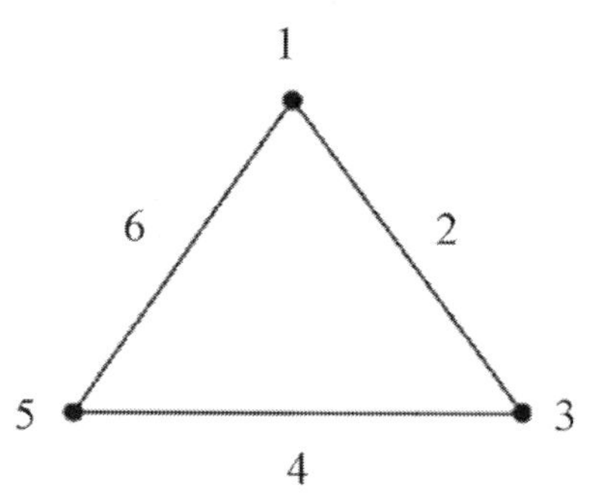

Figure 2. The second order element.

The physical interpretation of the second order equation is that the unknown quantities are assumed to only vary quadratic-ally in each direction [8].

The working approach of the chosen finite element simulation tool, is to define the vector potential, A, in each element. By evaluating the vector potential in each of the six points, see Figure 2, all six coefficients in the element's polynomial is found. Each element's potential can thereby be approximately calculated by Equation (12). When the vector potential in all elements is known, the magnetic flux in the generator can be calculated by [6]:

$$\Phi = \int (\nabla \times A) \mathrm{d}\boldsymbol{a} = \boldsymbol{A} \cdot \mathrm{d}\mathbf{s} \tag{13}$$

The current density in a conducting region, J_z, can be expressed as [6]:

$$\nabla^2 A_z = -\mu_0 J_z \tag{14}$$

With the unknown parameters defined, the simulations are straight forward.

THE MACHINE MODEL

To have a unique solution, the model must be limited by boundary conditions. These conditions may be describing the continuity or points, lines and surfaces in the model with known values of potential and field [8]. The single boundary condition the authors have chosen to work with, presented in Figure 3, states that there is no fringing of the magnetic field at infinity.

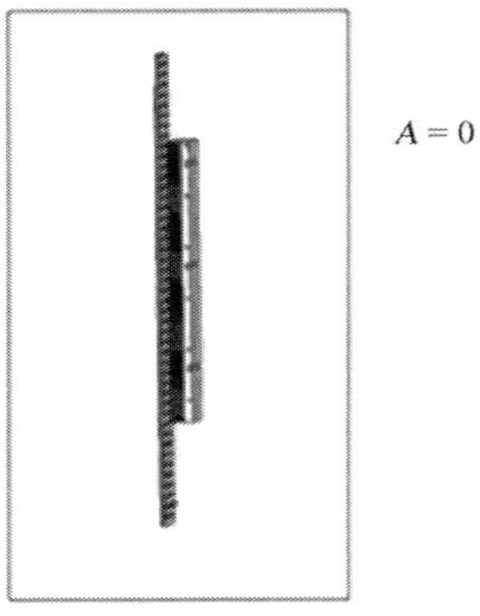

Figure 3. The chosen boundary condition.

Figure 4. a) Interval I, b) Interval II.

With a translator stroke length of 2 m, 1 m in each direction, the active area changes after half of the stroke length, and the generator's operation is divided in to two intervals. During Interval I, a magnet simultaneously slips in and out of the stator, resulting in a constant active stator area. During Interval II, presented in Figure 4(b), the active area decreases as the outmost magnet slips out of the stator. During Interval II, the total flux linkage in the machine is thereby not only varying due to the phenomena with changing amplitude of the flux linkage from the magnet trough the winding area, but also due to a varying size of the active area.

RESULTS

The calculated values, resulting from the analytical model, of the different parameters are presented in Table 2 together with a measurement on an existing prototype of the studied generator.

To investigate how the calculated values vary due to the relative motion between the stator and translator, finite element simulations were performed. The following presented results are from simulations, performed with a constant speed of the translator at 0.7 m/s with the generator operating at No Load. Figure 5 presents the inductance in the generator during Interval I. By Equation (11) the induced No Load voltage for phase a) was calculated, see Figure 6.

Table 2. The results from the analytic calculations and measured on the real machine.

Parameter	*Value*
I_{PM}	2380 A
$L_{PM,Calculated}$	10.3 mH
$L_{S,Calculated}$	19.5 mH
$L_{Mutual,Calculated}$	9.8 mH
$L_{S,Measured}$	18.5 mH

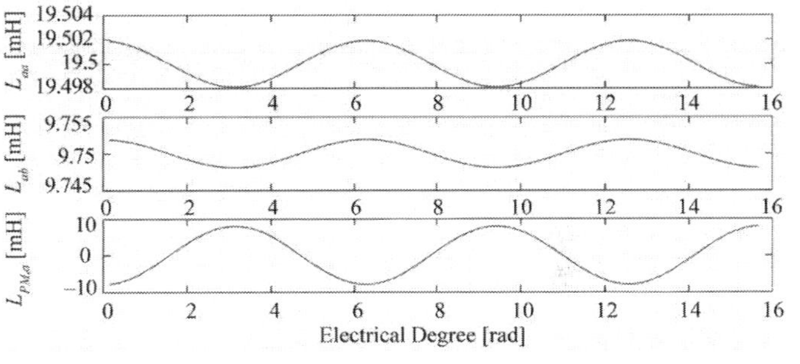

Figure 5. The upper graph presents the varying self inductance for one phase, the middle graph presents the varying mutual inductance each phase is exposed to from the two other phases, whereas the third graph presents the varying mutual inductance between the stator and translator, in terms of phase a.

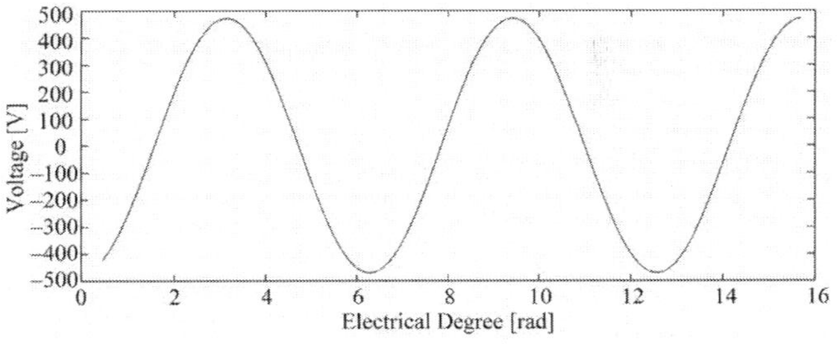

Figure 6. The induced No Load voltage for one phase at a constant speed of 0.7 m/s.

Let us take a look at what happens during Interval II. As the translator slips out of the stator, the active area and thereby the linked flux decreases. The part of the winding which is no longer covered by magnets loses the magnets flux linkage. Figure 7 presents the inductance terms in the generator during Interval II. By Equation (11) the induced No Load voltage for phase a) was calculated, and is presented in Figure 8.

DISCUSSION

The used surface mounted permanent magnets have a relative permeability close to one, i.e., the same as air. The length, see Equation (5), for the flux in low permeability material does thereby neither decrease nor increase if the air gap decrease or increase. The winding's selfand mutual inductance is thereby not affected by the air gap, which is strength by the presented result. The calculations, simulations and experimental measurement indicate together that the winding's selfand mutual inductance is highly not affected by the translator's position, i.e., the distance the flux transfers in the low permeability material is neither shorter or longer despite the change of the translator's position. However, the authors don't think this is the case if the low permeability permanent magnets are replaced with higher permeability-material, for example pole shoes of steel. The length for the flux in low permeability material can thereby be increased or decreased by increase or decrease the air gap. The designer can thereby by external parameters reach the required value, within values, on the inductance term.

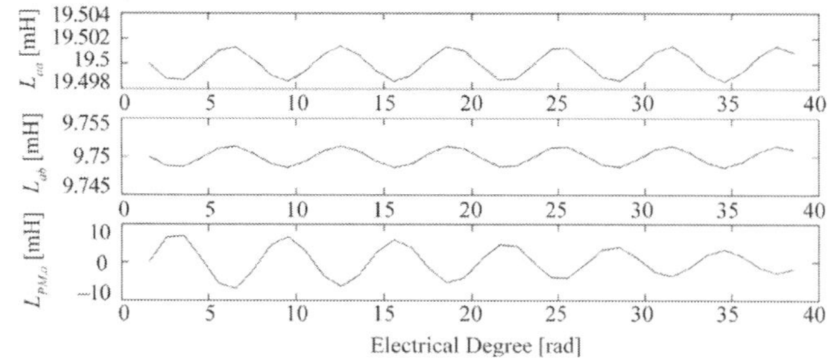

Figure 7. The upper graph presents the varying self inductance for one phase, the middle graph presents the varying mutual inductance each phase is exposed to from the two other phases, whereas the third graph presents the varying mutual inductance between the stator and translator.

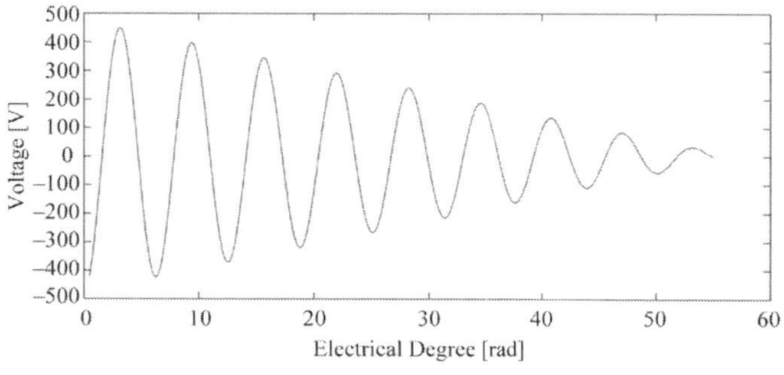

Figure 8. The calculated induced voltage for one phase.

The mutual inductance between the stator and translator changes drastically both during one electrical period and with the active area. The value is, as expected, highly affected by the decreasing active area. During this paper, the reluctance in the iron in the magnetic circuit is neglected, with the motivation of the iron's high permeability compared to air. Furthermore, the almost identical results of the measured and calculated value of the windings self-inductance, convinced the authors to neglect the winding's leakage inductance.

The decision to present the mutual inductance between the stator and translator as a signal oscillating around zero, Figures 5 and 7, was made to show the sign of the induced voltage. By this decision the direction of the coordinate system can be kept fixed during the full electric cycle. The physical meaning of a negative inductance does, of course, not exist, but the authors consider this representation to have pedagogic merits.

During the presented study the authors investigated different methods to examine the inductance terms of the machine, for example the Energy Perturbation Method, see [9,10]. However, the reduced opportunity to perform accurate analytical calculations were the reason the authors did not include the method in this paper.

CONCLUSION

An analysis of the varying inductance term in the ABC reference frame for a permanent magnet linear generator has been presented in this paper. The analysis is based on analytical calculations that have been compared to experimental measurements, and simulations in a finite element method program. The analytical results of the winding's self inductance showed very good agreement with the experimental measurements on the physical machine, thus serving as a reminder that time costly simulations are not always necessary. The numerical simulations were used to study the inductance terms during translator movement. As expected, it has been shown that the mutual inductance between the stator and translator is very much affected by the translator's position. On the other hand, the results show that the winding's selfand mutual inductance is barely affected by the air gap or the translator's position. This is attributed to the non-changing distance that the flux travels in the low permeability material despite changes of the external parameters. The conclusion is that only the design of the winding and stator affects the size of the different winding inductance terms.

ACKNOWLEDGEMENTS

This research was carried out as part of the Statkraft Ocean Energy Research Program, sponsored by Statkraft (www.statkraft.no). This support is gratefully acknowledged. The authors would also like to thank The Swedish Research Council for their financial support.

The authors would like to thank Cecilia Boström for her revision on spelling and grammar and a special thanks to Hjalmar Nyström for his contribution on the mechanical design and construction of the generator.

REFERENCES

1. L. Szabó, C. Oprea, I.-A. Viorel and K. Á. Biró, "Novel Permanent Magnet Tubular Linear Generator for Wave Energy Converters," Electric Machines & Drives Conference, IEMDC, IEEE International, Antalya, 3-5 May 2007, pp. 983-987. [Paper reference:1]
2. A. F. De O. Falcao, "Wave Energy Utilization: A Review of the Technologies," Renewable and Sustainable Energy Reviews, Vol. 14, No. 3, October 2009, pp. 899-918. [Paper reference:1]
3. M. Leijon, O. Danielsson, M. Eriksson, K. Thorburn, H. Bernhoff, J. Isberg, J. Sundberg, I. Ivanova, E. Sjöstedt, O. Ågren, K. E. Karlsspm and A. Wolfbrand, "An Electrical Approach to Wave Energy Conversion," Renewable Energy, Vol. 31, No. 9, July 2006, pp. 1309-1319. doi:10.1016/j.renene.2005.07.009 [Paper reference:1]
4. R. Escarela-Perez, E. Campero-Littlewood and A. L. Laureano-Cruces, "Steady-State Inductance Calculation of a Turbine Generator in the ABC Reference Frame," International Conference on Electrical and Electronics Engineering, Mexico City, 5-7 September 2007, pp. 233-236. [Paper reference:2]

5. F. Wu, X-P. Zhang, P. Ju and M. J. H. Sterling, "Modeling and Control of AWS-Based Wave Energy Conversion System Integrated Into Power Grid," IEEE Transactions on Power System, Vol. 23, No. 3, August 2008, pp. 1196-1204. [Paper reference:3]

6. K. Nilsson, O. Danielsson, and M. Leijon, "Electromagnetic Forces in the Air Gap of a Permanent Magnet Linear Generator at No Load," Journal of Applied Physics, Vol. 99, No. 3, 2 006, pp. 034505-034505-5. [Paper reference:3]

7. B. Bolund, "Electric Power Generation and Storage Using a High Voltage Approach," Ph.D. Dissertation, Acta Universitatis Upsaliensis, Uppsala, 2006. [Paper reference:1]

8. S.J. Salon, "Finite Element Analysis of Electric Machinery," IEEE Computer Applications in Power, Vol. 3, No. 2, 1990, pp. 29-32. doi:10.1109/67.53227 [Paper reference:3]

9. S. R. Shaudhry, S. Ahmed-Zaid and N. A. Demerdash, "Coupled Finite-Element/State Space Modeling of Turbogenerators in the ABC Frame of Reference—The No Load Case," IEEE Transactions on Energy Conversion, Vol. 10, No. 1, March 1995, pp. 56-62.

10. T. W. Nehl, F. A. Fouad, N. A. Demerdash, "Determination of Saturated Values of Rotating Machinery Incremental and Apparent Inductance by an Energy Perturbation Method," IEEE Transactions on Power Apparatus and System, Vol. PAS-101, No. 12, December 1982, pp. 4441-4451.

CHAPTER 17 DOUBLY FED INDUCTION GENERATOR

Ahmad M. Alkandari, S. A. Soliman, and Mansour H. Abdel-Rahman

Keywords: Asynchronous Operation, Double Feed Induction Generator (DFIG), Wind Power

INTRODUCTION

The growing demand of energy in industrialized world and environmental problems determined some important decision at political level that consider even more important to improve the percentage of energy produced by renewable sources. It is well-known the Kyoto Protocol and other important regulations that take care of environmental problems. Also the European Parliament approved the document 2001/77/CE on the "promotion of electrical energy produced by renewable sources" at-tending that before 2020 the 22% of total consumed energy should be produced by renewable sources. In this frame a lot of efforts are devoted to increase the efficiency of the generation systems based on wind, sun, hydro and biomass.

Wind power seems to be one of the most interesting technologies, especially considering the developments in the last decade. The electrical energy generation by wind depends on different factors, in particular the wind speed and the characteristics of the wind turbine generator [1]. The unpredictability of the wind power, that is variable with time, determines fluctuating power outputs of wind energy conversion systems. This variable power generation nature requires a careful analysis, design, and management of the generation system so that various approaches have been developed to study wind turbine generator behaviors.

There are different technical solutions that have been set up for different cases, for wind turbines in a range from less 1 kW to as large as 3 MW or more, to obtain the maximum efficiency and reliability. Traditionally the wind power generation has used fixed -speed induction generators that represent a simple and robust solution, then variable speed turbines have been considered be-cause they give higher energy, allow an extended control of both active and reactive power, and present less fluctuation in output power. The main solution adopted to realize wind generators can be divided, by electrical topology point of view and depending on the power, in the following categories [2]:

- Standard squirrel cage induction generator directly connected to the grid;
- Wound-rotor induction generator with variable rotor resistance;
- Doubly fed induction generator;
- Synchronous or induction generator with full-size power converter.

Both induction and synchronous generators can be used for wind turbine systems. Induction generators can be used in a fixed-speed system or a variable-speed system, while synchronous generators are normally used in power electronic interfaced variable-speed systems. Mainly, three types of induction generators are used in wind power conversion systems: cage rotor, wound rotor with slip control by changing rotor resistance, and DFIG. The cage rotor induction machine can be directly connected into an AC system and operates at a fixed speed or uses a full-rated power electronic system to operate at variable speed. The wound rotor generator with rotor resistanceslip control is normally directly connected to an AC system, but the slip control provides the ability of changing the operation speed in a certain range. The DFIG pro-vides a wide range of speed variation depending on the size of power electronic converter systems. In this chapter we first discuss the systems without power electronics except the thyristor soft starter, and then discuss the variable-speed wind turbine systems, including those with partially rated power electronics and the full-scale power electronic interfaced wind turbine systems.

In fixed -speed wind turbines, the generator is directly connected to the mains source grid. The frequency of the grid determines the rotational speed of the generator and thus of the rotor. The generator speed depends on the number of pole pairs and the frequency of the grid. The scheme consists of

Citation: A. Alkandari, S. Soliman and M. Abdel-Rahman. "Doubly Fed Induction Generator," Energy and Power Engineering, Vol. 3 No. 4, 2011, pp. 393-400. DOI: 10.4236/epe.2011.34050.

a squirrel-cage induction generator (SCIG), connected via a transformer to the grid. The wind turbine systems using cage rotor induction genera-tors almost operate at a fixed speed (variation of 1% - 2%). The power can be limited aerodynamically by stall control, active stall control, or by pitch control. The ad-vantage of wind turbines with induction generators is the simple and cheap construction. In addition, no synchronization device is required. These systems are attractive due to cost and reliability, but they are not fast enough (within a few ms) to control the active power. There are some other drawbacks also: the wind turbine has to operate at constant speed, it requires a stiff power grid to en-able stable operation, and it may require a more expensive mechanical construction in order to absorb high mechanical stress since wind gusts may cause torque pulsations in the drive train and the gearbox. Other dis-advantages with the induction generators are high starting currents and their demand for reactive power. They need a reactive power compensator to reduce (almost eliminate) the reactive power demand from the turbine generators to the grid. It is usually done by continuously switching capacitor banks following the production variation (5 - 25 steps) [2].

In variable-speed systems the generator is normally connected to the grid by a power electronic system. For synchronous generators and for induction generators without rotor windings, a full-rated power electronic system is connected between the stator of the generator and the grid, where the total power production must be fed through the power electronic system. For induction generators with rotor windings, the stator of the genera-tor is connected to the grid directly. Only the rotor of the generator is connected through a power electronic system. This gives the advantage that only a part of the power production is fed through the power electronic converter. This means the nominal power of the converter system can be less than the nominal power of the wind turbine. In general the nominal power of the converter may be 30% of the power rating of the wind turbine, enabling a rotor speed variation in the range of 30% of the nominal speed. By controlling the active power of the converter, it is possible to vary the rotational speed of the generator and thus of the rotor of the wind turbines [2].

Doubly-fed induction machines can be operated as a generator as well as a motor in both sub-synchronous and super synchronous speeds, thus giving four possible operating modes. Only the two generating modes at sub-synchronous and super-synchronous speeds are of interest for wind power generation.

In a DIFG the slip rings are making the electrical connection to the rotor. If the generator is running super-synchronously, electrical power is delivered to the grid through both the rotor and the stator. If the generator is running sub-synchronously, electrical power is delivered into the rotor from the grid. A speed variation of ±30% around synchronous speed can be obtained by the use of a power converter of 30% of nominal power. Furthermore, it is possible to control both active (P_{ref}) and reactive power (Q_{ref}), which gives a better grid performance, and the power electronics enables the wind turbine to act as a more dynamic power source to the grid. The DFIG system does not need either a soft starter or a reactive power compensator. The system is naturally a little bit more expensive compared to the classical systems However, it is possible to save money on the safety mar-gin of gear and reactive power compensation units, and it is also possible to capture more energy from the wind. [2]

In Reference [1] the case of variable speed dual-excited synchronous machine is investigated. For its control possibilities, this machine is particularly suitable for variable speed constant frequency applications like wind power generation systems. The mathematical model of the wind turbine and the generator is presented a control technique for output power maximization is proposed. It has been demonstrated that using the proposed control algorithm it is possible to achieve a significant increment on the power generated in comparison with that obtained using an unregulated induction machine.

Reference [3] presents a dynamic model of an important contemporary wind turbine concept namely a doubly fed (wound rotor) induction generator with a voltage source converter feeding the rotor. This wind turbine concept is equipped with rotor speed, pitch angle and terminal voltage controllers. It was shown that it is possible to develop a set of equations describing the behavior of the wind turbine. Furthermore, controllers for the rotor speed, the pitch angle and the terminal voltage were developed. The behavior of the system was investigated using two measured wind sequences.

Reference [4] presents the modal analysis of a grid connected to a DFIG. The change in modal properties for different system parameters, operating points, and grid strengths are computed and observed. The results offer a better understanding of the DFIG intrinsic dynamics, which can also be useful for control design and model justification.

Vector control is already applied to the DFIG control, which makes the DFIG gain good performance in the wind energy capturing operation. But in the two traditional vector control schemes, the stator magnetizing cur-rent is considered invariant in order to simplify the rotor current inner-loop controller. The two schemes are capable of performing very well when the grid is in normal condition [5]. However, when the grid disturbance, such as grid voltage dip or swell fault, occurs, the control performance will be getting worse, the rotor over current will occur, which seriously reduce the ride-through ability of the DFIG wind energy generation system. Based on the accurate model of the DFIG, the deficiency of the traditional vector control is deeply investigated. The improved control schemes of the two typical traditional vector control schemes used in DFIG is proposed, and the simulation study of the proposed and traditional control schemes is carried out. The validity of the proposed modified schemes to control the rotor current and to improve the ride-through ability of the DFIG wind energy generation system is proved in Reference [6] by the comparison study.

The growing integration of wind energy into power networks will have a significant impact on power system stability. With the development of Wind Turbine (WT) techniques, the DFI becomes the dominant WT type used in wind farms [6]. In this situation, DFIG should be modeled properly in power system stability analysis. A detailed model of the WT with DFIG and its associated controllers is presented, based on which the small signal stability model is derived. Small signal stability analysis shows that the DFIG control can significantly improve the stability of WT system. Applying a set of optimized controller parameters, the stability can be further enhanced [7].

This paper presents the steady state analysis of DFIG, where we assume that the three phase induction machine is connected to work as a generator for wind turbine farms, the machine is excited in the rotor circuit by a slip frequency current comes from an exciter mounted on the same shaft, with the main generator, so that the speed of the resulting flux in the air gap is always synchronous speed independent of the wind speed. Effects of the ex-citation voltage as well as the excitation voltage angle on the total active and reactive power, when the machine runs at constant speed are studied in this paper. Further-more, effects of the shaft speed on the active and reactive power at constant excitation voltage and phase angle are also studied.

STEADY STATE ANALYSIS OF DFIG

In the steady state analysis we assume a three-phase slip-ring induction machine, where the rotor is excited from a slip frequency exciter mounted on the same shaft of the rotating machine, and the stator is connected to the power grid having a constant voltage and constant frequency. The excitation voltage is expressed in a phasor form as $V_f Ðγ$. If the core resistance is neglected, then the equivalent circuit in p.u is given in Figure 1.

The loop voltage equations in per unit are given by:

$$V = I_s\left(R_1 + jX_1\right) + I_r\left(jX_m\right) \tag{1}$$

$$V_f\angle\gamma = I_s\left(jSX_m\right) + I_r\left(R_2 + jSX_1\right) \tag{2}$$

Solving Equations (1) and (2) yields

$$I_S = \frac{\left(VR_2 + V_f X_m \sin\gamma\right) + j\left(SVX_r - V_f X_m \cos\gamma\right)}{\left(R_s R_r - SX_s X_r + SX_m^2\right) + j\left(SR_s X_r + R_r X_s\right)} \tag{3}$$

$$I_r = \frac{(V_f R_s \cos\gamma - V_f X_s \sin\gamma) + j(V_f X_s \cos\gamma + R_s V_f \sin\gamma - SX_m V)}{(R_s R_r - SX_s X_r + SX_m^2) + j(SR_s X_r + R_r X_s)} \tag{4}$$

while the rotor current is given by Equation (4).

Let us define the parameters *a*, *b*, *c*, *d*, *e* and *g* as

$$\begin{aligned}
a &= \left(VR_r + V_f X_m \sin\gamma\right);\\
b &= \left(SVXr - V_f X_m \cos\gamma\right);\\
c &= \left(R_s R_r - SX_s X_r + SX_m^2\right);\\
d &= \left(SR_s X_r + R_r X_s\right)\\
e &= \left(V_f R_s \cos\gamma - V_f X_s \sin\gamma\right);\\
g &= \left(V_f X_s \cos\gamma + R_s V_f \sin\gamma - SX_m V\right).
\end{aligned} \tag{5}$$

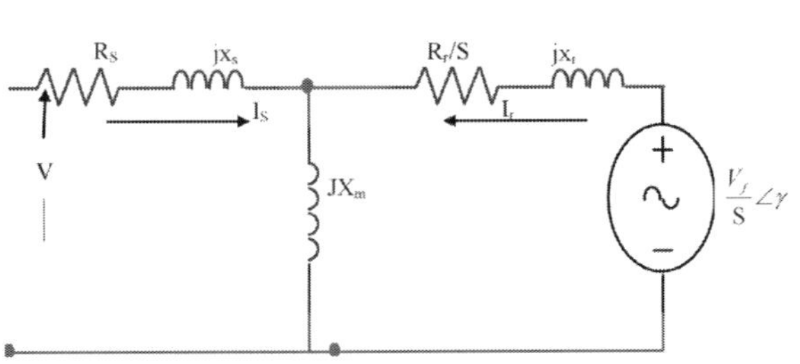

Figure 1. Steady state equivalent circuit of DFIG referred to the stator.

Then the stator and rotor currents can be written as:

$$I_s = \frac{a + jb}{c + jd}$$
$$I_r = \frac{e + jg}{c + jd} \tag{6}$$

The active and reactive power of the stator can be obtained as:

$$\begin{aligned}
S_S &= VI_S^* = P_S + JQ_S\\
P_S &= \text{Real}\left[VI_s^*\right]\\
Q_S &= \text{Im}\left[VI_S^*\right]
\end{aligned} \tag{7}$$

Substituting for I_s then the active and reactive power of the stator is n by:

$$\begin{aligned}
P_S &= \frac{V\left(ac - ad\right)}{c^2 + d^2}\\
Q_S &= \frac{-V\left(bc - ad\right)}{c^2 + d^2}
\end{aligned} \tag{8}$$

Note that the stator active power is positive for m toring operation and is negative for generating operation and vice versa for the reactive power.

The expressions for active and reactive power of the rotor (exciter) are given as:

$$\begin{aligned}
S_r &= V_f\angle\gamma I_r^* = P_r + JQ_r\\
P_r &= \text{Real}\left[V_f\angle\gamma I_r^*\right]\\
Q_r &= \text{Im}\left[V_f\angle\gamma I_r^*\right]
\end{aligned} \tag{9}$$

Substituting I_r into (9) we obtain

$$\begin{aligned}
P_r &= \frac{V_f\left(ec + gd\right)\cos\gamma + V_f\left(cg - ed\right)\sin\gamma}{c^2 + d^2}\\
Q_r &= \frac{V_f\left(ec + gd\right)\sin\gamma - V_f\left(cg - ed\right)\cos\gamma}{c^2 + d^2}
\end{aligned} \tag{10}$$

Note that, if the excitation voltage V_f equals zero, the machine performs in accordance with the basic operating characteristics similar to a conventional induction machine. Obviously, no power active or reactive is generated. Equation (10) describes the steady state operation of a doubly feed induction machine.

STEADY STATE CHARACTERISTICS

In studying the steady state characteristics we assume that the stator of the main machine is connected to an infinite bus source, and the primary circuit of the exciter is connected to the same source as such the output of the exciter is a source of a slip frequency and injects power to the slip rings of the rotor of the main machine. As such the current in the stator of the main machine has a constant frequency independent of the shaft speed.

Effects of Excitation Voltage Magnitude

In this study we assume a constant excitation voltage is applied to the machine rotor while the machine speed changes from zero to double the synchronous speed, *i.e.* S varies from 1 to −1. The parameters of the machine used in this study are given in Tables A.1 and A.2 [7].

Figure 2 gives the results obtained for the stator active power when the slip is changed from −1 to 1, *i.e.* speed changes to double the synchronous speed, while we changed the excitation voltage magnitude from 0 to 0.6 p.u in step of 0.2 p.u. Examining these curves reveals the following:

- At V_f = 0, induction motor operation, the machine delivers power to the network, via the stator windings, at speed greater synchronous speed, slip greater than −0.05. This is obviously correct, since the machines runs as an induction generator
- At excitation voltage magnitude of 0.2 and 0.4 p.u the machine delivers power to the source at this speed ranges.
- For excitation voltage magnitude of 0.6, the machine delivers power to the source at sub-synchronous speed, speed less than synchronous speed.

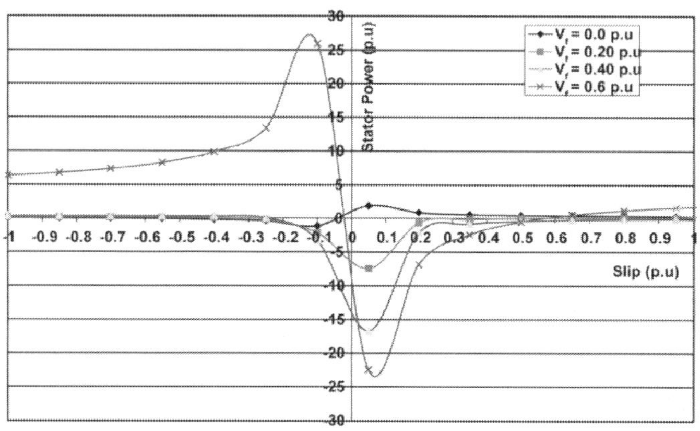

Figure 2. Stator active power versus speed at constant excitation voltage.

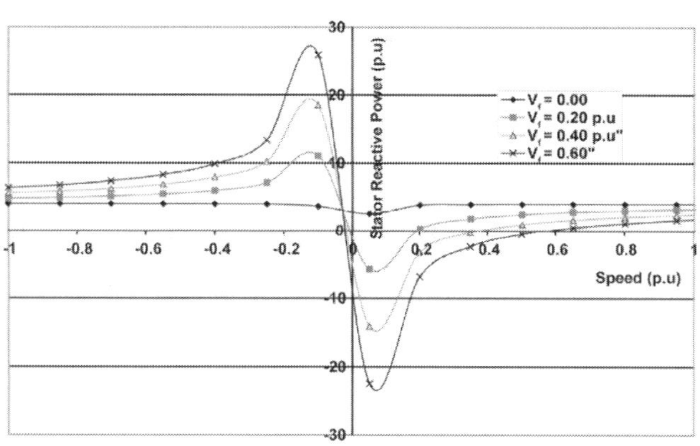

Figure 3. Stator reactive power versus the speed at constant excitation voltage.

Figure 3 gives the results obtained for the stator reactive power versus the speed at a constant excitation voltage magnitude ranging from V_f = 0.0 p.u to V_f = 0.60 p.u. Examining these curves reveals the following:

- •At sub-synchronous the machine absorbs reactive power from the network, while at super-synchronous speed the machine delivers reactive power to the network.
- The maximum absorbed or delivered reactive power to the network occurs at the same speed.
- At the same speed, the reactive power increases as the excitation voltage increases.

Figure 4 depicts the active power of the rotor at different excitation voltage magnitude when the machine speed varies from zero to double the synchronous speed.

Examining this table reveals that:

- The rotor is absorbed active power from the source at all speeds regards the excitation voltage magnitude.
- The maximum power absorbed at different excitation voltage magnitudes occurs at the same speed.
- When the excitation voltage is zero, the rotor is short circuited, no power is absorbed by the rotor (induction motor operation).

Figure 5 shows the variation of the rotor reactive power at different excitation voltage magnitudes, when the speed varies from zero to double the synchronous speed.

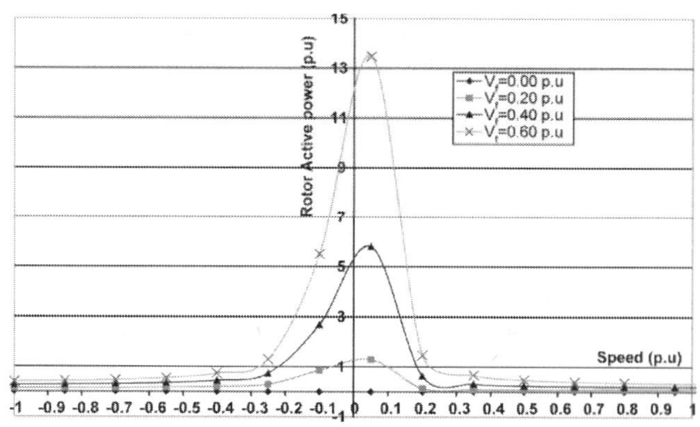

Figure 4. Rotor active power versus the speed at different excitation voltage.

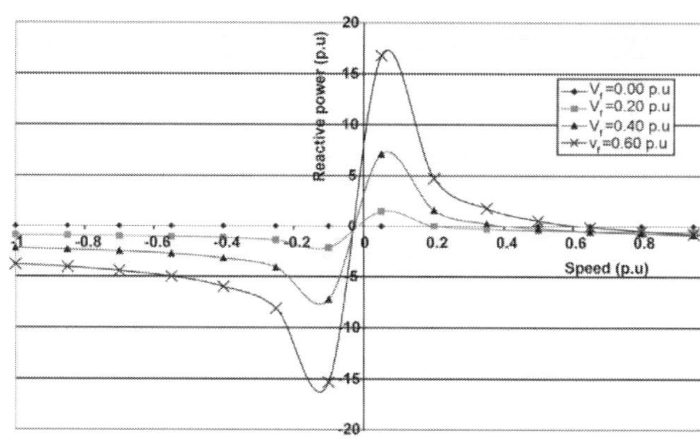

Figure 5. Rotor reactive power versus speed at different excitation voltage.

It can be noticed, from this figure, that:

- At sub-synchronous speed the machine absorbs reactive power from the source, while at super-synchronous speed the machine delivers reactive power to the network, (asynchronous capacitor) at different excitation voltage magnitudes, and the maximum of these reactive powers occurs at the same speed.

Effects of the Excitation Voltage Angle

In this section the characteristics of the machine is studied, where we assume that the machine run at super-synchronous speeds and we change the excitation voltage angle, the angle between the excitation voltage and the source voltage, from –90° to 90° while we kept excitation voltage magnitude constant at $V_f = 0.2$ p.u. Figure 6 shows the results obtained for the stator active power. Examining this curve reveals that:

- At synchronous speed, $s = 0.0$, the machine behaves as a synchronous machine, generator, and delivers power to the network, (P, δ,) characteristics.
- At super-synchronous speed the machine delivers power to the network for positive excitation voltage angle and absorbs power from the network for negative excitation voltage angle.
- The maximum delivers powers occur at different excitation voltage angles.

Figure 7 gives the same characteristics but the ma-chines runs above synchronous speed. Examining this curve reveals the following remarks:

- When the machine runs just speed 20% more than the synchronous speed, it delivers power to the network for the range of negative excitation voltage angle and little bit positive excitation voltage angle.
- The maximum delivered power occurs at different excitation voltage angle, negative angle, for every super-synchronous speed.
- The machine delivers power to the network at this rang of speeds, although the excitation voltage angle is zero.

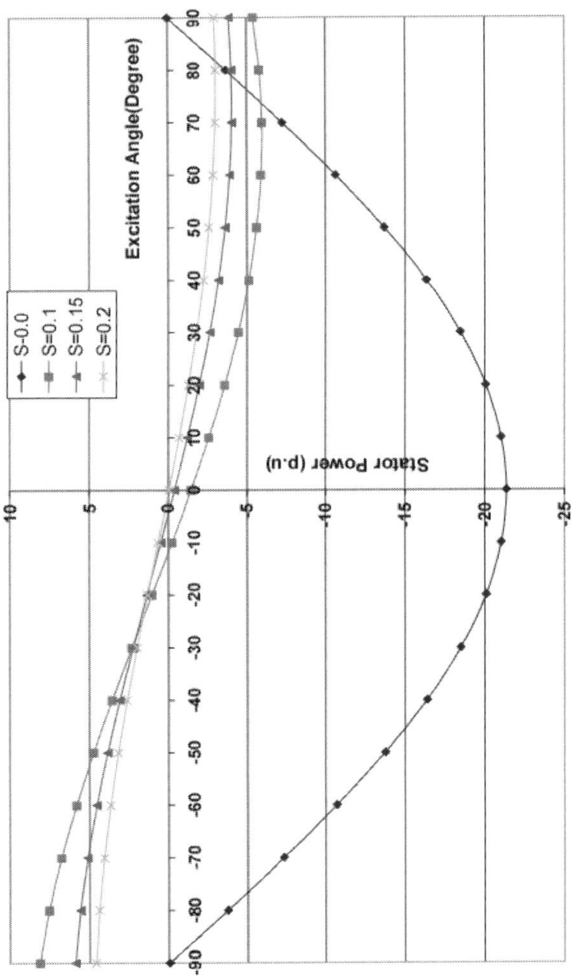

Figure 6. Stator power versus excitation angle at constant speed.

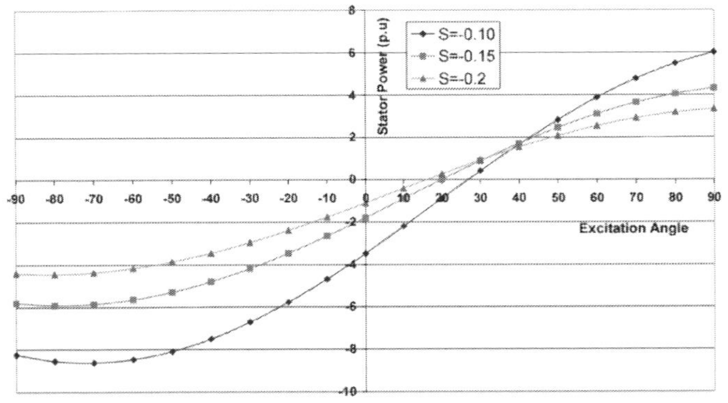

Figure 7. The stator power versus the excitation angle when the machine above synchronous speed, $V_f = 0.2$ p.u.

Figures 8 and 9 give the excitation power for sub-synchronous and super-synchronous speed operation, respectively, and the excitation voltage magnitude is kept constant at $V_f = 0.2$ p.u

Examining these curves reveals the following conclusions:

- At synchronous speed, $S = 0.0$, the excitation power is constant regardless the magnitude of the excitation voltage angle.
- The excitation power having the same trend regard-less the speed of the machine, sup or super synchronous speed.
- The machine delivers power to the network at almost negative excitation voltage angle and at sub-synchronous and super-synchronous speed, respectively.
- The maximum positive or maximum negative powers occur at different excitation voltage angle.

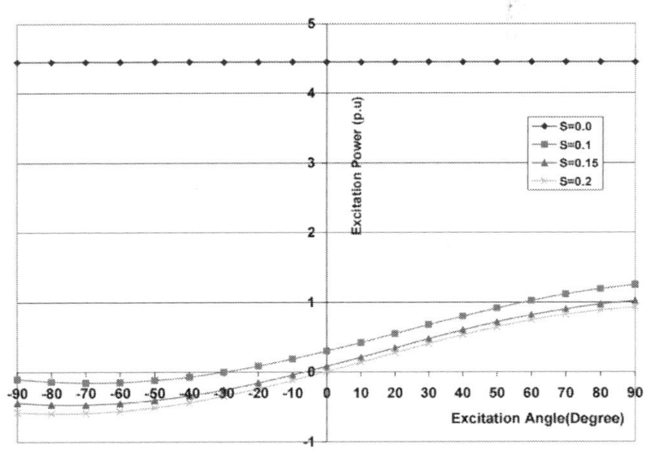

Figure 8. Excitation power versus excitation angle.

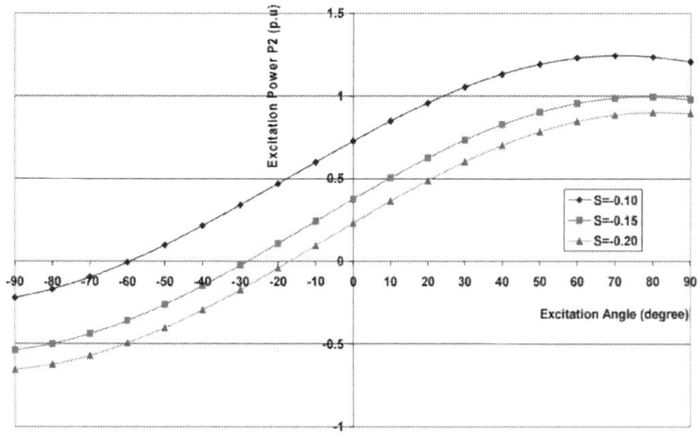

Figure 9. Excitation power versus excitation angle when the machine runs above synchronous speed, $V_f = 0.2$ p.u.

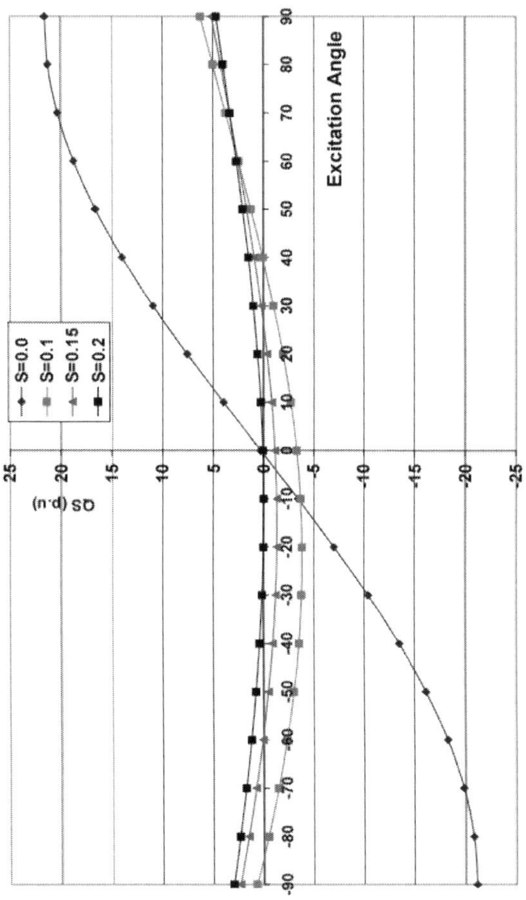

Figure 10. Stator reactive power versus excitation angle at $V_f = 0.2$ p.u.

Figures 10 and 11 give the stator reactive power variation with the excitation voltage angle, when the ma-chine runs at sub-synchronous and super-synchronous speeds, respectively, while the excitation voltage magnitude is kept constant at $V_f =$ 0.20 p.u. Examining these figures reveals that:

- At synchronous speed, $S = 0.0$, the machine be-haves as a non-salient pole synchronous machine and the excitation voltage angle becomes the rotor angle, *i.e.* the power angle.
- The machine absorbs and delivers reactive power to the network at different excitation voltage angle (Note that positive reactive power means absorption, inductive reactive power, while negative reactive power means delivering, capacitive reactive power).
- The maximum stator reactive power either positive or negative occurs at the same excitation voltage angle for every speed in the range.

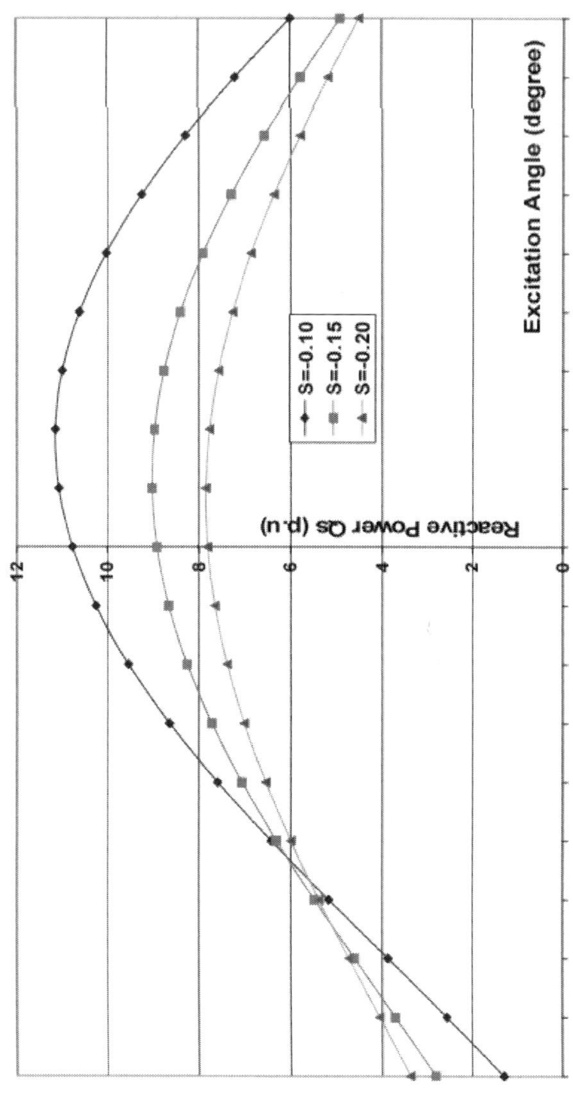

Figure 11. Stator reactive power when the machine runs above syn-chronous speed, $V_f = 0.2$ p.u.

Figure 12 shows the net active power delivered to the network versus the excitation voltage angle when the excitation voltage magnitude is kept constant, while the machine runs at super-synchronous speed. And the excitation voltage magnitude is kept constant at 0.2 p.u. Examining this figure reveals the following remarks:

When the machine runs at synchronous speed it de-livers more power to the network than the other speed.

At super-synchronous speeds the machine delivers power to the network for positive excitation voltage angle and absorbs power for negative excitation voltage angles.

At sup-synchronous speeds the machine delivers the maximum power at the same excitation voltage angle independent of the machine shaft speed.

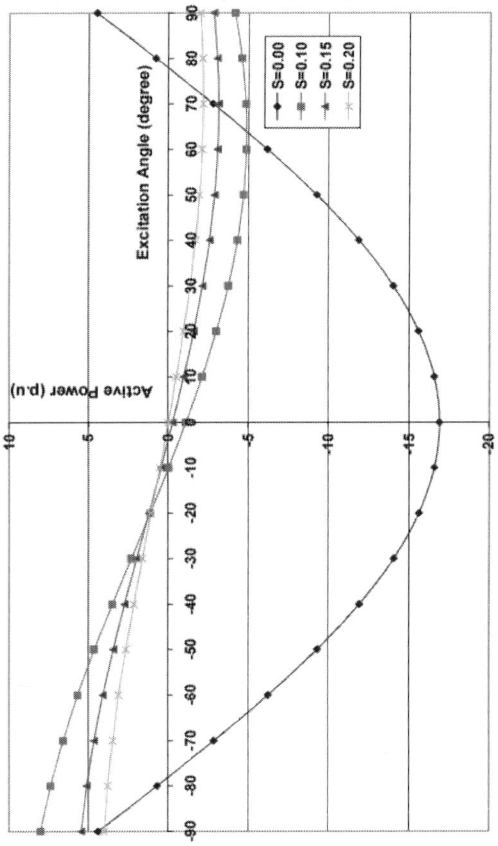

Figure 12. Total active power (p.u) versus excitation angle.

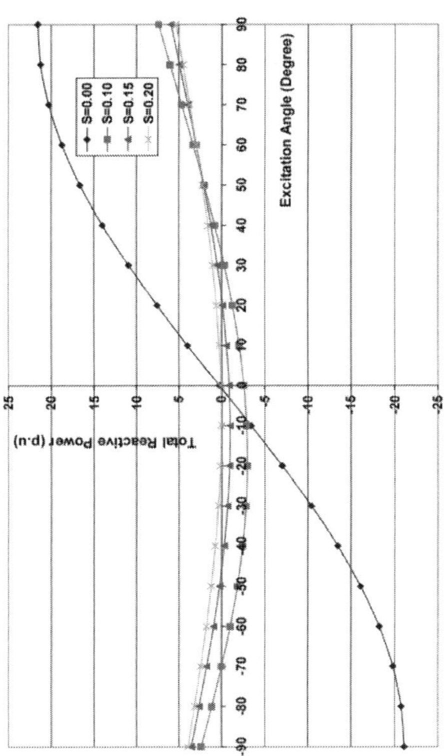

Figure 13. Total reactive power versus excitation angle.

Figure 13 depicts the variation of the total reactive power versus the excitation voltage angle, when the ex-citation voltage magnitude is kept constant at 0.2 p.u, while the machine runs at sup synchronous speed. It can be noticed from this figure that:

When the machine runs at synchronous speed, the machine delivers reactive power to the source, when the excitation voltage angle varies from zero to –90°, and absorbs the same magnitude of the reactive power when the excitation voltage angle varies from zero to 90°. *i.e.* the curve is inversely symmetrical at excitation voltage angle equals zero.

At speeds rather than the synchronous speed, the machine absorbs and delivers reactive power to the net-work for different modes of excitation voltage angles.

The maximum delivery of reactive power occurs at the same excitation voltage angle independent of the machine speed, except at synchronous speed.

CONCLUSIONS

In this paper the steady state analysis of DFIG is developed, we assume that the machine is excited on the rotor side by a slip-frequency current injected from an exciter mounted on the same shaft of the machine. The resulting rotating magnetic field rotates at synchronous speed. Effects of the excitation voltage magnitude and angle on both the active and reactive power when the machine runs at constant speed are investigated. It has been shown that controlling the excitation voltage magnitude and phase angle controls the mode of operation of the machine; motor or generator mode. Furthermore, effects of the shaft speed on the active and reactive power at constant excitation voltage magnitude and constant phase angle are also investigated. It can be concluded that the power angle stability of conventional synchronous machine has no meaning in this type of generators. Studying the control of the machine terminal voltage is under study and the results will appear in a forthcoming paper.

NOMENCLATURE

I_s, I_r are the stator and rotor current, respectively.

R_s and R_r are the stator and rotor resistance, respectively.

X_s & X_r are the total stator and rotor reactance, respectively.

$$X_s = x_s + X_m$$

$$X_r = x_r + X_m$$

X_m = Magnetizing reactance

x_s & x_r are Leakage reactance of the stator and rotor respectively.

S is the machine slip $\text{slip} = \frac{n_s - n}{n_s}$, *S* is positive at sub-synchronous speed $n < n_s$ and negative at super synchronous speed $n < n_s$ respectively.

V is the grid voltage a constant frequency f_s.

V_f is the excitation voltage magnitude, having constant frequency f_s.

g is the angle of the excitation voltage.

APPENDIX A.

Table A.1. Parameters of the induction machines.

Stator resistance	R_S	0.0022 Ω	0.010 p.u
Stator leakage Reactance	x_S	0.1200 mH	0.180 p.u
Rotor resistance	R_r	0.0018 Ω	0.009 p.u
Rotor leakage Reactance	x_r	0.05 mH	0.070 p.u
Magnetizing reactance	X_m	2.9 mH	4.400 p.u

Table A.2. Base values.

Base Voltage (Phase-neutral)	V_b	400 V
Base Current	I_b	1900 A
Base Frequency	ω_b	314 rad/s
Base Impedance	$Z_b = V_b/I_b$	0.21 Ω

REFERENCES

1. L. Piegari, R. Rizzo and P. Tricoli, "High Efficiency Wind Generators with Variable Speed Dual-Excited Synchronous Machines," *International Conference on Clean Electrical Power* 2007, Capri, 21-23 May 2007, pp. 795-800.
2. F. Blaabjerg and Z. Chen, "Power Electronics for Modern Wind Turbine," Morgan & Claypool Publisher, San Rafael, 2006.
3. J. G. Slootweg, H. Polinder and W. L. Kling, "Dynamic Modeling of a Wind Turbine with Doubly Fed Induction Generator," *IEEE Power Engineering Society Summer Meeting*, Vol. 1, 2001, pp. 644-649.
4. F. Mei and B. Pal, "Modal Analysis of Grid-Connected Doubly Fed Induction Generators," *IEEE Transactions on Energy Conversion*, Vol. 22, No. 3, 2007, pp. 728-736. doi:10.1109/TEC.2006.881080
5. Y. K. He, J. B. Hu and R. D. Zhao, "Modeling and Control of Wind-Turbine Used DFIG under Network Fault Conditions," *Proceedings of the 8th International Conference on Electrical Machines and Systems*, 2005, Nanjing, Vol. 2, 29 September 2005, pp. 986-991
6. F. Wu, X.-P. Zhang, K. Godfrey and P. Ju, "Modeling and Control of Wind Turbine with Doubly Fed Induction Generator," 2006 *IEEE PES Power Systems Conference and Exposition*, Atlanta, 29 October-1 November 2006.
7. A. Petersson, "Analysis, Modeling and Control of Doubly-Fed Induction Generators for Wind Turbines" Ph.D. Thesis, Division of Electric Power Engineering, Department of Energy and Environment, Chalmers University of Technology, Goteborg, 2005.

CHAPTER 18 HIGH EFFICIENCY DOUBLE–FED INDUCTION GENERATOR APPLIED TO WIND POWER GENERATOR TECHNICAL ANALYSES

Deng-Chern Sue

Keywords: Double-Fed Induction Generator, Cut-in Speed, Pushover Torque, Stall Regulator, Pitch Regulator, Inverter

INTRODUCTION

Taiwan is an island; with most of its energy imported. Because of the shortage of an indigenous energy supply in Taiwan, the international energy market has a strong effect on the local economy. Utilizing fossil fuels may cause global warming and temperature variation. The International Kyoto meeting requested to decrease the fossil fuel consumption and minimize the CO_2 emission to control the global warming effect. Therefore, American and European countries are concentrating on the study of renewable sources of energy, such as wind power, solar energy, ocean tides, hydro power biomass, etc., wherein the wind has a significant possibility of producing electric power.

The Taiwanese Government has set up "Sources of renewable energy policy" to increase the clean energy consumption and promote the renewable sources of energy.

This stable incentive policy creates the renewable sources of energy from the surrounding environment. It supports relative industrial development and promotes the renewable sources of energy for all applications.

Wind comes from the effects of solar and earth rotation. It causes airflow with kinetic energy. It is called wind power. Humanity has utilized wind-powered devices, such as sailboats, wind-powered water wheels, windmills, etc. In the late nineteenth century, a wind-powered machine was connected with a generator to produce power. Wind power becomes a main application for wind energy [1].

On Taiwan, the annual northeast monsoon produces strong winds along the coast, mountain and remote is-land areas. According to a recent investigation, over 2000 km^2 of land area has an annual average wind speed higher than 5 m/s. Wind energy has an excellent potential and the minimum recoverable wind energy is estimated at around 1000 MW. Taiwan has abundant wind energy.

Currently, the Peng-Hu Chung Tun Power Station and Taiwan Formosa Plastic Mai-Liao Power Station have installed wind-powered generators, which have operated successfully. However, Taiwan has a limited experience and few studies have been made on wind power. This study concentrates on the analysis and comparison of output characteristics for high efficiency Double-Fed Induction Generator and Traditional Generator.

WIND POWER EFFICIENCY

Rotation of the earth and solar radiation induce temperature differences at the earth's surface cause air circulation. Airflow in the horizontal direction produces wind. The wind velocity has the following units: m/s, km/hr, mph knots (nautical mile/hr), with the relationships as

$$1 \text{ m/s} = 3.6 \text{ km/hr} = 2.24 \text{ mph} = 1.94 \text{ knots}$$

In addition, the Beaufort scale can be used to identify the wind strength. The wind strength has 0 up to 17 grade (wind velocity from 0.3 m/s to 61.2 m/s). Wind pressure (F) is produced from airflow that is directly proportional to the cube (per formula) of its velocity.

Citation: D. Sue. "High Efficiency Double–Fed Induction Generator Applied to Wind Power Generator Technical Analyses," Energy and Power Engineering, Vol. 3 No. 3, 2011, pp. 253-261. DOI: 10.4236/epe.2011.33032.

$$F = \frac{1}{2} C_P \rho A U^2 \ \ (\text{kW}) \qquad (1)$$

where

A: cross-sectional area of wind stream (m^2)

ρ: air density (1.225 kg/m^3)

C_p: wind power coefficient

U: wind velocity (m/s)

The kinetic of wind pressure is called wind power (P), which is wind pressure (F) produced work per unit time,

$$P = \frac{1}{2} C_P \rho A U^3 \qquad (2)$$

For stall regulator, C_p is a l function and l is a ratio of the velocity at wind tip of regulator vs. wind velocity, i.e. l = $r\omega/V$ and r is the rotor radius of wind machine, w (rad/s) is angle velocity, Figure 1 is the C_p - l curve of various angles.

The power produced by a wind machine is directly proportional to its cross-sectional area, and the cube of the wind's velocity [2-4]. Therefore, the wind machine's location is the key factor of power production and power generating cost.

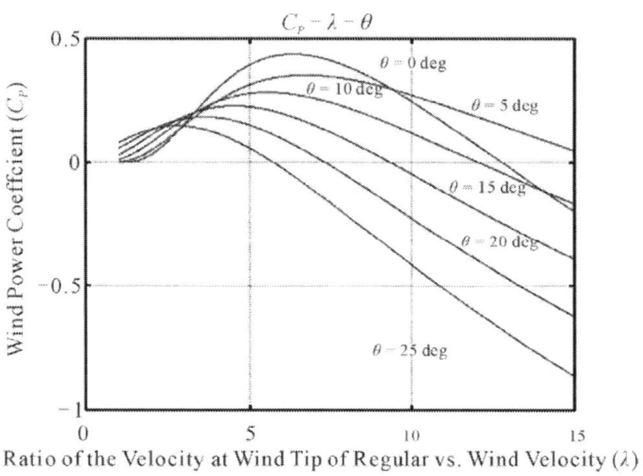

Figure 1. C_p-λ curve of various angles.

The wind machine operating at a pre-set range of wind velocities has to ensure the generator operates at rated speed. Means must be provided to shut down the genera-tor operating to protect the equipment if the wind velocity becomes excessive. But, the theoretical kinetic energy of the wind can only be partially utilized considering the overall efficiency of machine; only a part of wind power can be actually utilized.

CHARACTERISTICS OF WIND MACHINE

The kinetic energy of the moving air (wind) rotates the blades of the wind machine. The blades convert the wind energy to mechanical energy, which in turn, is converted to electrical energy by the generator. The power output of wind machine is affected by wind velocity, machine efficiency, blade design, blade pitch, cross-sectional area, etc. [1,2].

The wind machines type can be categorized into horizontal shaft and vertical shaft depends on shape and rotating shaft. The vertical shaft machine has a higher axial retardation factor, lower rotating speed, lower efficiency and requires more blade material, making it unsuitable for power generating use. The horizontal shaft machine can be categorized as facing upwind or downwind. The upwind machine uses a rear rudder (small machine) or universal mechanism to sense the wind direction and position the blades to produce maximum power. The downwind type uses the cone shape to minimize of the rotor to follow wind direction so the blades are facing the downwind side. Because the wind acts on the supporting structure of the wind machine and then on the blades, the resulting periodic fatigue loadings affects the machine operating life. In general, horizontal shaft wind machines are the upwind type, *i.e.*, blades are facing the wind. Figure 2 shows horizontal shaft wind machine type.

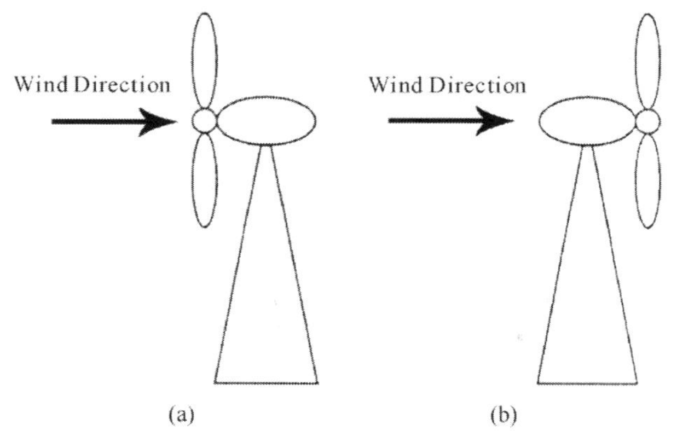

Figure 2. Horizontal shaft wind machine type. (a) Upwide; (b) Downwind.

The blade selection significantly affects the rotating speed and power output of machine. Currently, the major blades type is three-blade type. The three-blades has less loading variation more stable operation, decreased fatigue loadings, increased blade operating life compared to two blade and single blade units. In additional, the operation of three-blade is smooth and comfortable of vision, the two-blade units have an unstable rotation and single blade units are unbalance and uncomfortable. Therefore, the horizontal three-blades are selected for the current study of wind machines. Accordingly, the horizontal shaft, upwind and three-blade wind machines are evaluated as the best approach for an operating wind machine [1].

The power output is related to wind velocity or wind speed. When the wind speed matches the cut-in speed, the wind machine begins to produce power. From cut-in speed to the rated speed, the wind machine power output is directly proportional to the cube of the wind velocity. From the rated speed to cut-out speed, the power output remains at rated power. When the cut-out speed exceeded, the wind machine is shut down without power output to avoid any damage. The blades must be feathered to have no significant rotational velocity. Figure 3 shows wind velocity and power output of Vestas V47 -660 Type Wind Power Generator at rated power 660 kW, the cut-in speed, rated speed, and cut-out speed are 4, 15 and 25 m/s respectively. To maintain the rated power output for wind machine and without damaging the gearbox and/or generator, the power output of the wind machine has to be controlled. Control is separated into a stall regulator and a pitch regulator to control the amount of wind energy being absorbed and transferred to produce power [1,2]. Table 1 is the comparison of stall regulator and pitch regulator.

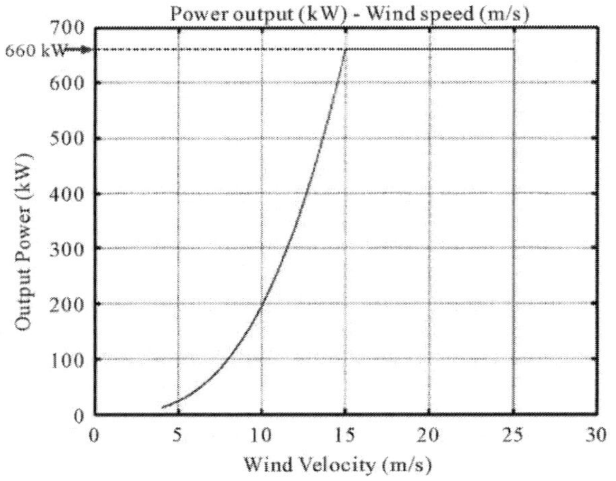

Figure 3. Wind velocity and power output of vestas V47-660 type wind power generator.

Table 1. Comparison of stall regulator and pitch regulator.

Item	Stall Regulator	Pitch Regulator
Output Characteristics	Has small range of output characteristic curve	Has large range of output characteristics curve
Constant Speed Control	Can meet control requirements	Has difficulty at high wind speed
Variable Speed Control	Can't meet	Has better power quality and lower transmission load compare to stall regulator
Start Wind Velocity	Higher wind velocity	Lower wind velocity
Safety	Needs breaker system to protect over speed	Without over speed problem
Structure	Structure and maintenance are easier	Structure and maintenance more complicated
Cost	Lower cost	Higher cost

TRADITIONAL WIND MACHINE OUTPUT CHARACTERISTICS

Currently, the major types of wind-powered generators are permanent magnet generator, synchronous generator, induction generator, and high efficiency double-fed generator [2,5]. The theory and output characteristics of the different types of wind-powered generators are explained as follows:

Permanent Magnet Generator

The rotor of the generator is made of permanently magnetic material and the rotor shaft is driven by the blades of the unit. When the rotating blades drive the magnetic rotor, a rotating magnetic field that interacts with the multiple windings of the stator. These rotating lines of magnetic force induce an alternating voltage in the windings. The alternating current induction voltage is

$$E_A = K\phi\omega \tag{3}$$

where

K: electrical coefficient of the generator

ϕ: rotor magnetic flux (Wb)

ω: angle velocity of rotor speed (rad/sec)

Because the rotating magnetic field is fixed and the rotor rotating speed varies with wind velocity, the output voltage and frequency are variable and cannot provide a stable power supply. The produced power must be run through a rectifier and an inverter to be compatible with a 60 Hz power system. Due to the small capacity and higher cost per kilowatt of output, this type is not suit-able for commercial applications. It is only suitable for remote areas without a grid power supply.

Synchronous Generator

A synchronous generator operates at the synchronous rotating speed of an alternating current system to which it is connected. A synchronous generator requires direct current to be supplied to the rotor winding via slipping to produce the rotor's magnetic flux. The prime mover (wind) drives the generator rotor forming a rotating magnetic field that induces a voltage in the stator windings of the unit. The windings of the stator are arranged so that a three-phase voltage is produced. Interactions of rotating magnetic field of synchronize rotation, which inducts a three-blade voltage. The induction voltage of alternating current is

$$E_A = K\phi\omega_{syn} \tag{4}$$

where

K: electrical coefficiency of generator

ϕ: rotor magnetic flux (Wb)

ω_{syn}: angular velocity of rotor rotating at synchronous speed

$\omega_{syn} = 2\pi f$ (rad/sec)

f: stator frequency, $f = n_m P/120$

n_m: rotor synchronize speed (rpm)

P: number of rotor poles

Output power (neglecting stator resistance) is

$$P_{out} = \frac{3V_T E_A}{X_s}\sin\delta \tag{5}$$

where,

V_T: voltage across generator terminals (volts)

X_s: synchronous reactance (Ω)

δ: phase angle difference between E_A and V_T (degree).

In general, the large power supply system needs to use two sets or more synchronous generators to combine into system for supply power. It has the advantages of in-creased system efficiency, decreased spare capacity, ease of unit maintenance, increased power supply reliability and meeting optimal dispatch, etc. Figure 4 shows out-put power and torque angle for raised pole and cylindrical rotor of synchronous generator.

Synchronous generator applies to wind-powered units when the rotating blades directly drive the rotor without any gearbox, avoiding the noise and wear of meshing drive gears. When the actual wind speed exceeds the design wind speed and the unit is operating at full load, a pitch control system is installed to adjust the blade angle to reduce the power extracted from the wind. Pitch control limits rotor speed and power output to eliminate overload/over speed damage.

The synchronous generator is same as the permanent magnet generator. The output voltage and frequency varies with the wind velocity. The wind machine normally rotates at 20 - 30 rpm compared to the maximum synchronous speed of 3600 rpm for a 60 Hz grid system.

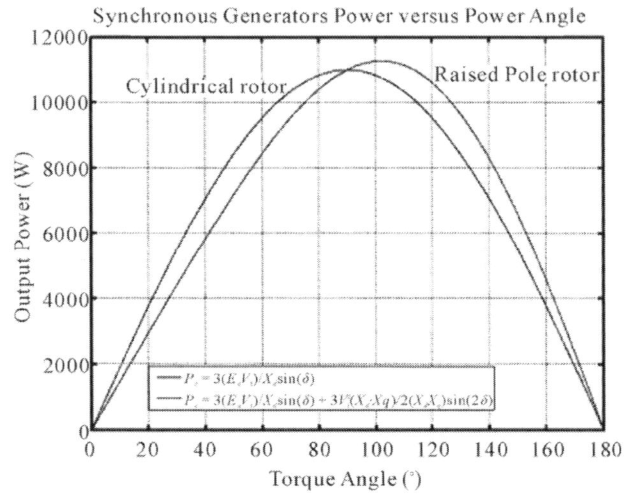

Figure 4. Output power characteristics of synchronous generator.

In order to make the wind machine's output compatible with the 60 Hz grid, the machine's output must pass through a rectifier to become a DC and then through an inverter to match the system frequency before being stepped-up via a transformer to feed into the power grid.

Induction Generator

The three-phase induction generator uses magnetic induction theory to transfer the electrical energy in the form of magnetic flux from stator to rotor, without any wire connection. Power from an external source energizes the stator, causing the rotor to turn, just like an induction motor. The rotating speed of the rotor is slightly lower than the rotating magnetic flux in the stator. This type generator is also called an asynchronous machine. When the rotor speed of an induction generator exceeds the speed rotating magnetic field in the stator, and the rotor direction is consistent with rotating magnetic field, the rotor will tend to pull the stator field faster. This action causes a reverse torque in the rotating direction, thus causing the induction generator to operate as a generator at the frequency and voltage of the initial power supply to the stator. In a properly designed machine, the magnetic link between the rotor and stator is strong enough to prevent the rotor going into over speed, regardless of the energy input from the blades. The definition of rotating difference ratio for an induction generator is

$$s = \frac{n_{syn} - n_m}{n_{syn}} \times 100\% \qquad (6)$$

where,

n_{syn}: synchronous speed of magnetic field (rpm)

n_m: rotor speed of rotor (rpm)

Figure 5 follows Kirchhoff's voltage law, the voltage loop equation for single-phase effective circuit of an in-duction generator is

$$V_s = \left[R_s + j\left(X_s + X_m\right)\right] I_S + jX_m I_r \qquad (7)$$

$$0 = jX_m I_s + \left\langle \frac{R_r}{s} + j\left[X_r + X_m\right] \right\rangle I_r \qquad (8)$$

In Equations (7) and (8), only the magnetic reactance, jX_m, is considered and the flux branch resistance of the core, R_c, is neglected. The conversion power and torque of an induction generator are represented as

$$P_{conv} = 3I_r^2 R_r \left(\frac{1-s}{s}\right) \qquad (9)$$

$$\tau_{ind} = \frac{P_{conv}}{\omega_m} \qquad (10)$$

Figure 6 shows the torque-speed characteristics curves of induction generators with differing rotor resistance values. The produced power is directly proportional to the torque applied to the stator by the blades and thus rotor, but if the input torque developed by the wind on the blades, is larger than pushover torque, the induction generator will over speed. Therefore, when applied to a wind-powered generator an adequate breaker, rotor resistance control, blade pitch control or similar devices to prevent high wind speed and/or excessive torque from inducing over speed of the blades and rotor [1].

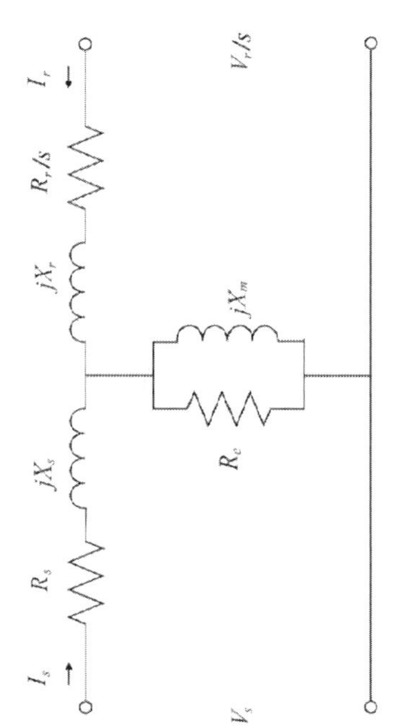

Figure 5. Effective circuit of a single phase of an induction generator.

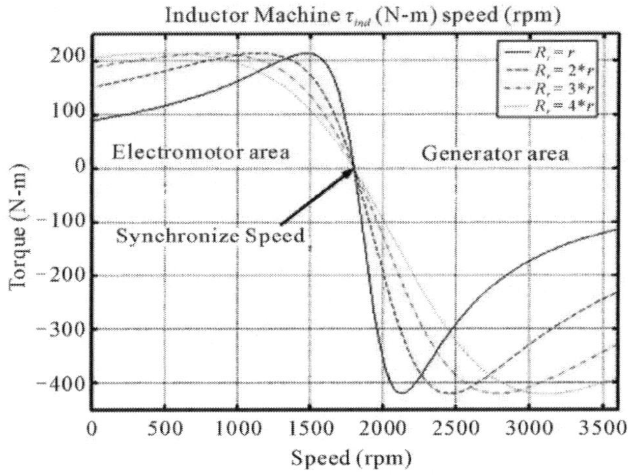

Figure 6. Torque-speed characteristic curve of an induction generator with different rotor resistance values.

The induction generator rotor is without a magnetic field circuit, to change the magnetic field to control the output voltage and is not self-exciting. The required reactive power for excitation needs is supplied by a connection from an external capacitor or from grid system. The excitation power controls the voltage at the generator terminals, which supply the power to grid system [5]. The advantages of induction generator are structural simplification; exciting field system is not required, unnecessary to synchronize operation, easy for operation and maintenance. When induction generator derives power from wind, the self-exciting type is used. A capacitor is installed across the terminals to provide the required reactive power for startup and improve the power factor during operation.

Comparison Synchronous Generator and Induction Generator

The stator construction is identical for a synchronous generator and an induction generator. The major difference is rotor design and construction. The winding of rotor of a synchronous generator need to have direct current (DC) for excitation to produce the rotor's magnetic field but the damper windings of the rotor of the induction generator need to be short-circuited to produce rotor magnetic field. Table 2 shows the comparison of synchronous and induction generators.

Table 2. Performance comparisons of synchronous and induction generators.

Items	Synchronous Generator	Induction Generator
Stator structure	Three-phase winding	Three-phase winding
Rotor structure	Has evidenced and controlled pole rotor winding need to connect to DC source	Has cage rotor, winding doesn't need to connect DC
Speed	Operating at synchronize speed	Operating at over-synchronous speed
Reactive power compensation	Not needed	From system or connect to capacitor
End voltage control	Cause exciting system control	Can't control
Converter device	Needed	Not needed
Maintenance	Complicate and difficult	Simple and easy
Cost	Expensive	Cheap

HIGH EFFICIENCY DOUBLE-FED INDUCTION GENERATOR

The disadvantages and limitations of the self-exciting induction generator include the difficulty to adjust the output voltage and frequency, the need to be operated at over-synchronous speed, the small range of output power, etc. Currently, new developments in the windings of the induction generator and vector control theory as applied to controlling the rotor input voltage and frequency can modulate the power input and output characteristics of the generator rotor. With the advantages of operating in an aubsynchronous speed range, output can exceed rated power; closer control of output voltage and frequency is obtained. Figure 7 shows a double-fed induction generator structure [2].

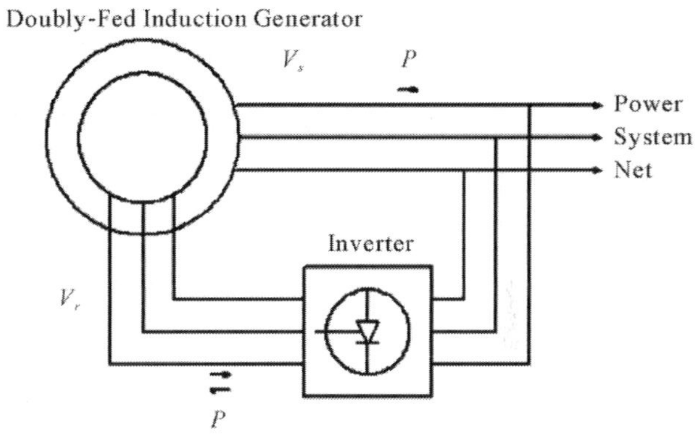

Figure 7. Structure of double-fed induction generator.

Model of Double-Fed Induction Generator

After transformation and d-q transformation of a three phase dynamic model for a double-fed induction generator [3,4], the equations of dynamic model can be written as:

$$V_{ds} = R_s I_{ds} + \frac{d\psi_{ds}}{dt} - \omega_s \psi_{qs}$$

$$V_{qs} = R_s I_{qs} + \frac{d\psi_{qs}}{dt} + \omega_s \psi_{ds} \quad (11)$$

$$V_{dr} = R_r I_{dr} + \frac{d\psi_{dr}}{dt} - \omega_r \psi_{qr}$$

$$V_{qr} = R_r I_{qr} + \frac{d\psi_{qr}}{dt} + \omega_r \psi_{dr}$$

$$\tau_m = \tau_e + J\frac{d\omega_r}{dt} + f\omega_r$$

$$\tau_e = \frac{M}{L_s}\left(\psi_{qs} I_{dr} - \psi_{ds} I_{qr}\right) \quad (12)$$

where

ω_r: angle speed of rotor ($\omega_r = s\omega_s$, rad sec)

$$\psi_{ds} = L_s I_{ds} + MI_{dr}$$

$$\psi_{qs} = L_s I_{qs} + MI_{qr}$$

$$\psi_{dr} = MI_{ds} + L_r I_{dr}$$

$$\psi_{qr} = MI_{qs} + L_r I_{qr} \quad (13)$$

Due to generator operating at steady condition, therefore, with further analysis, the term $\frac{d\psi}{dt}$ can be neglected, allowing the voltage of stator and rotor to be re-written as:

$$V_{ds} = R_s I_{ds} - \omega_s \left(L_s I_{qs} + M I_{qr} \right)$$
$$V_{qs} = R_s I_{qs} + \omega_s \left(L_s I_{ds} + M I_{dr} \right)$$
$$V_{dr} = R_r I_{dr} - \omega_r \left(L_r I_{qr} + M I_{qs} \right)$$
$$V_{qr} = R_r I_{qr} + \omega_r \left(L_r I_{dr} + M I_{ds} \right) \quad (14)$$

Related Equations of Vector Control

The output of double-fed induction generator is connected to power system; therefore, the voltage of shaft end is equal to the voltage of system. The vector control theory is to fix magnetic direction of stator to the vertical axis.

$$\psi_{ds} = \psi \text{ and } \omega_{qs} = 0$$
$$\tau_e = -\frac{M}{L_s} \psi_{ds} I_{qr} \quad (15)$$

The magnetic torque is only related with rotor current of cross-axis. Neglect wire resistance of stator, R_s the voltage of stator end and voltage of vertical axis-cross axis for double-fed induction generator can be written as:

$$V_{sn} \cong \frac{d\psi_{sn}}{dt}; \; n = a,b,c$$
$$V_{ds} = 0 \text{ and } V_{qs} = V_s \quad (16)$$

Use vertical axis-cross axis as reference base, rewrite the voltage, current and magnetic of stator end, as follow:

$$V_{ds} = 0$$
$$V_{qs} = V_s = \omega_s \psi_{ds}$$

$$\psi_{ds} = L_s I_{ds} + M I_{dr} = \psi_s$$
$$\psi_{qs} = L_s I_{qs} + M I_{qr} = 0$$

$$\psi_{dr} = M I_{ds} + L_r I_{dr}$$
$$\psi_{qr} = M I_{qs} + L_r I_{qr}$$

$$I_{ds} = \frac{1}{L_s}\left(\psi_s - M I_{dr} \right)$$

$$I_{qs} = -\frac{M}{L} I_{qr}$$

Derivative from the above equation, the magnetic field and voltage of rotor is

$$\psi_{dr} = \left[L_r - \frac{M^2}{L_s} \right] I_{dr} + \frac{M V_s}{\omega_s L_s}$$

$$\psi_{qr} = \left[L_r - \frac{M^2}{L_s} \right] I_{qr}$$

$$V_{dr} = R_r I_{dr} + \left[L_r - \frac{M^2}{L_s} \right] \frac{dI_{dr}}{dt} - s\omega_s \left[L_r - \frac{M^2}{L_s} \right] I_{qr}$$

$$V_{qr} = R_r I_{qr} + \left[L_r - \frac{M^2}{L_s} \right] \frac{dI_{qr}}{dt} + s\omega_s \left[L_r - \frac{M^2}{L_s} \right] I_{dr} + s\omega_s \frac{M V_s}{\omega_s L_s}$$

The real power and reactive power of the stator of a double-fed induction generator can be presented as:

$$P_s = V_{ds} I_{ds} + V_{qs} I_{qs} = -\frac{V_s M}{L_s} I_{qr}$$

$$Q_s = V_{qs} I_{ds} - V_{ds} I_{qs} = \frac{V_s \psi_s}{L_s} - \frac{V_s M}{L_s} I_{dr}$$

From the above derivative, the real power and reactive power of stator for the double-fed induction generator can be calculated. The current of vertical axis and cross axis for rotor can be separately controlled.

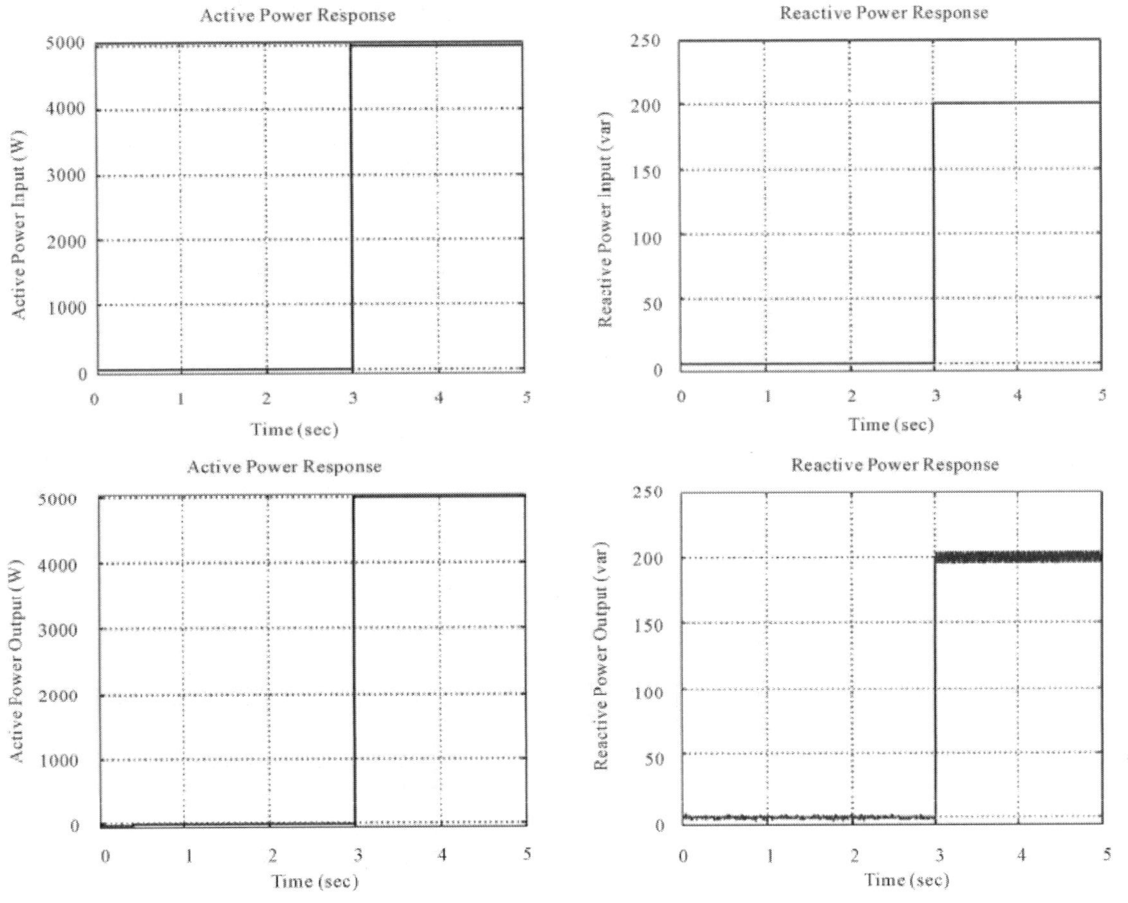

Figure 8. Simulation of response results for real power and reactive power.

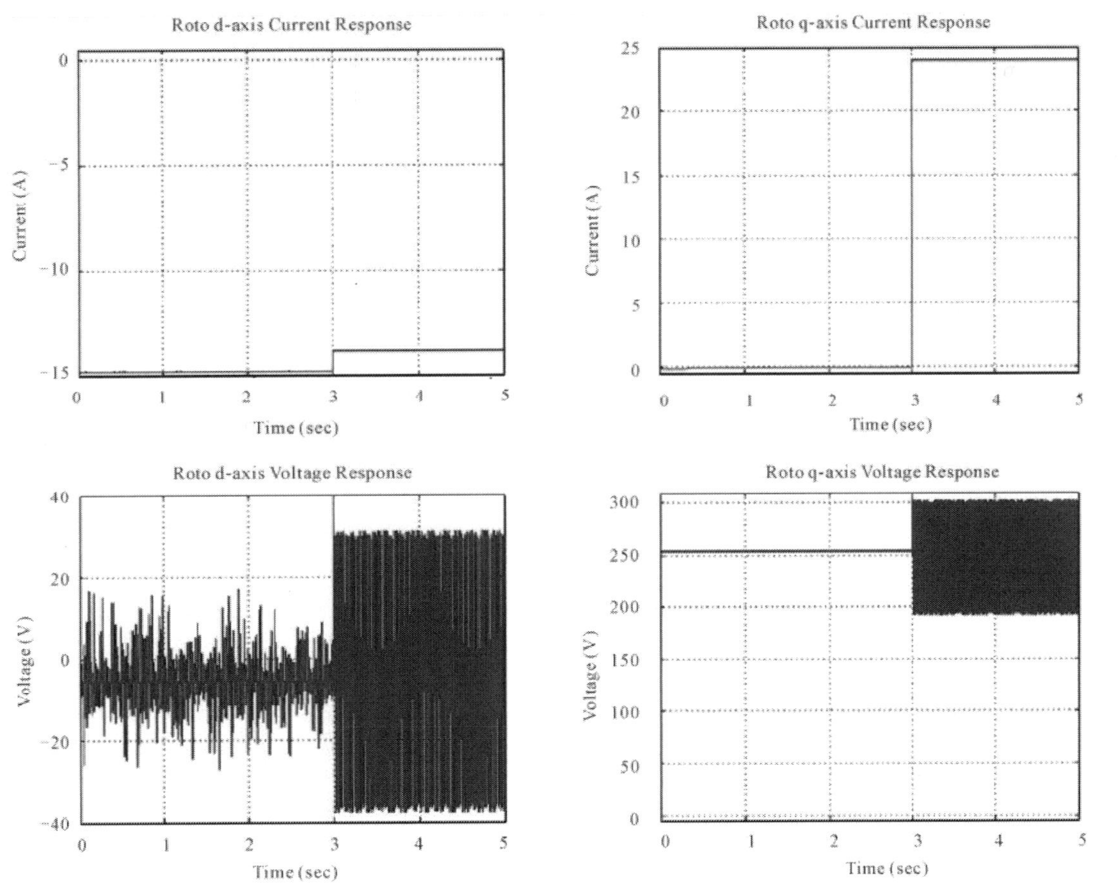

Figure 9. Simulation of response results of control current for rotor's vertical axis and cross axis.

Simulation Analysis

If the software MATLAB/Simulink is used to simulate the vector control response characteristic of Double-Fed Induction Generator, and the above theoretical derivative is solved using the following simulation parameters with the results shown on Figures 8 and 9:

$$R_s = 0.05\ \Omega,\ R_r = 0.05\ \Omega,\ M = 47.3\ \text{mH},$$
$$L_s = 50\ \text{mH},\ L_r = 50\ \text{mH}$$

The generator power output is 5000 W and speed is 3600 rpm. The power control setting of Double-Fed Induction Generator is 5000 W for real power and 200 VAR for reactive power, speed is 3420 rpm ($s = 0.05$).

From the above simulation results, the vector control of a high efficiency Double-Fed Induction Generator and the magnetic field of the stator is fixed on the vertical axis. The real power and reactive power components of the output only related to the relative strengths of the current of cross axis and the vertical axis of the rotor. The generator can operate at lower than synchronous speed, which does not limit to operate at over the synchronous speed. At over synchronous speeds, the stator and rotor produce power; therefore, the output is higher than the rated power and has an excellent characteristic of voltage and frequency adjustment.

CONCLUSIONS

From the above analysis and because suitable wind power is available in Taiwan, the following conclusions are derived:

- Use the vector control for the high efficiency Double-Fed Induction Generator, to control the output power of the wind-powered generator. Even if the wind velocity is unstable, it can rapidly adjust the current in the cross axis and vertical axis of the rotor to control the ratio of real power to reactive power to obtain the desired characteristics of the output power.
- The generator can be operated sub-synchronous speeds and output is higher than the rated power. It has excellent adjustment characteristics for voltage and frequency. Therefore, the high efficiency Double-Fed Induction Generator is an optimal selection for new wind powered generator applications.

REFERENCES

1. T. Burton, D. Sharpe, N. Jenkins and E. Bossanyi, "Wind Energy Handbook," John Wiley & Sons, Inc., Hoboken, 2001. doi:10.1002/0470846062
2. A. Petersson, "Analysis, Modeling and Control of Double-Fed Induction Generator," PhD Thesis, Chalmers University of Technology, Gothenburg, 2003.
3. L. Zhang and C. Watthanasarn, "A Matrix Converter Excited Double-Fed Induction Machine as a Wind Power Generator," 7th International Conference on Power Electronics and Variable Speed Drives, London, 21-23 September 1998, pp. 532-537. doi:10.1049/cp:19980583
4. J. G. Slootweg, H. Polinder and W. L. Kling, "Dynamic Modelling of a Wind Turbine with Double Fed Induction Generator," Power Engineering Society Summer Meeting, Vancouver, 2011, pp. 644-649. doi:10.1109/PESS.2001.970114
5. P. Kundur, "Power System Stability and Control," McGraw-Hill, Inc., New York, 1994.

CHAPTER 19

PERFORMANCE IMPROVEMENT SYNCHRONOUS MOTOR

V. Chandrasekaran, D. Nagarajan, and B. Nagaraj

Keywords: Double Winding Synchronous Motor (DWSyM), Double Winding Synchronous Generator (DWSyG), Performance Improvement, Energy Conservation, Twin Generator

INTRODUCTION

Synchronous generators are the major source of electrical power conversion to meet the existing demands. The excitation of synchronous generators is one of the main challenges in the power system. The field windings are energized through an Automatic Voltage Regulator (AVR) with control and measuring units which provides satisfactory operation with little disturbances like power swing. This can be overcome using additional control over the excitation system [1] [2]. Leakage flux can be reduced with double disc generator and by providing Permanent Magnets and field coils on the shaft [3] [4]. In order to eliminate the need of brushes, a three-phase rotor winding provides excitation current even from standstill. The control of the excitation current at different speeds is also made possible. Brushless excitation system and static excitation system are two methods employed to feed the field winding of synchronous machine [5] [6]. Double winding in the stator was recommended in the year 1930 for turbine alternators. In this machine, when stator is energized, the revolving magnetic field established in the air gap is used to develop three-phase EMF in one set of winding at the same time, and a mechanical power is also developed as conventional induction motor [7]. For energy conservation, induction machine was suggested with two sets of three-phase windings and these windings are energized based on shaft load thereby eddy current loss and copper losses are reduced. Improvement in the power factor is also observed due to this type of excitation [8]. Dual stator winding machines may be considered for various motor and generating applications with more flexible energy conversion. Both armature windings are of the same pole number, which may cause circulating currents due to unbalanced power supply [9]. Dual winding design modification can be carried out during rewinding process to reduce the cost of the machine in align with annual energy consumed [10]. From the above references, the various excitation methods of synchronous generators are studied. Energy conservation machines consist of two sets of three-phase windings in its stator. Based on the load demand, the windings are energized to minimize the losses.

In this paper an attempt is made in the design of synchronous machine in which armature consists of two set of three-phase windings. When operated as synchronous motor, the power factor and efficiency are slightly decreased for a given excitation. In the suggested model, since armature has two windings, a three-phase EMF is developed in the second set of winding, by connecting an external electrical load the overall performance of the motor is improved for the given excitation. While operated as synchronous generator, due to the two windings in the stator, the power and lighting loads can be separated, due to which the interruption to the lighting circuit can be minimized. In this mode of operation, the machine can be considered as "Twin Generator" due to two distinct three-phase outputs.

CONSTRUCTIONAL DETAILS

Construction of Armature

Stationary armature rotating field type of synchronous machine is presented in this discussion. The armature consists two double layers three-phase winding distributed in the stator core. Rotating field consists of excitation winding and slip ring arrangement. For the maximum utilization of developed power, it is suggested that, the two set of windings may be arranged with different phase angles, however for the optimum power output, shaft angle of zero degree or 60 degree may be considered. In the proposed machine zero degree displacement is provided between the two sets of windings. The stator

Citation: Chandrasekaran, V., Nagarajan, D. and Nagaraj, B. (2016). Performance Improvement Synchronous Motor. Circuits and Systems, 7, 2865-2874. DOI: 10.4236/cs.2016.710245.

windings of a double winding motor can be arranged with different shift angles between them. In double winding induction motor, shift angle of 60 degrees or zero degrees are the best choice [11]. For As a proof for the discussion, a 3 kW, 415 V, 4-pole, 3-phase Double Winding Synchronous Motor (DWSyM) has been considered and tested.

Design Details

Armature of DWSyM consists of two sets of three-phase windings. By adjusting electrical and mechanical loads, current flow in the machine is maintained such a way that the machine does not exceed its thermal capacity. This of design is suitable for lower capacity synchronous machines. For conventional three-phase winding the slot utility factor is around 25% where is in DWSyM it is about 43.3% which ensures the better utility of slots. The armature and field construction of DWSyM is shown in Figure 1.

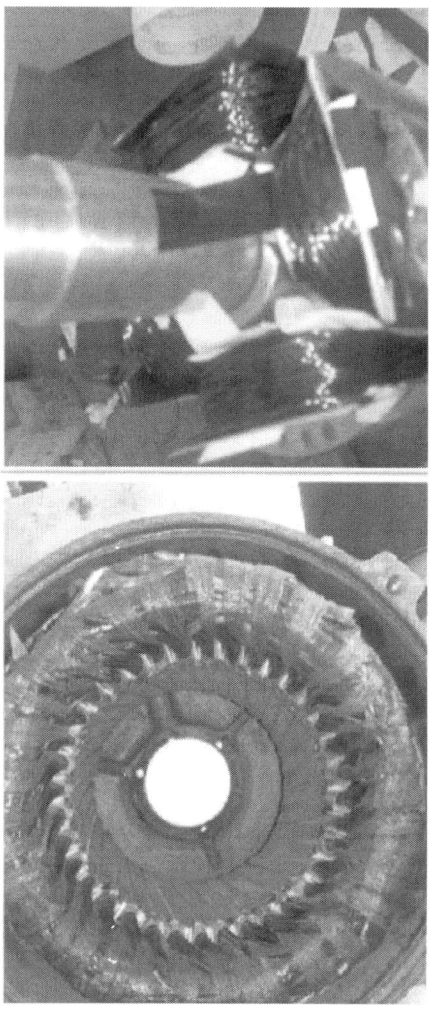

Figure 1. Double winding synchronous motor.

Design details of DWSyM

Armature design

Number of poles = 4

Synchronous speed = 1500 rpm

Diameter of the core = 0.139 m

Length of the core = 0.11 m

Number of turns per phase = 215

Electrical loading = 18,000 A/m^2

Magnetic loading = 0.44 wb/m^2

Stator consists of two sets of double layers windings with the same number of turns and cross sectional area.

Current density chosen is 6.0 A/mm^2.

Field Design

Filed MMF = 1254 A

Excitation voltage = 60 V

Number of turns = 1127

Current density = 6.0 A/mm^2

EXPERIMENTAL INVESTIGATIONS

The current and flux density waveforms help to calculate iron losses of electrical machine. Numerical method using finite element analysis is used for calculation. This type of testing is suitable for synchronous and DC machines [12]. Control strategy with vector controller helps to compensate core loss component ripple measurement [13]. The testing arrangement of DWSyM is shown in Figure 2. Experimental set up shows the mechanical loading through brake drum arrangement and electrical loading with lamp load and loading rheostats. The main focus of the testing is to observe the performance of DWSyM for the various excitation conditions by loading both the windings.

Initially load test was carried out using brake drum arrangement for the various filed currents while second set of winding is left free. Torque, mechanical output and efficiency are observed. Another set of testing was carried out in which for different excitations in which second winding is also loaded using loading rheostat and lamp loads, the performance of the machine is observed.

Table 1 shows the reading observed with rated field excitation with 60 V and field current of 0.85 A. It is observed that the maximum power factor is 0.75, efficiency is 78% for the load current of 6 A. The minimum power factor is observed as 0.64 for the load current of 3 A, the efficiency of the machine for this load is 83%. The efficiency and power factor characteristics are shown in Figure 3& Figure 4.

Table 2 shows the reading observed with rated field excitation with 50 V and field current of 0.72 A. It is observed that the maximum power factor is 0.66 and efficiency is 84.3% for the load current of 4.5 A. The minimum power factor is observed as 0.5 for the load current of 3 A, the efficiency of the machine for this load is 83%. The efficiency and power factor characteristics are shown in Figure 5& Figure 6. When the excitation is reduced, it is observed that the power factor of the machine decreases.

Figure 2. Experimental setup.

Table 1. Conventional brake test with excitation 60 V, field current 0.85 A.

Input Voltage (V)	Line Current (A)	Input Power (W)	Output Power (W)	% Efficiency	Power Factor
415	2.4	240	0	0	0.13
415	3.0	1400	1162	83.0	0.64
415	3.5	1840	1540	83.6	0.73
415	4.0	2080	1727	83.0	0.72
415	4.5	2400	2025	84.3	0.74
415	5.0	2680	2250	83.1	0.74
415	5.5	2840	2371	83.4	0.71
415	6.0	3240	2556	78.0	0.75

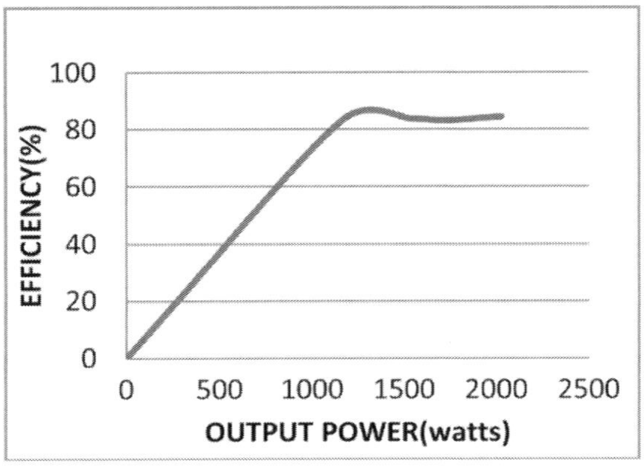

Figure 3. Efficiency characteristics.

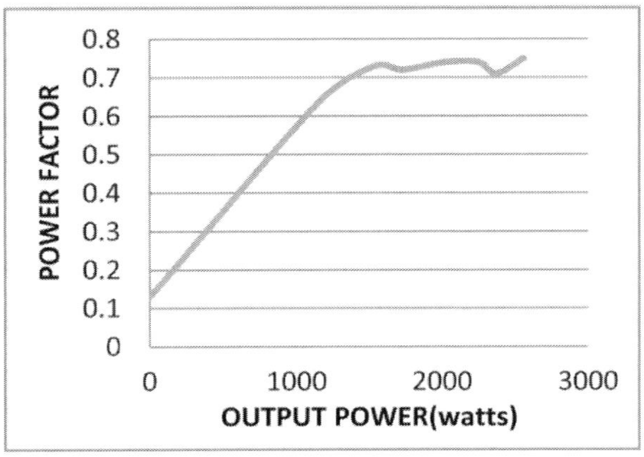

Figure 4. Power factor characteristics.

Table 2. Conventional brake test with excitation 50 V, field current 0.72 A.

Input Voltage (V)	Line Current (A)	Input Power (W)	Output Power (W)	% Efficiency	Power Factor
415	2.5	440	0	0`	0.24
415	3.0	1080	618	83.0	0.50
415	3.5	1520	1047	83.6	0.60
415	4.0	1840	1384	83.0	0.63
415	4.5	2160	1816	84.3	0.66
415	5.0	2400	1954	83.1	0.66
415	5.5	2600	2033	83.4	0.65
415	6.0	2800	2264	78.0	0.64

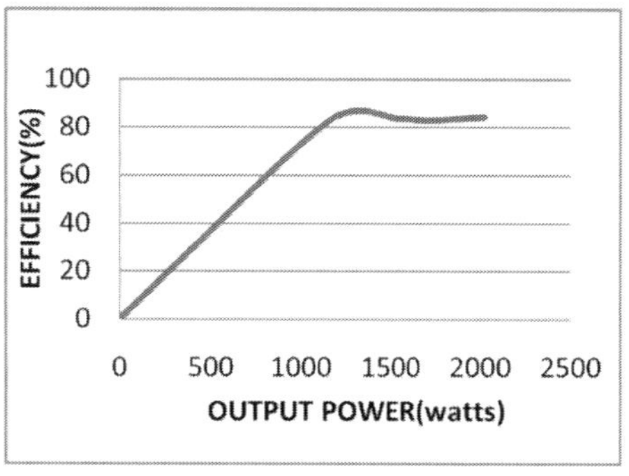

Figure 5. Efficiency characteristics.

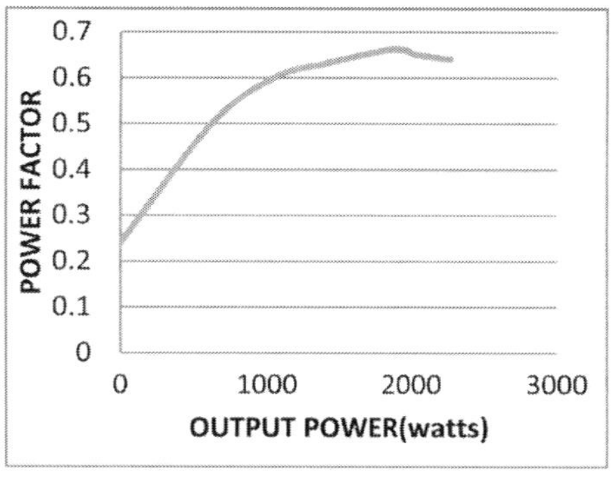

Figure 6. Power factor characteristics.

Table 3 shows the reading observed with rated field excitation with 40 V and field current of 0.62 A. It is observed that the maximum power factor is 0.64 and efficiency is 79.6% for the load current of 5.0 A. The minimum power factor is observed as 0.55 for the load current of 3.5 A, the efficiency of the machine for this load is 55 %. The efficiency and power factor characteristics are shown inFigure 7 & Figure 8. When the excitation is reduced, it is observed that the power factor of the machine decreases.

Table 4 shows the reading observed with rated field excitation with 60 V and field current of 0.85 A. The second winding is connected to a load of 1 A. It is observed that the maximum power factor is 0.95 and efficiency is 85.5% for the load current of 6.0 A. The minimum power factor is observed as 0.71 for the load current of 3.5 A, the efficiency of the machine for this load is 83.1%. The efficiency and power factor characteristics are shown in Figure 9 & Figure 10. When the load is applied on the second set of winding, power factor and efficiency is improved for the same load current.

Table 5 shows the reading observed with field excitation with 50 V and field current of 0.72 A. The second winding is connected to a load of 1 A. It is observed that the maximum power factor is 0.74 and efficiency is 92.3% for the load current of 5.0 A. The minimum power factor is observed as 0.68 for the load current of 3.5 A, the efficiency of the machine for this load is 78.3%. The efficiency and power factor characteristics are shown in Figure 11 & Figure 12.

Table 3. Brake test with excitation 40 V, current 0.62 A.

Input Voltage (V)	Line Current (A)	Input Power (W)	Output Power (W)	% Efficiency	Power Factor
415	2.9	480	0	0	0.23
415	3.5	1400	777	55.0	0.55
415	4.0	1680	1384	82.0	0.58
415	4.5	1960	1520	77.0	0.60
415	5.0	2320	1848	79.6	0.64
415	5.5	2520	1970	78.0	0.63
415	6.0	2760	2094	75.0	0.63

Table 4. Excitation 60 V, field current 0.85 A with 1 A electrical load in second winding.

Line Current (A)	Mechanical Output (W)	Output Voltage (V)	Electrical Output (W)	Total Output (W)	Overall % Efficiency	Overall Power Factor
3.0	0	390	580	580	0	0.46
3.5	777	380	720	1497	83.1	0.71
4.0	1016	375	720	1736	80.0	0.75
4.5	1382	375	720	2102	84.7	0.76
5.0	1663	370	720	2383	82.7	0.80
5.5	1934	365	720	2654	86.1	0.77
6.0	2187	365	720	2907	85.5	0.95

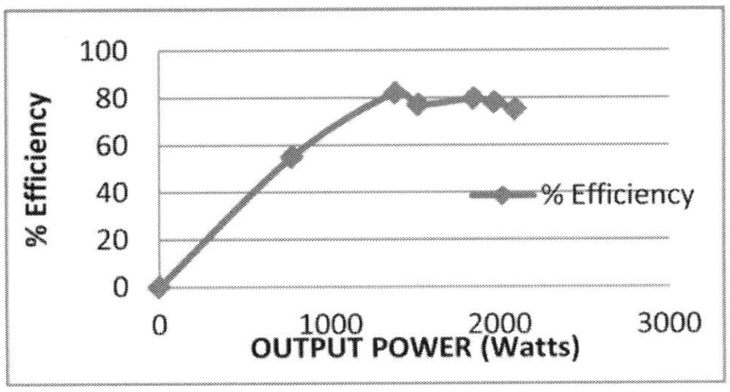

Figure 7. Efficiency characteristics.

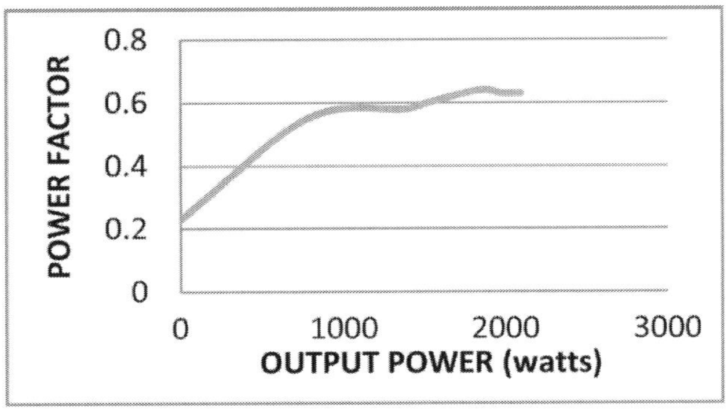

Figure 8. Power factor characteristics.

Table 5. Excitation 50 V, Field current 0.72 A, with 1 A electrical load in second winding.

Line Current (A)	Mechanical Output (W)	Output Voltage (V)	Electrical Output (W)	Total Output (W)	Overall % Efficiency	Overall Power Factor
3.0	0	385	780	780	0	0.46
3.5	647	380	700	1347	78.3	0.68
4.0	1047	375	700	1747	83.3	0.72
4.5	1323	370	720	2043	85.1	0.74
5.0	1775	370	720	2495	92.3	0.74
5.5	1786	365	660	2446	83.7	0.73
6.0	2278	360	640	2918	91.1	0.74

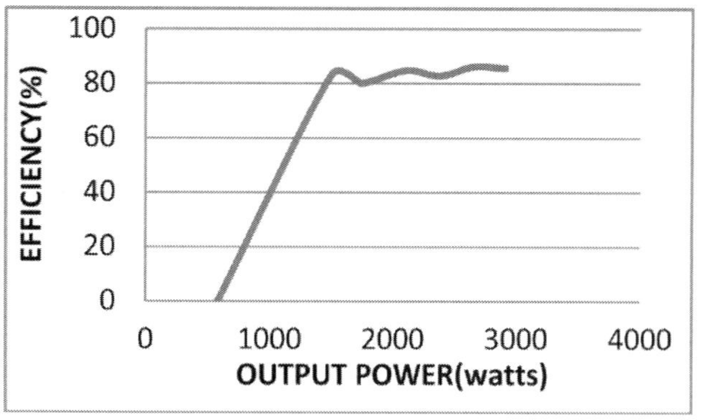

Figure 9. Efficiency characteristics.

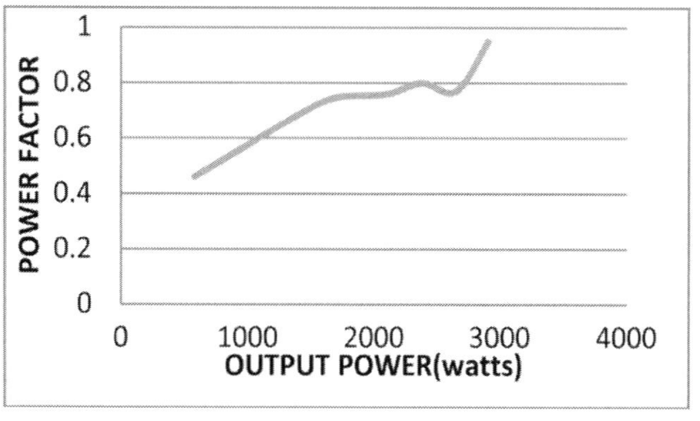

Figure 10. Power factor characteristics.

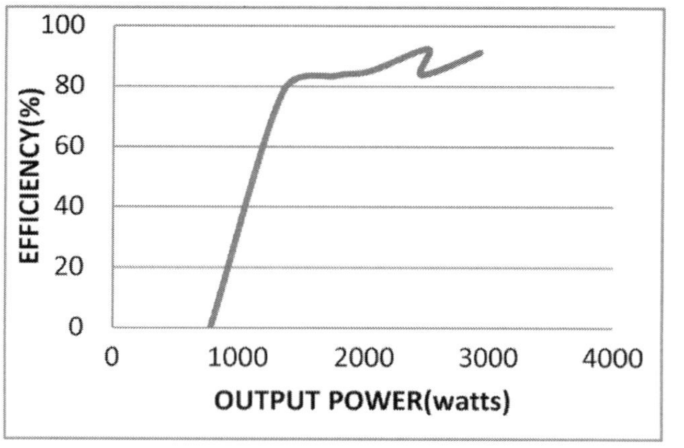

Figure 11. Efficiency characteristics.

Table 6 shows the reading observed with field excitation with 40 V and field current of 0.52 A. The second winding is connected to a load of 1 A. It is observed that the maximum power factor is 0.70 and efficiency is 82.5 % for the load current of 4.8 A. The minimum power factor is observed as 0.66 for the load current of 3.8 A, the efficiency of the machine for this load is 77.6%. The efficiency and power factor characteristics are shown in Figure 13 & Figure 14.

Table 6. Excitation 40V, Field current 0.52A, with 1A electrical load in second winding.

Line Current (A)	Mechanical Output (W)	Output Voltage (V)	Electrical Output (W)	Total Output (W)	Overall % Efficiency	Overall Power Factor
3.3	0	380	720	720	0	0.60
3.8	584	375	720	1304	77.6	0.66
4.3	892	375	720	1612	79.0	0.70
4.8	1294	370	720	2014	82.5	0.70
5.4	1509	365	680	2189	80.4	0.70
6.0	1771	360	680	2451	80.6	0.70

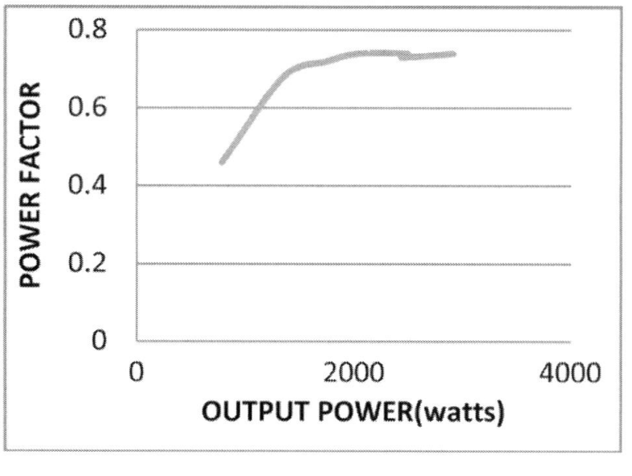

Figure 12. Power factor characteristics.

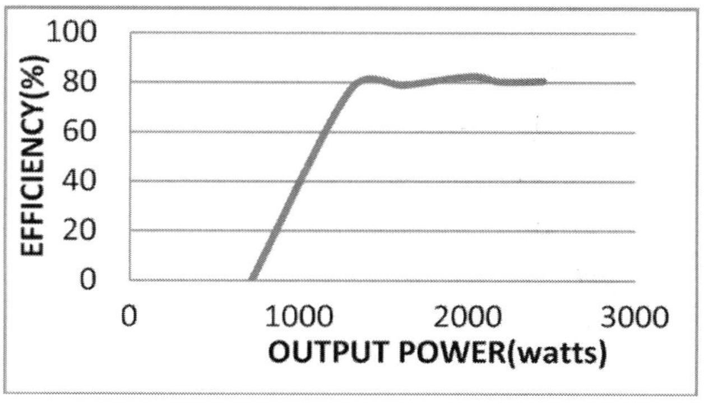

Figure 13. Efficiency characteristics.

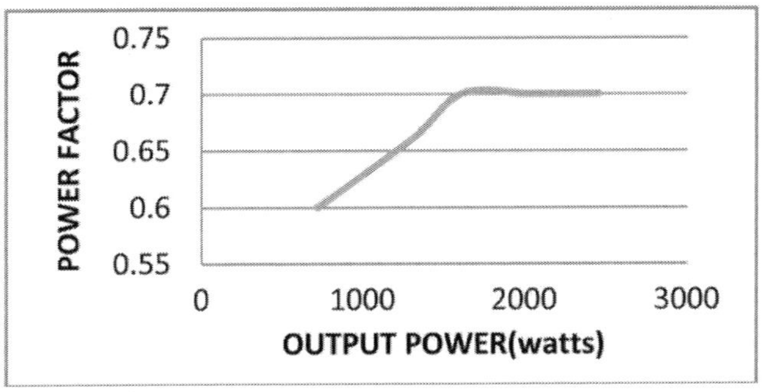

Figure 14. Power factor characteristics.

PERFORMANCE COMPARISON AND INFERENCES

In conventional synchronous motor, for the given load, when the excitation is decreased, the efficiency and power factor also reduces. The main focus of this paper is to improve the performance of the machine at reduced excitation and minimum load. A series of testing was carried out to study the performance of DWSyM. Tables 1-3 refer the reading observed during brake test with various excitations. Figures 3-7 refer the efficiency and power factor characteristics for various excitations. Tables 4-6 refer the reading observed with both mechanical and electrical load. A 3-phase lamp load and a three-phase loading rheostat are used for loading the second set of winding. Figures 4-8 refer the efficiency and power factor characteristics for various excitations with electrical load in the second set of winding. The comparison of performance analysis is shown in Table 7. For the rated excitation, for the load current of 6 A, efficiency is 78% and power factor 0.78, the same is improved to 85.5% and 0.95 when electrical load is connected in the second of winding. When the excitation is reduced, for the load current of 6 A, the efficiency improves from 78% to 91.1% and power factor improves from 0.64 to 0.74. When the excitation is further reduced for the load current of 3.5A, the efficiency is improved from 55% to 77.6% and power factor from 0.55 to 0.66. In general, dual winding machine provides opportunity to improve its performance for the reduced shaft load while in the conventional machine it is not possible.

Industrial Application

DWSyM is coupled with centrifugal pump and tested for its performance. Lamp load is connected across the second set of winding. By adjusting the mechanical and electrical loads the machine can be operated simultaneously. The load connected to second set of winding is not depending on separate supply. The designed DWSyM is tested in an industry and the testing set up is shown in Figure 15. DWSyM is coupled with a pump, a lamp load is connected in the second winding. Mechanical and electrical outputs are made available in the same machine. Both the loads are adjusted within its thermal limit such that motor is not over loaded.

Energy Conservation

DWSyM consists of two sets of stator winding. The load connected to the second set of winding is depend on separate supply. When the machine is not over loaded, by loading both the windings, the machine can be used to rated capacity. For example, if electrical load of 1 A is connected in second winding, power developed in this winding is 700 Watts, if it is operated for 10 hours a day, then the energy conserved is about 210 units per month. When the second winding is loaded, it is also observed that there is slight increase in the input current, however, when compared to the load current of second winding, this is minimum and also accounted.

Table 7. Performance analysis.

Input	Conventional Brake Test Method			With External Load in Second Winding		
	Line Current I_L	% Efficiency	Power Factor	Line Current I_L	% Efficiency	Power Factor
Armature:415 V, 50 Hz, 3-Phase Field: DC 60 V, 0.85 A	5.0	83.1	0.74	5.0	82.7	0.77
	6.0	78.0	0.75	6.0	85.5	0.95
Armature:415 V, 50 Hz, 3-Phase Field: DC 50 V, 0.72 A	4.0	83.0	0.63	4.0	83.0	0.74
	6.0	78.0	0.64	6.0	91.1	0.74
Armature:415 V, 50 Hz, 3-Phase Field: DC 40 V, 0.62 A	3.5	55.0	0.55	3.8	77.6	0.66
	6.0	75.0	0.63	6.0	80.6	0.70

Figure 15. Industrial application.

CONCLUSION

A 3 kW, 3-phase, 4-pole, 1500 rpm, 415 V Double Winding Synchronous Motor was fabricated to verify the performance. The main focus is to improve the performance of synchronous motor during lightly loaded condition; by connecting load on the second winding the overall performance is improved. When operated as conventional motor for the load current of 6 A, the efficiency is 78% and power factor is 0.75, when second winding is connected to a load of 1 A, for the same excitation, the efficiency is improved 85.5% and power factor to 0.95. For the reduced excitation, when operated as conventional motor for a load current of 3.5 A, the efficiency is 55% and power factor is 0.55, when a load of 1 A is connected to the second winding, for the same excitation, the efficiency is improved to 77.6% and power factor to 0.66. This result ensures that there is performance improvement by loading the second set of winding. Another major observation is when the excitation is reduced for the given load torque, the power factor and efficiency of the synchronous motor is reduced; by adjusting load on the second set of winding, the overall performance is improved. Just for a load of 1 A, 210 units is conserved in one month. The Double Winding Synchronous Motor can be used where it is necessary to run continuously like in textile industries and product manufacturing units. The output power from the second winding can be used for charging and supplying lighting loads.

REFERENCES

1. Tesnjak, S., Erceg, G., Erceg, R., Klarin, D. and Komericki, Z. (2000) Excitation System Models of Synchronous Generator. https://bib.irb.hr/datoteka/475822.Final_paper_-_SiP2010_Jerkovic.pdf [Paper reference:1]
2. Machowski, J., Bialek, W., Robak, S. and Bumby, J.R. (1998) Excitation Control System for Use with Synchronous Generators. IEE Proceedings—Generation, Transmission and Distribution, 145, 537-546. http://dx.doi.org/10.1049/ip-gtd:19982182 [Paper reference:1]
3. Naoe, N. and Fukami, T. (2001) Trial Production of a Hybrid Excitation Type Synchronous Machine. IEEE International Electric Machines and Drives Conference, Cambridge, 17-20 Jun 2001, 545-547. http://dx.doi.org/10.1109/IEMDC.2001.939362 [Paper reference:1]
4. Eastham, J.F., Evans, P.D. and Coles, P.C. (1992) Double Disc Alternators with Hybrid Excitation. IEEE Transactions on Magnetics, 28, 3039-3041. http://dx.doi.org/10.1109/20.179710 [Paper reference:1]
5. (2007) IEEE Standard Definitions for Excitation Systems for Synchronous Machines. IEEE Std 421.12007 (Revision of IEEE Std 421.1-1986), 1-33. [Paper reference:1]
6. Klempner, G. and Kerszenbaum, I. (2004) Operation and Maintenance of Large Turbo-Generators. IEEE Press Series on Power Engineering, John Wiley and Sons, Inc. http://dx.doi.org/10.1002/0471683388 [Paper reference:1]
7. Alger, P.L., Freiburghouse, E.H. and Chase, D.D. (1930) Double Windings for Flexible Alternators. Transactions of the American Institute of Electrical Engineers, 49, 226-244. [Paper reference:1]
8. Weber, H.J. (1989) A.C. Induction Motor Energy Conserving Power Control Method and Apparatus. US Patent, Publication Number: US4806838A. [Paper reference:1]
9. Bu, L.L., Li, C., Krukowski, J., Xu, W. and Liu, X. (2003) A New Energy Recovery Double Winding Cage-Rotor Induction Machine. IEEE Transactions on Energy Conversion, 18, 315- 320. [Paper reference:1]
10. Xu, L., Xu, X., Lipo, T.A. and Novotny, D.W. (1991) Vector Control of a Synchronous Reluctance Motor Including Saturation and Iron Loss. IEEE Transactions on Industry Applications, 27, 977-985. http://dx.doi.org/10.1109/28.90356 [Paper reference:1]
11. Razik, H., Rezzoug, A. and Hadiouche, D. (2005) Modelling and Analysis of Dual – Stator Induction Motors. IEEJ Transcations Industrial Applications, 125, 1093-1104. [Paper reference:1]
12. Leonardi, F., Matsuo, T. and Lipo, T.A. (1996) Iron Loss Calculation for Synchronous Reluctance Machines. Proceedings of the 1996 International Conference on Power Electronics, Drives and Energy Systems for Industrial Growth, New Delhi, 8-11 January 1996, 307-312. [Paper reference:1]
13. Fletcher, J.E., Williams, B.W. and Green, T.C. (1995) Efficiency Aspects of Vector Control Applied to Synchronous Reluctance Motors. Proceedings of IAS Annual Meeting, Orlando, 8-12 October 1995, 294-300. [Paper reference:1]

CHAPTER 20

PHASOR CONTROL OF CONVERTER OUTPUT VOLTAGE FOR FREQUENCY REGULATION

Masahide Hojo, Ryo Ikeshita, Yoshinobu Ueda, and Toshihisa Funabashi

Keywords: Self-Commutated Power Converter, Synchronous Generator, Frequency Control, Voltage Phasor

INTRODUCTION

A conventional power system has been supplied by a lot of synchronous generators. They have kept a good power balance among them by their synchronizing torques. On the other hand, a distributed generator injects its maximum output power by observing the phase angle of the voltages at its coupling point. The power balance can be kept successfully while the total amount of power generated by the distributed generators is negligible compared to the synchronous generators.

Today, the synchronizing torque of a power system may be weakened by increasing installation of static power converters accompanied by renewable energy resources with a growing demand of high penetration of renewable energies. In a micro-grid, batteries are often installed with the renewable energy resources to compensate their fluctuating output power. As a result, a lot of static power converters will be normally installed with traditional synchronous generators. In case that the total rated power of the converters exceeds the one of synchronous generators, it is afraid that the synchronizing torque of the micro-grid may be spoiled. In the circumstances, there is a proposal of a Virtual Synchronous Machine (VSM) [1] -[3], where the converter behaves like a traditional synchronous machine with the inertia. In recent literatures, the effectiveness of the VSM is evaluated as compared to the traditional droop controller [4] and its dynamic performance is also analyzed [5]. Assuming that the converter is supported by an ideal dc voltage source and consists of ideal semiconductor devices, the VSM can be realized to be equipped with a swing equation onto its controller. The effectiveness is promised by the designed swing equation under an assumed system condition. However, it may be difficult to achieve the same effect when the power system condition is continuously modified according to the increasing installation of the renewable energy resources.

In the circumstances, unlike the VSM, authors proposed a phase-shifting method of the output voltage phasor [6], which simply only emulates an active power exchange between the power converter and the power system to enhance the synchronizing torque of the power system. In the traditional converter control, the output voltage phasor is determined to be synchronized with the terminal voltage at the point of common coupling as described above. On the other hand, the infinite bus voltage can be considered as a voltage phasor with the frequency fixed to its rated value. In the proposed control strategy, each converter independently determines its output voltage phase angle between the two phasors using single control gain and the amplitude of the output voltage is decided by a reactive power control in order to reduce inverter output currents. The control gain can regulate its contribution level, according to its surplus capacity of the converter or energy storage. This is the reasons why the proposed control can be an autonomous control system and it has robustness where the converters continue to be installed increasingly. Therefore, it can provide an effective support to regulate the system frequency where the total rated power of the converter-based generators increases as much as one of the conventional rotating generators like a micro-grid.

However, it can be hardly estimated where the available converters are installed and how many inverters join the frequency regulation. Therefore, such control strategy is required to be useful under uncertain conditions. In the proposed control, each converter can share the task automatically in accordance with the electrical distance among the generation units and, if it is overtasked, it can reduce the task by modulating the control gain. This paper explains the behaviors of multiple power converters equipped with the proposed control and verifies its operating characteristics discussed above by simulation study based on instantaneous

Citation: Hojo, M., Ikeshita, R., Ueda, Y. and Funabashi, T. (2014). Phasor Control of Converter Output Voltage for Frequency Regulation. Journal of Power and Energy Engineering, 2, 19-27. DOI: 10.4236/jpee.2014.27003.

values. Especially, it is confirmed that multiple static power converters employing the proposed controller can properly cooperate with each other when they are distributed or they use other control parameters.

CONVERTER CONTROL FOR COOPERATION WITH A SYNCHRONOUS GENERATOR

Phase Angle Regulation for Active Power Control

In order to consider an active power interaction between a synchronous generator and a static power converter, a simple system model is used in this paper, as shown in Figure 1. For the simplicity, the synchronous generators and the converters are represented by single ones, respectively. V_g and V_c represent the fundamental components of terminal voltages at the synchronous generator and the converter and their phase angles are represented by θ_g and θ_c, respectively. The terminal voltage of the synchronous generator is regulated by a traditional Automatic Voltage Regulator equipped with the generator.

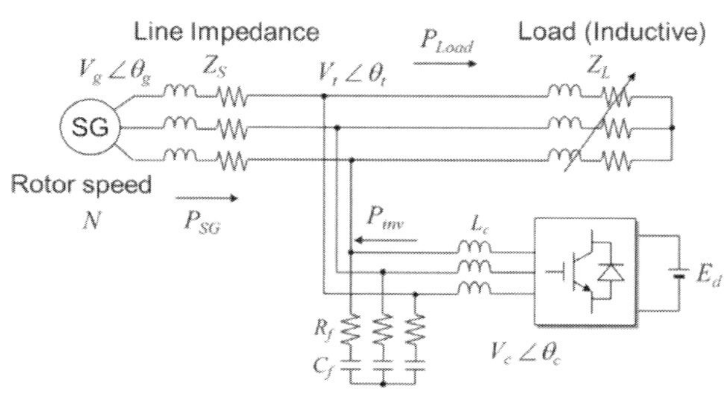

Figure 1. A model with a synchronous generator and a self-commutated static power converter.

Generally, the self-commutated static power converter regulates its output voltage based on the phase angle of the terminal voltage at the point of common coupling. It detects the phase angle of the terminal voltage at the coupling point θ_t and decides its output voltage. Therefore, the converter follows the frequency regulated by the synchronous generator and trades its favorable output power. Assume that the initial value of the converter phase angle is θ_{c0}, the converter phase angle θ_c can be determined as:

$$\theta_c = \theta_{c0} + \int \omega_g \mathrm{d}t \,, \tag{1}$$

where w_g is the angular velocity of the synchronous generator. In this case, the converter can be considered as a synchronous machine whose inertia is zero.

On the other hand, the infinite bus has an infinite inertia and a constant voltage. If there is an ideal self-commutated converter with an ideal energy source, it can realize the infinite bus and it is quite stable all through the operation of the micro-grid. In this case, the converter phase angle can be represented by the nominal value of the angular velocity w_0, as

$$\theta_c = \theta_{c0} + \int \omega_0 \mathrm{d}t \,. \tag{2}$$

The synchronous generators trade their power after a disturbance and keep a stable cooperation in a power system. The power trade is realized by the difference of the phase angle between them. From this viewpoint, the power trade can be realized by modulating the phase angle of the converter.

Figure 2 summarizes relationships among voltage phasors. Assume that a small disturbance makes the phase angle of the synchronous generator θ_g lag behind the reference angle θ_0 by $\Delta\theta$. When the converter is operated in the traditional operating mode, the voltage phasors can be indicated as Figure 2(a). On the other hand, the converter is operated as shown in Figure 2(b) in case that the converter acts as an infinite bus voltage. Therefore, this paper proposes that the output voltage phasor is placed as shown in Figure 2(c), in order enhance the synchronizing torque of the power system.

To set the voltage phasor as shown in Figure 2(c), the converter phase angle θ_c is determined as

$$\theta_c = \theta_{c0} + \theta_g + k_c\left(\theta_0 - \theta_g\right), \tag{3}$$

$$0 \le k_c < 1 \,. \tag{4}$$

The phase angle of the converter θ_c can be increased to lead θ_0 by $\Delta\theta$ as shown in Figure 2(d). In this case, the angular velocities can be formulated by

$$\int \left(\omega_c - \omega_0\right) \mathrm{d}t = \int \left(\omega_0 - \omega_g\right) \mathrm{d}t \,. \tag{5}$$

Therefore, the phase angle θ_c can be derived as

$$\theta_c = \theta_{c0} + \int \left(2\omega_0 - \omega_g\right) \mathrm{d}t \,. \tag{6}$$

However, Equation (6) can be represented by substituting 2.0 to k_c of Equation (3). Therefore, all operating modes can be represented as Figure 2 with different value of k_c.

Amplitude Modification for Reactive Power Control

In order to reduce the converter rated current, reducing the reactive power output is effective. Therefore, the voltage phasor amplitude is regulated in addition to the phase angle control.

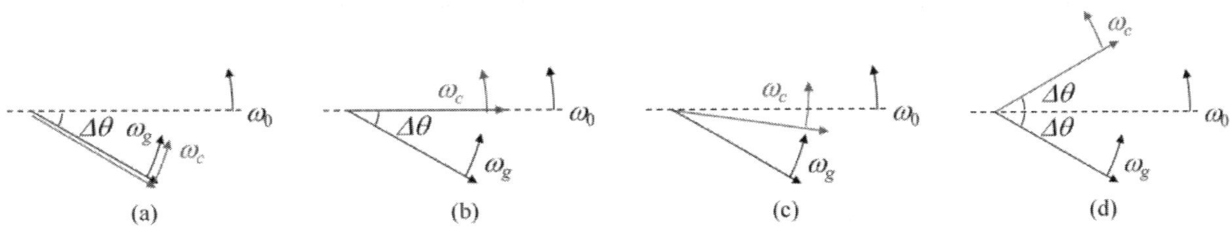

Figure 2. Various placements of the converter output voltage phasors. (a) $k_c = 0.0$, (b) $k_c = 1.0$, (c) $0.0 < k_c < 1.0$, (d) $k_c = 2.0$.

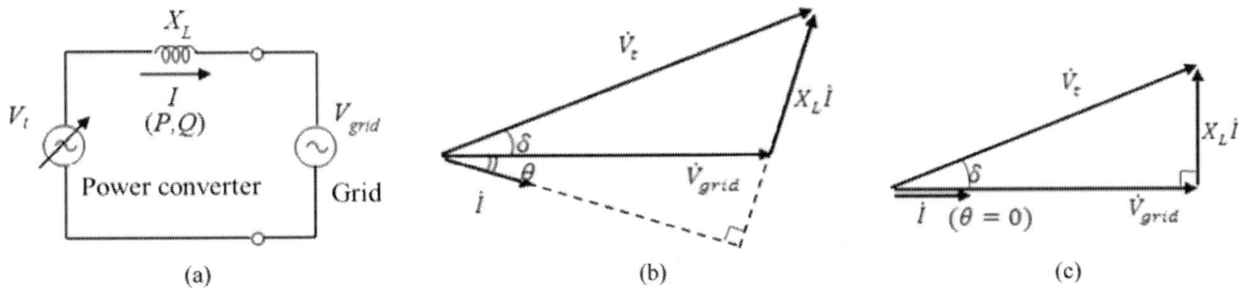

Figure 3. Reactive power regulation of the converter. (a) A simplified equivalent circuit, (b) Phase-shifted output voltage phasors, (c) Regulated output voltage phasor.

Figure 3(a) is a simple equivalent circuit when a self-commutated power converter is interconnected to a power system with a linkage inductance X_L. V_t represents the output voltage of the converter and V_{grid} is the voltage at the coupling point. By the phase angle difference δ between V_t and V_{grid}, voltage phasors can be indicated as shown in Figure 3(b). In this case, the converter trades active and reactive power as:

$$P = V_{grid} I \cos\theta \,, \tag{7}$$

$$Q = V_{grid} I \sin\theta \,. \tag{8}$$

Considering the relations among the voltage phasors as shown in Figure 3(b),

$$X_L I \cos\theta = V_t \sin\delta \,, \tag{9}$$

$$X_L I \sin\theta = V_t \cos\delta - V_{grid} \,. \tag{10}$$

Therefore, the active and reactive power can be formulated as

$$P = \frac{V_t V_{grid}}{X_L} \sin\theta \,, \tag{11}$$

$$Q = \frac{\left(V_t V_{grid} \cos\delta - V_{grid}^2\right)}{X_L} \,. \tag{12}$$

Assumed that the reactive power Q is equal to zero, the following equation can be derived from Equation (12);

$$V_t = \frac{V_{grid}}{\cos\delta} \,. \tag{13}$$

By detecting the phase angle of the voltage V_{grid}, δcan be achieved. Then, the reactive power can be equal to zero by keeping the relation of Equation (13).

Proposed Voltage Phasor Controller

The output voltage controller consists of the above two concept as shown in Figure 4. First, the phase angle of the voltage is determined and then its amplitude is modified based on the reactive power control. The proposed control is equipped with a sinusoidal wave generator and the well-known triangular-wave Pulse Width Modulation of a three-phase voltage sourced converter, as shown in Figure 4.

It should be noted that the reference angle is ordinary determined by the traditional method such as normal PLL. When the converter controller detects a large frequency variation beyond its prescribed threshold value, the proposed controller slightly shift its reference angle by this method.

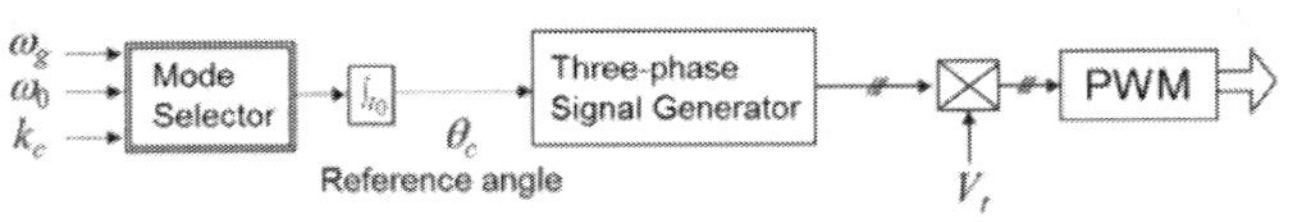

Figure 4. Output voltage controller of the converter.

SIMULATION RESULTS WITH SINGLE INVERTER UNIT

The proposed control is verified by simulation study. A system model shown in Figure 1 is used with the parameters summarized in Table1 The dc side voltage of the converter is assumed to be supported by ideal dc voltage source. On the other hand, the traditional rotating synchronous machine is represented by a Motor-andGenerator combination which employs a speed governor and an automatic voltage regulator.

In the actual system, it may be difficult to detect the frequency of the center of inertia. For the simplicity, the referential frequency is assumed to be detected by the rotor speed of the rotating generator in this paper. When $k_c = 0.5$, Figure 5 shows how the frequency and active power output of synchronous generator and power converter behaves after the load variation. It can be confirmed in the upper figures that the controller set the frequency reference as a half of system frequency.

If the control gain is adjusted, the effectiveness of the controller can be modified as shown in Figure 6. When k_c = 0.0, the active power output of the converter immediately recovered to the initial value and it does not contribute the frequency regulation. However, increasing k_c enhances the control effect as shown in Figure 6.

From these results, the proposed controller can adjust its control effect by simply modifying the single control gain k_c.

CONTROL EFFECT BY MULTIPLE CONVERTERS

When the proposed controller is applied to multiple power converters, their parameters can be designed independently. Figure 7 shows a model system with two power converters which are equipped with the proposed control. The control gains K_1 and K_2 are defined to adjust their contributions to frequency regulation. Table 2 summarizes simulation conditions.

First, same circuit parameters are given to the two converters, which are indicated as the Case 1 inTable 2, but different control gains K_1 and K_2 are applied to them. Figure 8 displays simulation results, comparing with the different control gains. As shown in Figure 8(a), same control gains result in the same behaviors, where the same active powers are injected and both reactive powers are regulated as zero.

On the other hand, different control gains introduce different behaviors as shown in Figure 8(b). There is no difference in the frequency variations of the two cases because the values of $K_1 + K_2$ are same. However, two converters show different contributions after the load variation. The converter 2 with the larger control gain output high active power, resulting in the high contributions to the frequency regulation.

Therefore, it is confirmed that the contribution to the frequency regulation can be simply adjusted by the control gain.

Figure 9 shows simulation results in Case 2, where the different circuit parameters are used as indicated in Table2 The different linkage inductance implies that they are not installed at the same point but distributed in a power network. However, as the proposed control directly modifies the phase angle of the output voltage of the converter, the output active power is influenced by the interconnecting inductance. Therefore, although the same control gains are used in the proposed controller in the case of Figure 9(a), the output active power of the converter 2 is larger than the one of the converter 1. In addition, if the control gain of the converter 2 is changed to the larger one, the output active power is increased as shown in Figure 9(b).

The two converters are connected at the same coupling point as indicated in Figure 7, but the interconnecting inductance means the electrical distance from the coupling point. Therefore, Figure 9(a) also indicates that the converter which is near the generator contributes well to the frequency regulation.

Table 1. Conditions for simulation study.

Microgrid	
Operating frequency f_0 Rated RMS voltage (3ϕ, line-to-line) Line impedance Z_s Load (inductive) Z_L(t < 2.5 s) (2.5 s < t)	60 Hz 200 V 0.10 Ω, 6.0 mH 2.4 Ω, 2.1 mH 1.6 Ω, 2.1 mH
Converter	
Dc voltage source E_d Interconnecting Inductance L_C Ac filter capacitor C_f Switching frequency	400 V 5.0 mH 6.0 μF 3 kHz

Table 2. Conditions for simulation with two converters.

System parameters		
Rated voltage (3φ, line-to-line) Operating frequency f_0 Line impedance Z_S Load (inductive) Z_L(t < 2.5 s) (t > 2.5 s)	200 V 60 Hz 0.10 Ω, 5.0 mH 2.4 Ω, 2.1 mH 1.6 Ω, 2.1 mH	
Converter 1, 2		
Dc voltage source E_{d1}, E_{d2}	Case 1	Case 2
linkage inductance L_{C1}, L_{C2} Ac filter capacitor C_{f1}, C_{f2} Switching frequency f_{s1}, f_{s2}	400 V 5.0 mH 6 μF 3 kHz	400 V 5.0 mH, 3.0 mH 6 μF, 11 μF 3 kHz

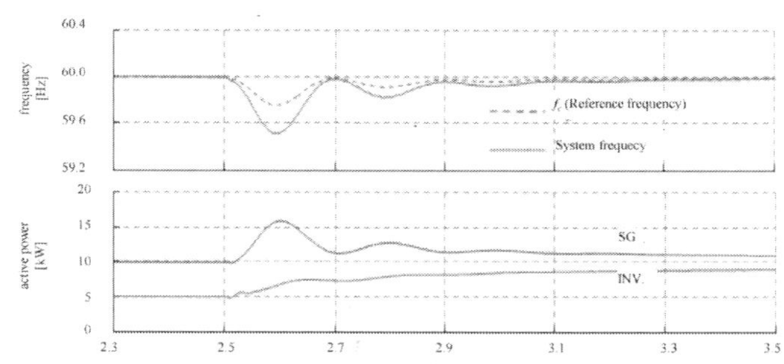

Figure 5. Frequency and active powers after a load variation when k_c = 0.5.

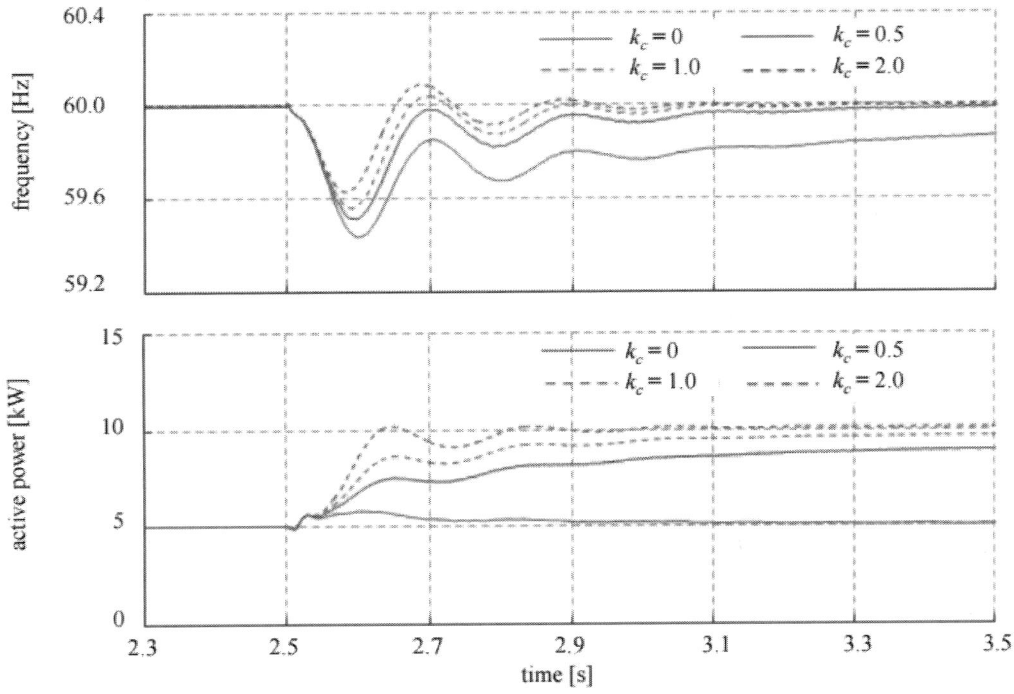

Figure 6. Comparisons of frequency and active powers with different gain k_c.

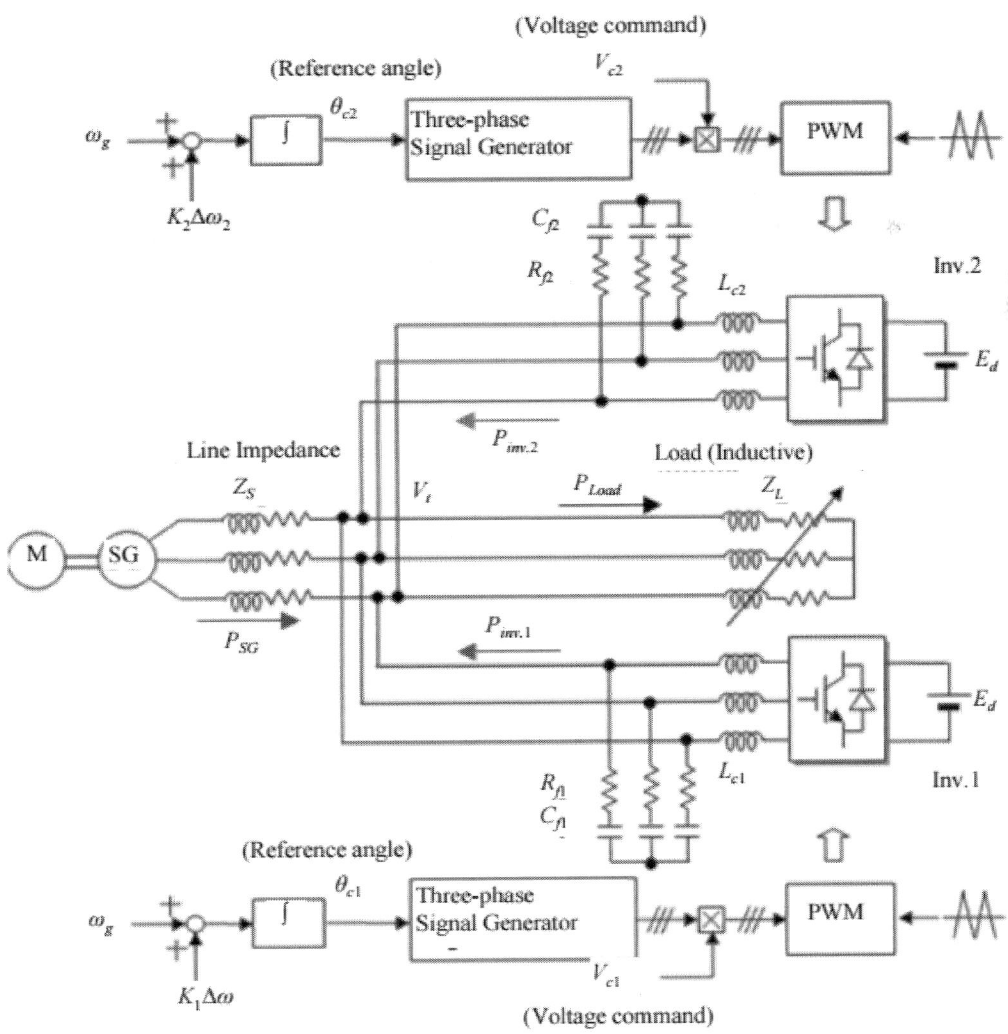

Figure 7. System model with two converter system.

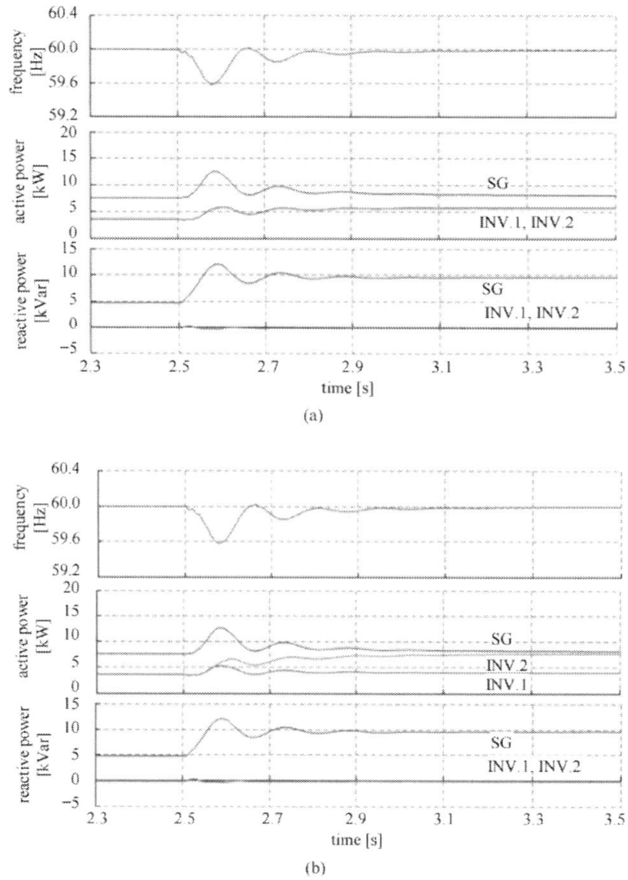

Figure 8. Simulation results when the same linkage inductances and filter capacitors are used (Case 1). (a) $K_1 = K_2 = 1/2$, (b) $K_1 = 1/4$, $K_2 = 3/4$.

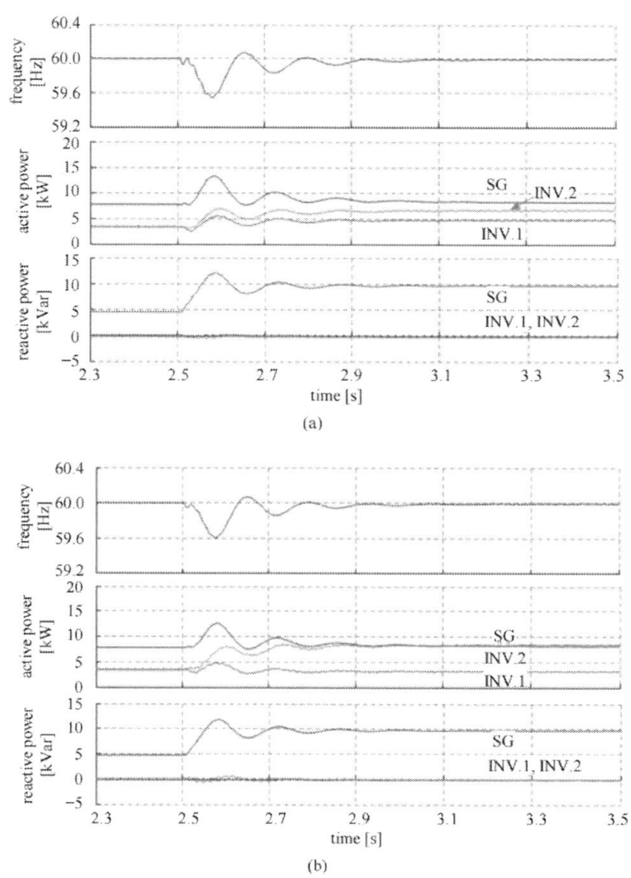

Figure 9. Simulation results when different linkage inductances and filter capacitors are used (Case 2). (a) $K_1 = K_2 = 1/2$, (b) $K_1 = 1/4$, $K_2 = 3/4$.

CONCLUSION

This paper proposed a simple converter controller to contribute to the system frequency regulation. The converter control emulating the power trade between the synchronous machines is developed and equipped to the static power converter. The effectiveness of the controller is tested by simulation study in case of a small disturbance like a load variation. In the simulation study, it is confirmed that the contribution of the converter can be easily adjusted by the single control gain. In addition, the controller can be also tested when multiple converters employ it. By the detailed simulation, the two converters share the role of the frequency regulation, according to their control gains or the electrical distance among the generation units. From these results, the proposed controller can be easily applied to the power converter such as being used for renewable energy resources with some amount of energy storage, resulting in enhancement of frequency regulations.

ACKNOWLEDGEMENTS

This work was supported by Grant-in-Aid for Scientific Research (C) of Japan Society for the Promotion of Science (JSPS), as JSPS KAKENHI Grant Number 25420258.

REFERENCES

1. Beck, H.P. and Hesse, R. (2007) Virtual Synchronous Machine. International Conference on Electrical Power Quality and Utilization, Barcelona, 9-11 October 2007, 6 p. [Paper reference:1]
2. Visscher, K. and Haan, S.W.H. (2008) Virtual Synchronous Machines (VSG's) for Frequency Stabilization in Future Grids with a Significant Share of Decentralized Generation. CIRED Seminar 2008: Smart-Grids for Distribution, Frankfurt, 23-24 June 2008, 0118.
3. Sakimoto, K., Sugimoto, K. and Shindo, Y. (2013) Low Voltage Ride through Capability of a Grid Connected Inverter Based on the Virtual Synchronous Generator. Proceedings of the IEEE 10th International Conference on Power Electronics and Drive Systems, Kitakyushu, 22-25 April 2013, 1066-1071. [Paper reference:1]
4. Alipoor, J., Miura, Y. and Ise, T. (2014) Voltage Sag Ride-Through Performance of Virtual Synchronous Generator. Proceedings of the 2014 International Power Electronics Conference—ECCE, Hiroshima, 18-21 May 2014, 3298-3305. [Paper reference:1]
5. Liu, J., Miura, Y. and Ise, T. (2014) Dynamic Characteristics and Stability Comparisons between Virtual Synchronous Generator and Droop Control in Inverter-Based Distributed Generators. Proceedings of the 2014 International Power Electronics Conference—ECCE, Hiroshima, 18-21 May 2014, 1536-1543. [Paper reference:1]
6. Hojo, M., Ikeshita, R., Terauchi, T., Ueda, Y. and Funabashi, T. (2011) A Converter Controller of Virtual Synchronous Machine for Stable Operation of Microgrid. Conference Proceedings of 21th International Conference and Exhibition on Electricity Distribution (CIRED), Frankfurt, 6-9 June 2011, 0535. [Paper reference:1]

CHAPTER 21

PMSG WIND ENERGY CONVERSION SYSTEM: MODELING AND CONTROL

Omessaad Elbeji, Mouna Ben Hamed, and Lassaad Sbita

Keywords: Wind Turbine, Wind Energy Conversion System, Permanent Magnet Synchronous Generator, Speed Control, Current Control

INTRODUCTION

Renewable energy was strongly encouraged. It doesn't produce emissions either provided by the sun, wind, waterfalls or plant growth. They participate in the fight against the greenhouse effect and CO_2 emissions in the atmosphere, facilitate rational management of local resources, and also create jobs. Solar (solar photovoltaic, solar thermal), hydro, wind, biomass, geothermal energy are inexhaustible resource of energy versus "energy stock" from fossil fuels deposits such us scarce oil, carbon, natural gas.

In this paper, study focuses on wind energy conversion systems. Indeed, wind energy has become a major producer of renewable electric energy.

A wind turbine generator system (WTGS) transforms the wind energy into electrical energy. In fact, wind turbines generate mechanical forces such as windmills of the past. Through their blades, wind turbine captures the wind kinetic energy and transforms it into mechanical one.

Then this later was transformed into electric energy by a generator.

There are many types of generator available for wind energy conversion; as example induction generator in all its forms like wound rotor asynchronous generator, dual stator induction generator, MADA, etc. The permanent magnet synchronous generator is selected for many reasons. A permanent magnet synchronous generator is characterized by the absence of gearbox and reduced active weight, besides having a high power density and a high efficiency (disappearing of the copper losses in rotor).

Citation: Elbeji, O., Hamed, M. and Sbita, L. (2014). PMSG Wind Energy Conversion System: Modeling and Control.International Journal of Modern Nonlinear Theory and Application, 3, 88-97. DOI: 10.4236/ijmnta.2014.33011.

Generally the wind turbine generator based on rotational speed can be splited into two types: fixed and variable speed WTGS. Fixed speed turbines are easier to interface with the electrical grid. However, variable speed turbines are able to extract more energy from the wind and are the design preferred by the wind industry.

This paper interested to a variable speed WTGS. It has higher efficiency, especially at low wind speeds and also its power variations are lower than fixed speed turbines.

The paper analyzes a complete model of a variable wind turbine equipped with a permanent magnet synchronous generator it also proposes a vector control strategy to control the wind turbine generator. This strategy includes a speed controller and two current controllers.

The wind conversion system model and the control schemes were verified using Matlab/Simulink. Simulations results are selected, dicussed and come to prove the obtained performances of the used controllers.

WIND TURBINE MODEL

The wind energy captured by the blades was transformed by the wind turbine into mechanic energy.

The model studied is illustrated by Figure 1 that contains the wind model, an aerodynamic part and a mechanical model.

Wind Model

The wind speed model requires wind climate and geographical data of the concerned site and the period of the concerned year by the study. The wind model is given by a Fourier series representation of the wind which has as a signal consisting of a superposition of several harmonics. It is given by:

$$v_v(t) = A + \sum_{k=1}^{i} a_k \sin(\omega_k t) \quad (1)$$

where A is the average value of the wind speed, a_k is harmonic amplitude of the order k, ω_k is pulse harmonic of order k.

Aerodynamic Model

The aerodynamic energy of the wind can be represented as [1]:

$$P_w = \frac{1}{2}\rho A v_w^3 \quad (2)$$

where A is the circular area, $\rho$is the air density, and v_w is the wind speed.

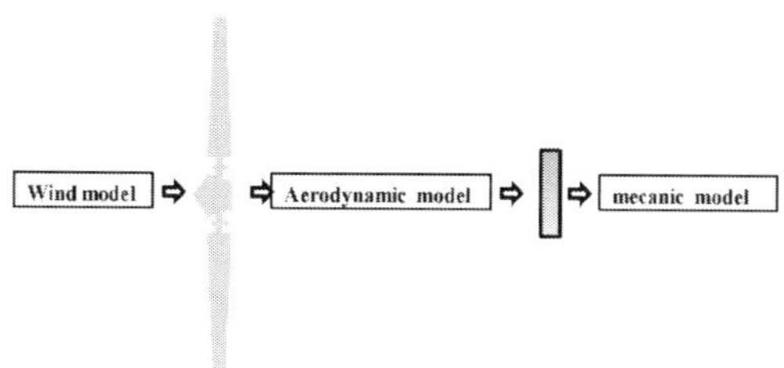

Figure 1. Simplified scheme of wind turbine.

Using the wind aerodynamic energy, aerodynamic power can be produced by the turbine. It can be expressed by [2]:

$$P_t = \frac{1}{2}\rho A v_w^3 C_p(\beta,\lambda) \quad (3)$$

where C_p is the power coefficient, it depending on the pitch angle β and the tip speed ratio λ that given by [3]:

$$\lambda = \frac{\Omega R}{v_w} \quad (4)$$

where Ω is the turbine rotor speed, R is the turbine radius. Then the power coefficient C_p can be expressed by [4]:

$$C_p(\beta,\theta) = 0.5176\left(\frac{116}{\lambda_i} - 0.4\beta - 5\right)\exp^{-\frac{21}{\lambda_i}} + 0.006795\lambda_i \quad (5)$$

where λ_i is given by:

$$\lambda_i = \frac{1}{\dfrac{1}{\lambda + 0.08\beta} - \dfrac{0.035}{\beta^3 + 1}} \quad (6)$$

C_p is depending on λ andβ, the C_p family of curve is obtained with different value of λ and with changing β value. This dependence is clearly visually in Figure 2.

However, the power coefficient is maximal when $\beta = 0$ that given by Figure 3.

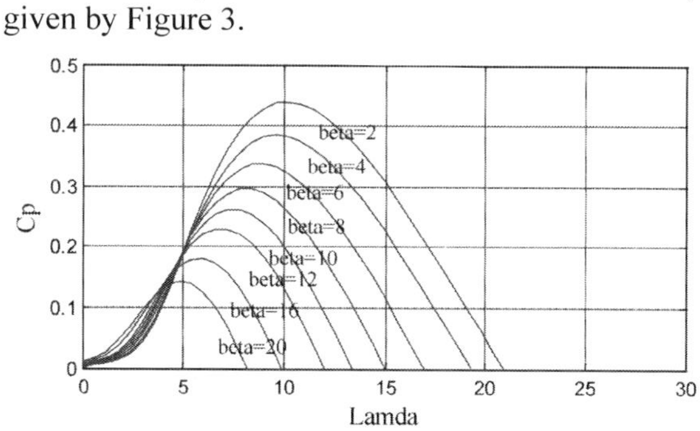

Figure 2. Power coefficient curve family versus β and λ.

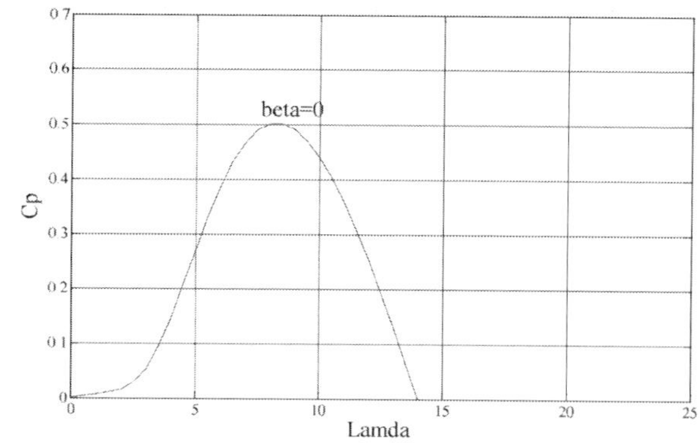

Figure 3. Power coefficient with different values of λ (with $\beta = 0$).

To have best results, next simulations are carried with $\beta = 0$.

The aerodynamic torque is determined by [5]:

$$T_t = \frac{P_t}{\Omega} \quad (7)$$

$$T_t = \frac{0.5\rho\pi R^3 v^2 C_p}{\lambda} \quad (8)$$

The aerodynamic turbine power curves family with varying the turbine rotor speed for different value of wind speed illustrated in Figure 4.

With varying the turbine rotor speed for different wind values, the curves family of the aerodynamic torque speed is given by Figure 5.

Mechanic Model

The fundamental dynamic equation is described with the following equation [2]:

$$J\frac{d\Omega}{dt} = T_t - T_{em} - f\Omega \quad (9)$$

where T_{em} is the electromagnetic torque, f is the turbine rotor friction.

Then, the wind turbine generator drive that represents the mechanical bloc can be given by:

$$T_t - T_{em} = J\frac{d\Omega}{dt} + f\Omega \quad (10)$$

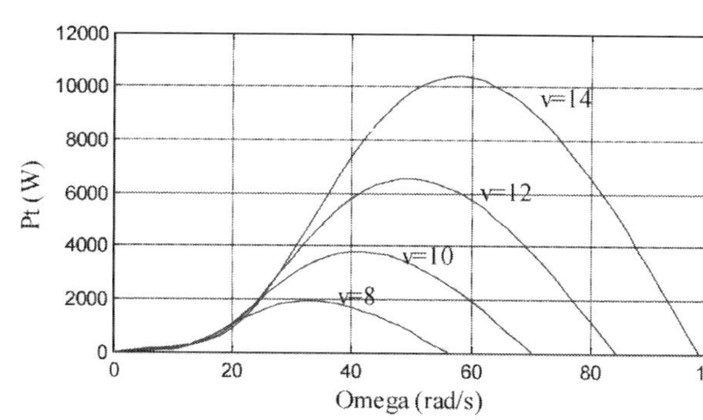

Figure 4. Curve family of turbine power versus ω and v.

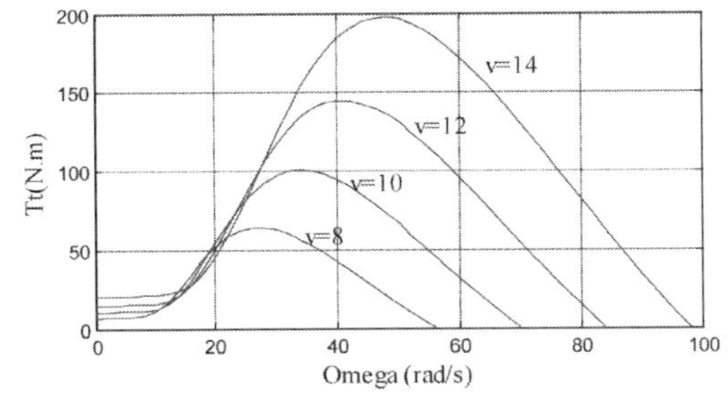

Figure 5. Curve family of aeodynamic torque versus ω and v.

PMSG MODEL

The PMSG model can be written, in the d-q synchronously rotating reference frame, by the following equation system [1] [6]

$$\begin{cases} \dfrac{di_d}{dt} = -\dfrac{R_a}{L_d} i_d + \dfrac{L_q}{L_d} p\omega i_q + \dfrac{1}{L_d} u_d \\ \dfrac{di_q}{dt} = -\dfrac{R_a}{L_q} i_q - \dfrac{L_d}{L_q} p\omega i_d - \dfrac{1}{L_q} p\omega \phi_m + \dfrac{1}{L_q} u_q \end{cases} \quad (11)$$

where d, q are the synchronous rotating reference frame; R_a is the armature resistance; L_d, L_q are the generator inductance on the d-q axis; i_d, i_q are, respectively, the d and q-axis components of current; u_q, u_d are the d and q-axis voltage components, respectively, p is the pole pairs number and ϕ_m is the permanent magnet flux.

In the d-q synchronously rotating reference frame, the electromagnetic torque is represented by[1]:

$$T_{em} = \frac{3}{2} n\left(\left(L_d - L_q\right) i_d i_q + \phi_m i_q\right). \quad (12)$$

CONTROL STRATEGY

Vector control strategy is used to have more preferment results, to control the wind turbine.

As given in Equation (11) we have a problem of coupling between d and q-axis, represented by the terms $p\omega i_q$ and $p\omega i_d$. Vector control concept is recommended, in order to overcome this problem of coupling

Vector Control Strategy

Vector control strategy is based on the field orientation, $i_d = 0$. Then according to (12), since ϕ_m is constant, the electromagnetic torque is directly proportional to i_q [7].

Two inputs v_d and v_q are defined to compensate the cross-coupling terms [1] [8],

$$v_d = p\Omega L_q i_q + u_d \quad (13)$$

$$v_q = -p\Omega L_d i_d - e_q + u_q \quad (14)$$

where

$$e_q = \Omega \phi_m \quad (15)$$

$$v_d = L_d \frac{di_d}{dt} + R_a i_d \quad (16)$$

$$v_q = L_q \frac{di_q}{dt} + R_a i_q \quad (17)$$

Current Regulators

According to (16) and (17), two separate first-order models in the d-q axis [8] [9]. Thus,

$$\frac{i_d}{v_d} = \frac{1}{\left(sL_d + R_a\right)} \quad (18)$$

$$\frac{i_q}{v_q} = \frac{1}{\left(sL_q + R_a\right)} \quad (19)$$

Therefore, we obtain two similar PI regulators witch used in two independent current loops that one of them controls the q-axis component and the second controls d-component as described inFigure 6.

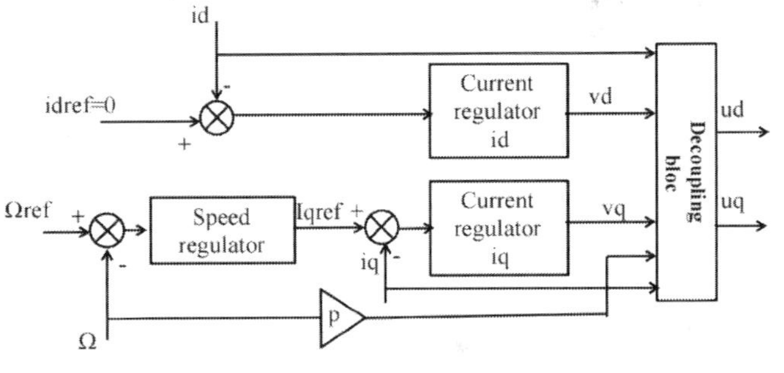

Figure 6. Control schemes.

Speed Regulator

Using the mechanical equation of wind turbine (10) the transfer function of the wind speed is written by:

$$\Omega = \frac{\left(T_t - T_{em}\right)}{Js + f} \quad (20)$$

Therefore, based on the last model the wind speed regulator is designed by a PI controller. Figure 6illustrates the overall schemes of the wind turbine control strategy. In fact, the speed controller takes as input the error between the reference speed Ω_{ref} and the actual rotational speed. Where Ω_{ref} is obtained by λ expression as shown in (4), it can be represented by:

$$\Omega_{ref} = \lambda \frac{v_w}{R} \quad (21)$$

Then, the speed controller output presents i_{qref} that is subtracted by the actual i_q and the result is the input of the current regulator of the quadrature current component. The direct component has as input the results of the subtraction between the reference direct current i_{dref} that it is null in our strategy and the actual one i_q. The two controllers have as outputs the two voltage v_d and v_q that presents the inputs of the decoupling bloc. This later has as outputs the voltage components u_q and u_d which present also the outputs of overall control schemes [8][10].

SIMULATIONS RESULTS

Simulations are carried out, in order to prove the wind turbine model and the effectiveness of the proposed control strategy. The block simulated is represented by Figure 7. It include, the different parts of the Wind Energy Conversion System [1] [11] such as the wind turbine model that has as input the electromagnetic torque T_{em} and the mechanic rotational speed Ω as output. The permanent magnet synchronous generator model which has three inputs; the two voltage components u_q and u_d and the electric rotational speed of the wind turbine ω. The control model take the quadrature current i_q, the direct current i_d and the reference electric wind turbine rotational speed ω_{ref} as inputs.

The complete WECS bloc given by Figure 7 was built in Simulink and its simulation results are given by Figures 8-19. The wind turbine parameters and the PMSG one used in simulation are illustrated in the Appendix.

Simulations are carried with the following model:

$$v(t) = 10 + 0.2\sin(0.1047t) + 2\sin(0.2665t) + \sin(1.293t) + 0.2\sin(3.6645t)$$

(22)

Simulation results demonstrate the performances of the wind turbine model and the vector control strategy. In fact, the output mechanical rotational speed Ω is assumed at the reference one Ω_{ref}; i_d is null as desired, T_{em} is proportional to i_q. Therefore the value of the power coefficient can be better than the current one as shown by Figure 19 C_p isn't the maximum one so the wind turbine power also isn't in the max and the tip speed ratio λ isn't optimal as given by Figure 18. The real λ = 12 but the optimum one is that illustrated by Figure 19 when C_p is maximum λ= 8.877.

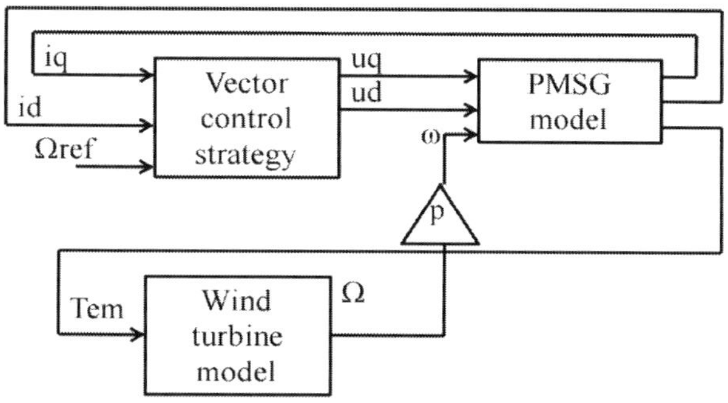

Figure 7. Complete model of the wind energy conversion system.

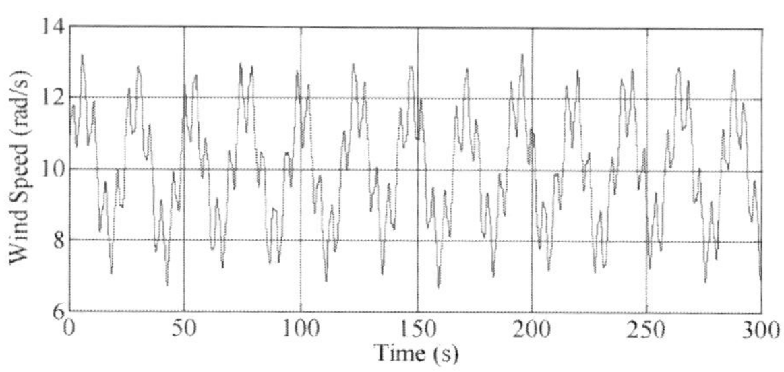

Figure 8. Wind speed profile.

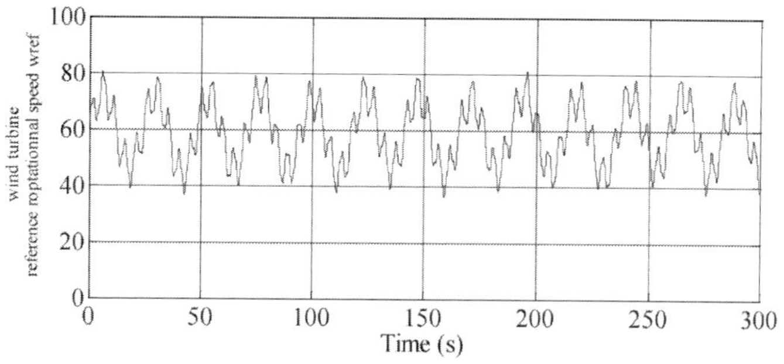

Figure 9. Wind turbine reference rotationnal speed.

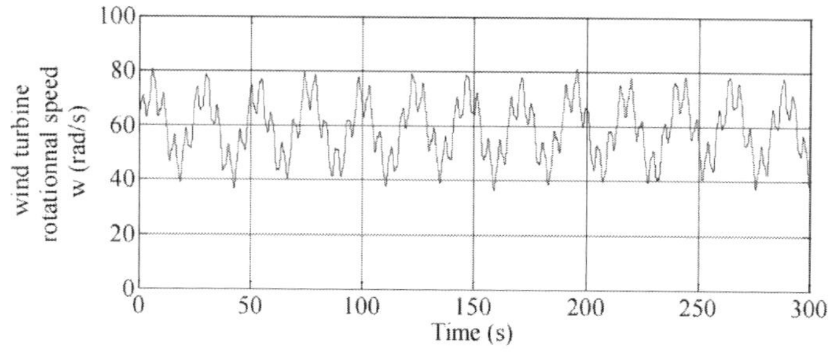

Figure 10. Wind turbine actual rotationnal speed.

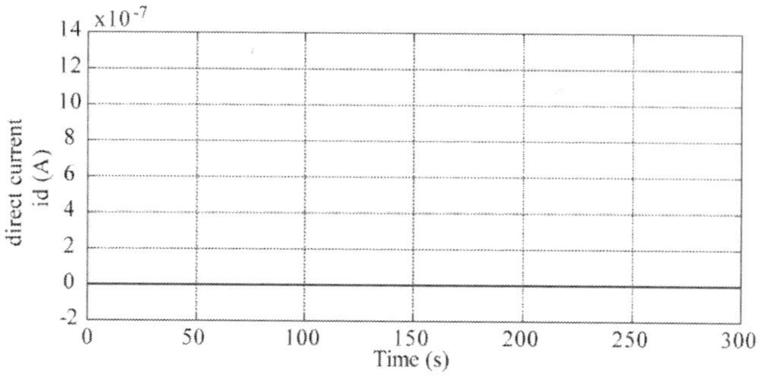

Figure 11. Current component i_d.

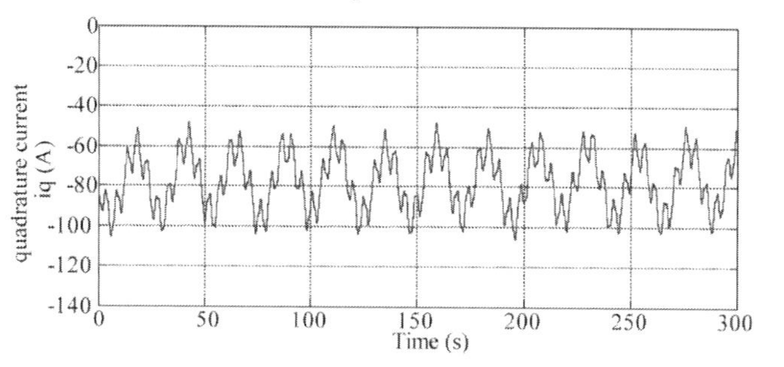

Figure 12. Current component i_q.

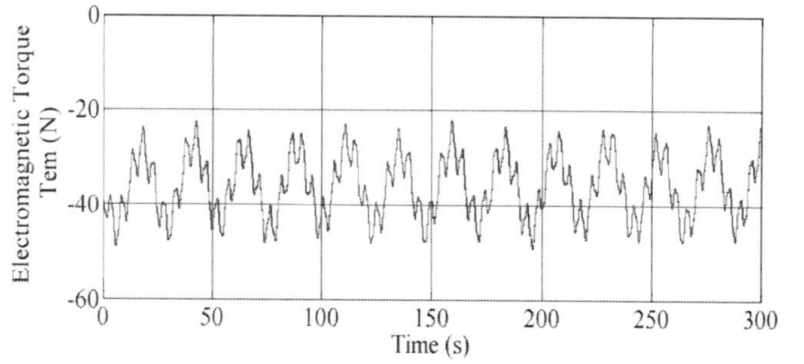

Figure 13. Electromagnetic torque T_{em}.

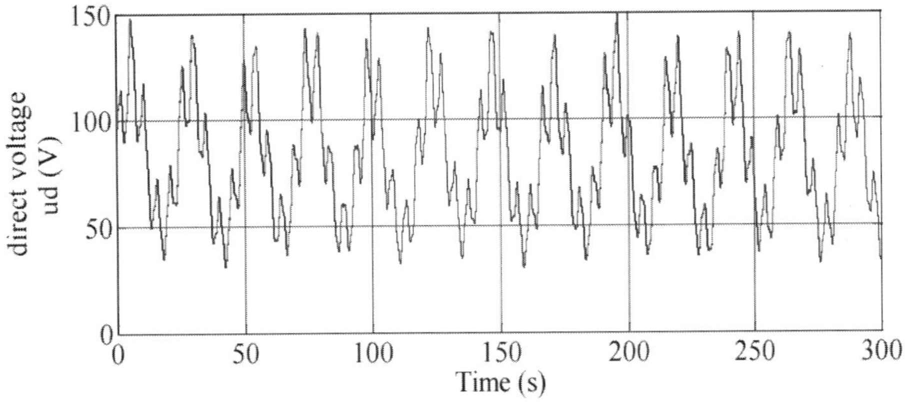

Figure 14. Voltage component u_d.

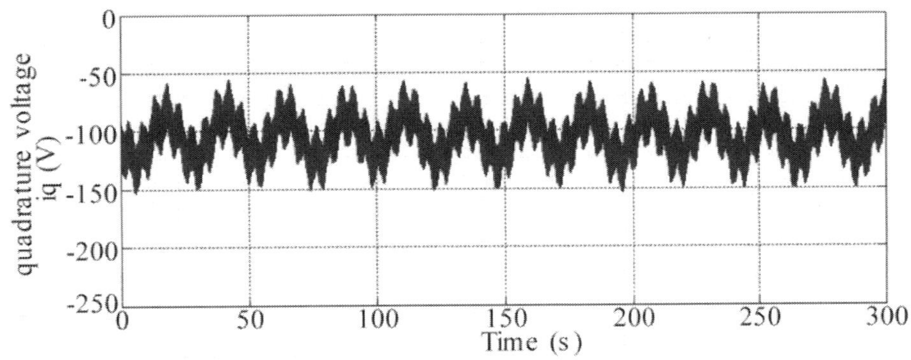

Figure 15. Voltage component u_q.

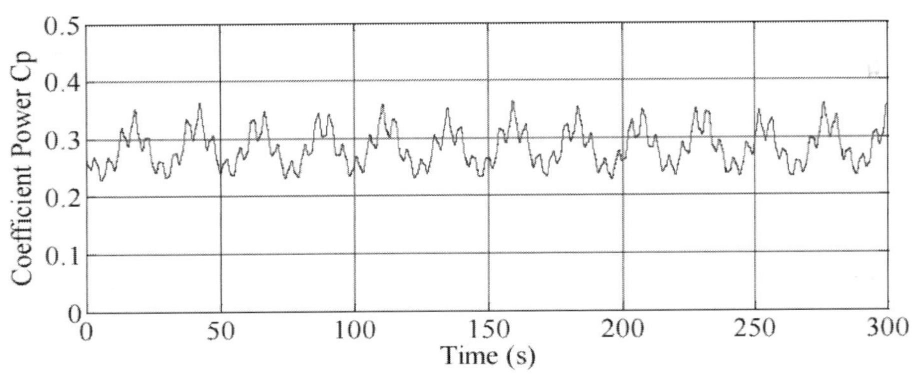

Figure 16. Coefficient power C_p.

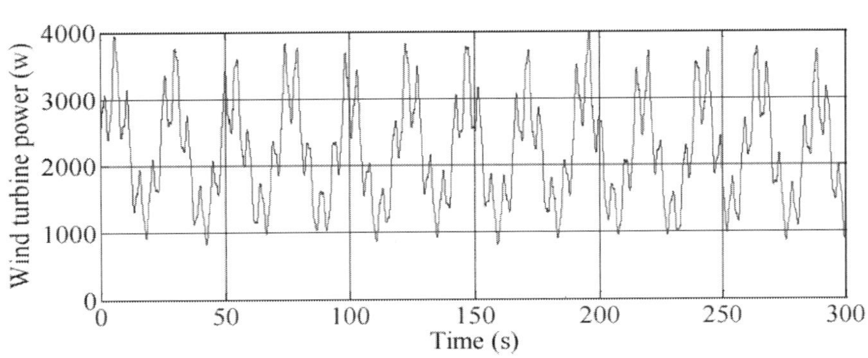

Figure 17. Wind turbine power P_t.

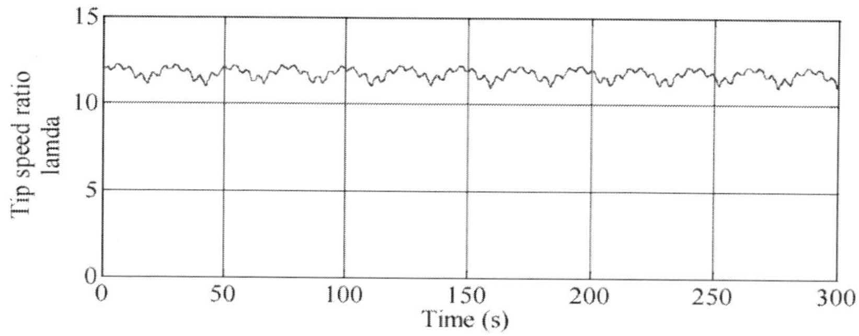

Figure 18. Tip speed ratio λ.

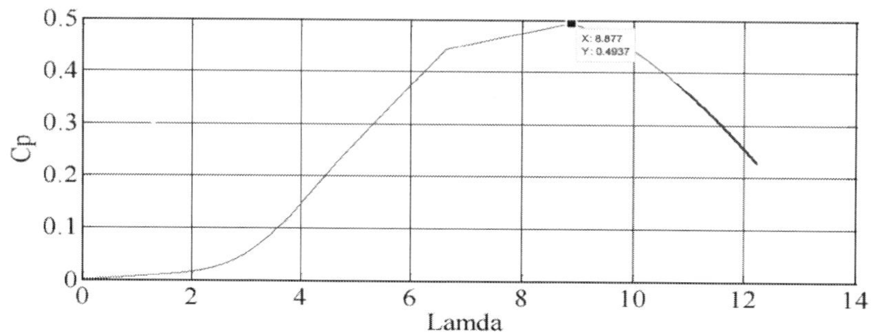

Figure 19. Coefficient power C_p.

CONCLUSION

In this paper, the different bloc of the wind energy conversion system was studied and modeled. In fact, the model of the wind turbine has been presented, the permanent magnet synchronous generator using in variable speed wind turbine has been modeled, controlled and simulated. Using Matlab/Simulink, simulation results were carried when the proposed WECS model was confirmed and the performances of vector control strategy are substantiating.

We faced a number of problems, when making the different component model of the wind energy conversion system. First, the performances of the WECS are relied to the choice of pitch angle. Then, the aerodynamic power is related to the wind speed. And as assumed by simulation results, the coefficient power isn't in its maximum, the speed ratio isn't in the optimum one and the turbine power isn't maximal. So, as results, trying to overcome these problems, we propose for the continuation of this chain of conversion energy controlling the pitch angle, studying and research of the maximum power point tracking to have more preferment results.

APPENDIX

Appendix Wind turbine parameters	PMSG parameters
ρ= 1.2 kg/m³, R = 2 Ω J = 0.002, f = 7E−4 m²	L_d = 0.0066 H, L_q = 0.0058 H R_a = 1.4 Ω, ϕ_m = 0.1546 Wb, p = 3

REFERENCES

1. Yin, M., Li, G., Zhou, M. and Zhao, C. (2007) Modeling of the Wind Turbine with a Permanent Magnet Synchronous Generator for Integration. IEEE, Power Engineering Society General Meeting, Tampa, 24-28 June 2007, 1-6.
2. Carranza, O., Figueres, E., Garcera, G. and Gonzalez-Medina, R. (2013) Analysis of the Control Structure of Wind Energy Generation Systems Based on a Permanent Magnet Synchronous Generator. Applied Energy, 103, 522-538. http://dx.doi.org/10.1016/j.apenergy.2012.10.015
3. Qiu, Z., Zhou, K. and Li, Y. (2011) Modeling and Control of Diode Rectifier Fed PMSG Based Wind Turbine. DRPT: 4th International Conference on Electric Utility Deregulation and Restructuring, and Power Technologies, Weihai, 6-9 July 2011, 1384-1388.
4. Thongam, J.S., Bouchard, P., Ezzaidi, H. and Ouhrouche, M. (2009) Wind Speed Sensorless Maximum Power Point Traking Control of Variable Speed Wind Energy Conversion System Electric Machines and Drives Conference, Miami, 3-6 May 2009, 1832-1837.
5. Nath, S. and Rana, S. (2011) The Modeling and Simulation Wind Energy Based Power System Using MATLAB. International Journal of Power System Operation and Energy Management, 1, 12-17.
6. Chen, X.F., Shu, Z.B. and Zhao, Y.K. (2005) Mathematic Model and Performance Analysis of PMSM, Based Servo System. Journal of Mechanism and Electronics, 1, 1-43.
7. Sanchez, A.G., Molina, M.G. and Lede, A.M. (2012) Dynamic Model of Wind Energy Conversion Systems

with PMSGBased Variable-Speed Wind Turbines for Power System Studies. International Journal of Hydrogen Energy, 37, 10064-10069. http://dx.doi.org/10.1016/j.ijhydene.2011.12.077

8. Bourourou, F., Louar, F. and Bourass, L. (2013) Simulation of Wind Conversion Chain Model with PI Voltage Regulation. Clearn Electric Power (ICCEP), Alghero, 11-13 June 2013, 707-711.

9. Abdal Rassul Abbs, F. and Abdulla Abdulsada, M. (2010) Simulation of Wind-Turbine Speed Control by MATLAB. International Journal of Comùputer and Electrical Engineering, 2, 912-915.

10. El Magri, A., Giri, F., Besançon, G., El fadili, A., Dugard, L. and Chaoui, F.Z. (2013) Sensorless Adaptative Output Feedback Control of Wind Energy Systems with PMS Generators. Control Engineering Practise, 21, 530-543.

11. Muyeen, S.M., Al-Durra, A. and Tamura, J. (2011) Variable Speed Wind Turbine Generator System with Current Controlled Voltage Source Inverter. Energy Conversion and Management, 52, 2688-2694. http://dx.doi.org/10.1016/j.enconman.2011.02.001

INDEX